Tutor in a Book Algebra

Table of Contents

Pre-Algebra - Integers

Objective: Add, Subtract, Multiply and Divide Positive and Negative Numbers.

The ability to work comfortably with negative numbers is essential to success in algebra. For this reason we will do a quick review of adding, subtracting, multiplying and dividing of integers. **Integers** are all the positive whole numbers, zero, and their opposites (negatives). As this is intended to be a review of integers, the descriptions and examples will not be as detailed as a normal lesson.

World View Note: The first set of rules for working with negative numbers was written out by the Indian mathematician Brahmagupta.

When adding integers we have two cases to consider. The first is if the signs match, both positive or both negative. If the signs match we will add the numbers together and keep the sign. This is illustrated in the following examples

Example 1.

$$-5 + (-3) \quad \text{Same sign, add } 5+3, \text{keep the negative}$$
$$-8 \quad \text{Our Solution}$$

Example 2.

$$-7 + (-5) \quad \text{Same sign, add } 7+5, \text{keep the negative}$$
$$-12 \quad \text{Our Solution}$$

If the signs don't match, one positive and one negative number, we will subtract the numbers (as if they were all positive) and then use the sign from the larger number. This means if the larger number is positive, the answer is positive. If the larger number is negative, the answer is negative. This is shown in the following examples.

Example 3.

$$-7 + 2 \quad \text{Different signs, subtract } 7-2, \text{use sign from bigger number, negative}$$
$$-5 \quad \text{Our Solution}$$

Example 4.

$$-4 + 6 \quad \text{Different signs, subtract } 6-4, \text{use sign from bigger number, positive}$$
$$2 \quad \text{Our Solution}$$

Example 5.

$4 + (-3)$ Different signs, subtract $4 - 3$, use sign from bigger number, positive

    1   Our Solution

Example 6.

$7 + (-10)$ Different signs, subtract $10 - 7$, use sign from bigger number, negative

   -3 Our Solution

For subtraction of negatives we will change the problem to an addition problem which we can then solve using the above methods. The way we change a subtraction to an addition is to add the opposite of the number after the subtraction sign. Often this method is refered to as "add the opposite." This is illustrated in the following examples.

Example 7.

$8 - 3$ Add the opposite of 3

$8 + (-3)$ Different signs, subtract $8 - 3$, use sign from bigger number, positive

   5   Our Solution

Example 8.

$-4 - 6$ Add the opposite of 6

$-4 + (-6)$ Same sign, add $4 + 6$, keep the negative

  -10 Our Solution

Example 9.

$9 - (-4)$ Add the opposite of -4

$9 + 4$ Same sign, add $9 + 4$, keep the positive

  13   Our Solution

Example 10.

$-6 - (-2)$ Add the opposite of -2

$-6 + 2$ Different sign, subtract $6 - 2$, use sign from bigger number, negative

  -4 Our Solution

Multiplication and division of integers both work in a very similar pattern. The short description of the process is we multiply and divide like normal, if the signs match (both positive or both negative) the answer is positive. If the signs don't match (one positive and one negative) then the answer is negative. This is shown in the following examples

Example 11.

$$(4)(-6) \quad \text{Signs do not match, answer is negative}$$
$$-24 \quad \text{Our Solution}$$

Example 12.

$$\frac{-36}{-9} \quad \text{Signs match, answer is positive}$$

$$4 \quad \text{Our Solution}$$

Example 13.

$$-2(-6) \quad \text{Signs match, answer is positive}$$
$$12 \quad \text{Our Solution}$$

Example 14.

$$\frac{15}{-3} \quad \text{Signs do not match, answer is negative}$$

$$-5 \quad \text{Our Solution}$$

A few things to be careful of when working with integers. First be sure not to confuse a problem like $-3-8$ with $-3(-8)$. The second problem is a multiplication problem because there is nothing between the 3 and the parenthesis. If there is no operation written in between the parts, then we assume that means we are multiplying. The $-3-8$ problem, is subtraction because the subtraction separates the 3 from what comes after it. Another item to watch out for is to be careful not to mix up the pattern for adding and subtracting integers with the pattern for multiplying and dividing integers. They can look very similar, for example if the signs match on addition, the we keep the negative, $-3+(-7)=-10$, but if the signs match on multiplication, the answer is positive, $(-3)(-7)=21$.

0.1 Practice - Integers

Evaluate each expression.

1) $1 - 3$

2) $4 - (-1)$

3) $(-6) - (-8)$

4) $(-6) + 8$

5) $(-3) - 3$

6) $(-8) - (-3)$

7) $3 - (-5)$

8) $7 - 7$

9) $(-7) - (-5)$

10) $(-4) + (-1)$

11) $3 - (-1)$

12) $(-1) + (-6)$

13) $6 - 3$

14) $(-8) + (-1)$

15) $(-5) + 3$

16) $(-1) - 8$

17) $2 - 3$

18) $5 - 7$

19) $(-8) - (-5)$

20) $(-5) + 7$

21) $(-2) + (-5)$

22) $1 + (-1)$

23) $5 - (-6)$

24) $8 - (-1)$

25) $(-6) + 3$

26) $(-3) + (-1)$

27) $4 - 7$

28) $7 - 3$

29) $(-7) + 7$

30) $(-3) + (-5)$

Find each product.

31) $(4)(-1)$

32) $(7)(-5)$

33) $(10)(-8)$

34) $(-7)(-2)$

35) $(-4)(-2)$

36) $(-6)(-1)$

37) $(-7)(8)$

38) $(6)(-1)$

39) $(9)(-4)$

40) $(-9)(-7)$

41) $(-5)(2)$

42) $(-2)(-2)$

43) $(-5)(4)$

44) $(-3)(-9)$

45) $(4)(-6)$

Find each quotient.

46) $\frac{30}{-10}$

47) $\frac{49}{-7}$

48) $\frac{12}{4}$

49) $\frac{2}{-1}$

50) $\frac{30}{6}$

51) $\frac{20}{10}$

52) $\frac{27}{3}$

53) $\frac{-35}{5}$

54) $\frac{80}{-8}$

55) $\frac{-8}{-2}$

56) $\frac{50}{5}$

57) $\frac{-16}{2}$

58) $\frac{48}{8}$

59) $\frac{60}{10}$

60) $\frac{54}{6}$

Pre-Algebra - Fractions

Objective: Reduce, add, subtract, multiply, and divide with fractions.

Working with fractions is a very important foundation to algebra. Here we will briefly review reducing, multiplying, dividing, adding, and subtracting fractions. As this is a review, concepts will not be explained in detail as other lessons are.

World View Note: The earliest known use of fraction comes from the Middle Kingdom of Egypt around 2000 BC!

We always like our final answers when working with fractions to be reduced. Reducing fractions is simply done by dividing both the numerator and denominator by the same number. This is shown in the following example

Example 15.

$$\frac{36}{84} \qquad \text{Both numerator and denominator are divisible by 4}$$

$$\frac{36 \div 4}{84 \div 4} = \frac{9}{21} \qquad \text{Both numerator and denominator are still divisible by 3}$$

$$\frac{9 \div 3}{21 \div 3} = \frac{3}{7} \quad \text{Our Soultion}$$

The previous example could have been done in one step by dividing both numerator and denominator by 12. We also could have divided by 2 twice and then divided by 3 once (in any order). It is not important which method we use as long as we continue reducing our fraction until it cannot be reduced any further.

The easiest operation with fractions is multiplication. We can multiply fractions by multiplying straight across, multiplying numerators together and denominators together.

Example 16.

$$\frac{6}{7} \cdot \frac{3}{5} \quad \text{Multiply numerators across and denominators across}$$

$$\frac{18}{35} \quad \text{Our Solution}$$

When multiplying we can reduce our fractions before we multiply. We can either reduce vertically with a single fraction, or diagonally with several fractions, as long as we use one number from the numerator and one number from the denominator.

Example 17.

$$\frac{25}{24} \cdot \frac{32}{55} \quad \text{Reduce 25 and 55 by dividing by 5. Reduce 32 and 24 by dividing by 8}$$

$$\frac{5}{3} \cdot \frac{4}{11} \quad \text{Multiply numerators across and denominators across}$$

$$\frac{20}{33} \quad \text{Our Solution}$$

Dividing fractions is very similar to multiplying with one extra step. Dividing fractions requires us to first take the reciprocal of the second fraction and multiply. Once we do this, the multiplication problem solves just as the previous problem.

Example 18.

$$\frac{21}{16} \div \frac{28}{6}$$ Multiply by the reciprocal

$$\frac{21}{16} \cdot \frac{6}{28}$$ Reduce 21 and 28 by dividing by 7. Reduce 6 and 16 by dividing by 2

$$\frac{3}{8} \cdot \frac{3}{4}$$ Multiply numerators across and denominators across

$$\frac{9}{32}$$ Our Soultion

To add and subtract fractions we will first have to find the least common denominator (LCD). There are several ways to find an LCD. One way is to find the smallest multiple of the largest denominator that you can also divide the small denomiator by.

Example 19.

$$\text{Find the LCD of 8 and 12}$$ Test multiples of 12

$$12? \; \frac{12}{8}$$ Can't divide 12 by 8

$$24? \; \frac{24}{8} = 3$$ Yes! We can divide 24 by 8!

$$24$$ Our Soultion

Adding and subtracting fractions is identical in process. If both fractions already have a common denominator we just add or subtract the numerators and keep the denominator.

Example 20.

$$\frac{7}{8} + \frac{3}{8}$$ Same denominator, add numerators $7 + 3$

$$\frac{10}{8}$$ Reduce answer, dividing by 2

$$\frac{5}{4}$$ Our Solution

While $\frac{5}{4}$ can be written as the mixed number $1\frac{1}{4}$, in algebra we will almost never use mixed numbers. For this reason we will always use the improper fraction, not the mixed number.

Example 21.

$$\frac{13}{6} - \frac{9}{6} \qquad \text{Same denominator, subtract numerators } 13 - 9$$

$$\frac{4}{6} \qquad \text{Reduce answer, dividing by 2}$$

$$\frac{2}{3} \qquad \text{Our Solution}$$

If the denominators do not match we will first have to identify the LCD and build up each fraction by multiplying the numerators and denominators by the same number so the denominator is built up to the LCD.

Example 22.

$$\frac{5}{6} + \frac{4}{9} \qquad \text{LCD is } 18.$$

$$\frac{3 \cdot 5}{3 \cdot 6} + \frac{4 \cdot 2}{9 \cdot 2} \qquad \text{Multiply first fraction by 3 and the second by 2}$$

$$\frac{15}{18} + \frac{8}{18} \qquad \text{Same denominator, add numerators, } 15 + 8$$

$$\frac{23}{18} \qquad \text{Our Solution}$$

Example 23.

$$\frac{2}{3} - \frac{1}{6} \qquad \text{LCD is } 6$$

$$\frac{2 \cdot 2}{2 \cdot 3} - \frac{1}{6} \qquad \text{Multiply first fraction by 2, the second already has a denominator of 6}$$

$$\frac{4}{6} - \frac{1}{6} \qquad \text{Same denominator, subtract numerators, } 4 - 1$$

$$\frac{3}{6} \qquad \text{Reduce answer, dividing by 3}$$

$$\frac{1}{2} \qquad \text{Our Solution}$$

0.2 Practice - Fractions

Simplify each. Leave your answer as an improper fraction.

1) $\frac{42}{12}$

2) $\frac{25}{20}$

3) $\frac{35}{25}$

4) $\frac{24}{9}$

5) $\frac{54}{36}$

6) $\frac{30}{24}$

7) $\frac{45}{36}$

8) $\frac{36}{27}$

9) $\frac{27}{18}$

10) $\frac{48}{18}$

11) $\frac{40}{16}$

12) $\frac{48}{42}$

13) $\frac{63}{18}$

14) $\frac{16}{12}$

15) $\frac{80}{60}$

16) $\frac{72}{48}$

17) $\frac{72}{60}$

18) $\frac{126}{108}$

19) $\frac{36}{24}$

20) $\frac{160}{140}$

Find each product.

21) $(9)(\frac{8}{9})$

22) $(-2)(-\frac{5}{6})$

23) $(2)(-\frac{2}{9})$

24) $(-2)(\frac{1}{3})$

25) $(-2)(\frac{13}{8})$

26) $(\frac{3}{2})(\frac{1}{2})$

27) $(-\frac{6}{5})(-\frac{11}{8})$

28) $(-\frac{3}{7})(-\frac{11}{8})$

29) $(8)(\frac{1}{2})$

30) $(-2)(-\frac{9}{7})$

31) $(\frac{2}{3})(\frac{3}{4})$

32) $(-\frac{17}{9})(-\frac{3}{5})$

33 $(2)(\frac{3}{2})$

34) $(\frac{17}{9})(-\frac{3}{5})$

35) $(\frac{1}{2})(-\frac{7}{5})$

36) $(\frac{1}{2})(\frac{5}{7})$

Find each quotient.

37) $-2 \div \frac{7}{4}$

38) $\frac{-12}{7} \div \frac{-9}{5}$

39) $\frac{1}{9} \div \frac{1}{2}$

40) $-2 \div \frac{3}{2}$

41) $\frac{3}{2} \div \frac{13}{7}$

42) $\frac{5}{3} \div \frac{7}{5}$

43) $-1 \div \frac{2}{3}$

44) $\frac{10}{9} \div -6$

45) $\frac{8}{9} \div \frac{1}{5}$

46) $\frac{1}{6} \div \frac{5}{3}$

47) $\frac{-9}{7} \div \frac{1}{5}$

48) $\frac{-13}{8} \div \frac{-15}{8}$

49) $\frac{2}{9} \div \frac{3}{2}$

50) $\frac{4}{5} \div \frac{13}{8}$

51) $\frac{1}{10} \div \frac{3}{2}$

52) $\frac{5}{3} \div \frac{5}{3}$

Evaluate each expression.

53) $\frac{1}{3} + \left(-\frac{4}{3}\right)$

54) $\frac{1}{7} + \left(-\frac{11}{7}\right)$

55) $\frac{3}{7} - \frac{1}{7}$

56) $\frac{1}{3} + \frac{5}{3}$

57) $\frac{11}{6} + \frac{7}{6}$

58) $(-2) + \left(-\frac{15}{8}\right)$

59) $\frac{3}{5} + \frac{5}{4}$

60) $(-1) - \frac{2}{3}$

61) $\frac{2}{5} + \frac{5}{4}$

62) $\frac{12}{7} - \frac{9}{7}$

63) $\frac{9}{8} + \left(-\frac{2}{7}\right)$

64) $(-2) + \frac{5}{6}$

65) $1 + \left(-\frac{1}{3}\right)$

66) $\frac{1}{2} - \frac{11}{6}$

67) $\left(-\frac{1}{2}\right) + \frac{3}{2}$

68) $\frac{11}{8} - \frac{1}{2}$

69) $\frac{1}{5} + \frac{3}{4}$

70) $\frac{6}{5} - \frac{8}{5}$

71) $\left(-\frac{5}{7}\right) - \frac{15}{8}$

72) $\left(-\frac{1}{3}\right) + \left(-\frac{8}{5}\right)$

73) $6 - \frac{8}{7}$

74) $(-6) + \left(-\frac{5}{3}\right)$

75) $\frac{3}{2} - \frac{15}{8}$

76) $(-1) - \left(-\frac{1}{3}\right)$

77) $\left(-\frac{15}{8}\right) + \frac{5}{3}$

78) $\frac{3}{2} + \frac{9}{7}$

79) $(-1) - \left(-\frac{1}{6}\right)$

80) $\left(-\frac{1}{2}\right) - \left(-\frac{3}{5}\right)$

81) $\frac{5}{3} - \left(-\frac{1}{3}\right)$

82) $\frac{9}{7} - \left(-\frac{5}{3}\right)$

Pre-Algebra - Order of Operations

Objective: Evaluate expressions using the order of operations, including the use of absolute value.

When simplifying expressions it is important that we simplify them in the correct order. Consider the following problem done two different ways:

Example 24.

$2 + 5 \cdot 3$	Add First	$2 + 5 \cdot 3$	Multiply
$7 \cdot 3$	Multiply	$2 + 15$	Add
21	Solution	17	Solution

The previous example illustrates that if the same problem is done two different ways we will arrive at two different solutions. However, only one method can be correct. It turns out the second method, 17, is the correct method. The order of operations ends with the most basic of operations, addition (or subtraction). Before addition is completed we must do repeated addition or multiplication (or division). Before multiplication is completed we must do repeated multiplication or exponents. When we want to do something out of order and make it come first we will put it in parenthesis (or grouping symbols). This list then is our order of operations we will use to simplify expressions.

Order of Operations:

Parenthesis (Grouping)

Exponents

Multiply and Divide (Left to Right)

Add and Subtract (Left to Right)

Multiply and Divide are on the same level because they are the same operation (division is just multiplying by the reciprocal). This means they must be done left to right, so some problems we will divide first, others we will multiply first. The same is true for adding and subtracting (subtracting is just adding the opposite).

Often students use the word PEMDAS to remember the order of operations, as the first letter of each operation creates the word PEMDAS. However, it is the author's suggestion to think about PEMDAS as a vertical word written as:

P
E
MD
AS

so we don't forget that multiplication and division are done left to right (same with addition and subtraction). Another way students remember the order of operations is to think of a phrase such as "Please Excuse My Dear Aunt Sally" where each word starts with the same letters as the order of operations start with.

World View Note: The first use of grouping symbols are found in 1646 in the Dutch mathematician, Franciscus van Schooten's text, Vieta. He used a bar over the expression that is to be evaluated first. So problems like $2(3+5)$ were written as $2 \cdot \overline{3+5}$.

Example 25.

$$
\begin{array}{ll}
2 + 3(\underline{9-4})^2 & \text{Parenthesis first} \\
2 + 3(\underline{5})^2 & \text{Exponents} \\
2 + 3(\underline{25}) & \text{Multiply} \\
\underline{2 + 75} & \text{Add} \\
77 & \text{Our Solution}
\end{array}
$$

It is very important to remember to multiply and divide from from left to right!

Example 26.

$$
\begin{array}{ll}
\underline{30 \div 3} \cdot 2 & \text{Divide first (left to right!)} \\
\underline{10 \cdot 2} & \text{Multiply} \\
20 & \text{Our Solution}
\end{array}
$$

In the previous example, if we had multiplied first, five would have been the answer which is incorrect.

If there are several parenthesis in a problem we will start with the inner most parenthesis and work our way out. Inside each parenthesis we simplify using the order of operations as well. To make it easier to know which parenthesis goes with which parenthesis, different types of parenthesis will be used such as $\{ \}$ and $[\,]$ and $(\,)$, these parenthesis all mean the same thing, they are parenthesis and must be evaluated first.

Example 27.

$$
\begin{array}{ll}
2\{8^2 - 7[32 - 4(\underline{3}^2 + 1)](-1)\} & \text{Inner most parenthesis, exponents first} \\
2\{8^2 - 7[32 - 4(\underline{9+1})](-1)\} & \text{Add inside those parenthesis} \\
2\{8^2 - 7[32 - \underline{4(10)}](-1)\} & \text{Multiply inside inner most parenthesis} \\
2\{8^2 - 7[\underline{32 - 40}](-1)\} & \text{Subtract inside those parenthesis} \\
2\{\underline{8}^2 - 7[-8](-1)\} & \text{Exponents next} \\
2\{\underline{64 - 7}[-8](-1)\} & \text{Multiply left to right, sign with the number} \\
2\{\underline{64 + 56}(-1)\} & \text{Finish multiplying} \\
2\{\underline{64 - 56}\} & \text{Subtract inside parenthesis} \\
\underline{2\{8\}} & \text{Multiply} \\
16 & \text{Our Solution}
\end{array}
$$

As the above example illustrates, it can take several steps to complete a problem. The key to successfully solve order of operations problems is to take the time to show your work and do one step at a time. This will reduce the chance of making a mistake along the way.

There are several types of grouping symbols that can be used besides parenthesis. One type is a fraction bar. If we have a fraction, the entire numerator and the entire denominator must be evaluated before we reduce the fraction. In these cases we can simplify in both the numerator and denominator at the same time.

Example 28.

$$\frac{\tilde{2}^4 - (-8)\cdot 3}{15 \div 5 - 1}$$ Exponent in the numerator, divide in denominator

$$\frac{16 - \overbrace{(-8)\cdot 3}}{3 - 1}$$ Multiply in the numerator, subtract in denominator

$$\frac{\overbrace{16 - (-24)}}{2}$$ Add the opposite to simplify numerator, denominator is done.

$$\frac{40}{2}$$ Reduce, divide

$$20$$ Our Solution

Another type of grouping symbol that also has an operation with it, absolute value. When we have absolute value we will evaluate everything inside the absolute value, just as if it were a normal parenthesis. Then once the inside is completed we will take the absolute value, or distance from zero, to make the number positive.

Example 29.

$$1 + 3|-4^2 - (-8)| + 2|3 + (-5)^2|$$ Evaluate absolute values first, exponents
$$1 + 3|-16 - (-8)| + 2|3 + 25|$$ Add inside absolute values
$$1 + 3|-8| + 2|28|$$ Evaluate absolute values
$$1 + 3(8) + 2(28)$$ Multiply left to right
$$1 + 24 + 2(28)$$ Finish multiplying
$$1 + 24 + 56$$ Add left to right
$$25 + 56$$ Add
$$81$$ Our Solution

The above example also illustrates an important point about exponents. Exponents only are considered to be on the number they are attached to. This means when we see -4^2, only the 4 is squared, giving us $-(4^2)$ or -16. But when the negative is in parentheses, such as $(-5)^2$ the negative is part of the number and is also squared giving us a positive solution, 25.

0.3 Practice - Order of Operation

Solve.

1) $-6 \cdot 4(-1)$

2) $(-6 \div 6)^3$

3) $3 + (8) \div |4|$

4) $5(-5+6) \cdot 6^2$

5) $8 \div 4 \cdot 2$

6) $7 - 5 + 6$

7) $[-9 - (2-5)] \div (-6)$

8) $(-2 \cdot 2^3 \cdot 2) \div (-4)$

9) $-6 + (-3-3)^2 \div |3|$

10) $(-7-5) \div [-2-2-(-6)]$

11) $4 - 2|3^2 - 16|$

12) $\frac{10 \quad 6}{(\quad 2)^2} - 5$

13) $[-1-(-5)]|3+2|$

14) $-3 - \{3 - [-3(2+4) - (-2)]\}$

15) $\frac{2+4\ 7+2^2}{4 \cdot 2 + 5 \cdot 3}$

16) $-4 - [2 + 4(-6) - 4 - |2^2 - 5 \cdot 2|]$

17) $[6 \cdot 2 + 2 - (-6)](-5 + \left|\frac{-18}{6}\right|)$

18) $2 \cdot (-3) + 3 - 6[-2 - (-1-3)]$

19) $\frac{13 \quad 2}{2 \ (\quad 1)^3 + (\quad 6) \ [\quad 1 \ (\quad 3)]}$

20) $\frac{-5^2 + (-5)^2}{|4^2 - 2^5| - 2 \cdot 3}$

21) $6 \cdot \frac{-8 - 4 + (-4) - [-4 - (-3)]}{(4^2 + 3^2) \div 5}$

22) $\frac{-9 \cdot 2 - (3-6)}{1 \ (\quad 2+1) \ (\quad 3)}$

23) $\frac{2^3 + 4}{18 \ 6 + (\quad 4) \ [\quad 5(\quad 1)(\quad 5)]}$

24) $\frac{13 + (\quad 3)^2 + 4(\quad 3) + 1 \ [\quad 10 \ (\quad 6)]}{\{[4+5] : [4^2 \ 3^2(4 \ 3) \ 8]\} + 12}$

25) $\frac{5 + 3^2 \ 24 : 6 \cdot 2}{|5 + 3(2^2 - 5) + 2^3 - 5|^2}$

Pre-Algebra - Properties of Algebra

Objective: Simplify algebraic expressions by substituting given values, distributing, and combining like terms

In algebra we will often need to simplify an expression to make it easier to use. There are three basic forms of simplifying which we will review here.

World View Note: The term "Algebra" comes from the Arabic word al-jabr which means "reunion". It was first used in Iraq in 830 AD by Mohammad ibn-Musa al-Khwarizmi.

The first form of simplifying expressions is used when we know what number each variable in the expression represents. If we know what they represent we can replace each variable with the equivalent number and simplify what remains using order of operations.

Example 30.

$$p(q+6) \text{ when } p = 3 \text{ and } q = 5 \qquad \text{Replace } p \text{ with 3 and } q \text{ with 5}$$
$$(3)((5) + 6) \qquad \text{Evaluate parenthesis}$$
$$(3)(11) \qquad \text{Multiply}$$
$$33 \qquad \text{Our Solution}$$

Whenever a variable is replaced with something, we will put the new number inside a set of parenthesis. Notice the 3 and 5 in the previous example are in parenthesis. This is to preserve operations that are sometimes lost in a simple replacement. Sometimes the parenthesis won't make a difference, but it is a good habbit to always use them to prevent problems later.

Example 31.

$$x + zx(3-z)\left(\frac{x}{3}\right) \text{ when } x = -6 \text{ and } z = -2 \qquad \text{Replace all } x's \text{ with 6 and } z's \text{ with 2}$$
$$(-6) + (-2)(-6)(3-(-2))\left(\frac{(-6)}{3}\right) \qquad \text{Evaluate parenthesis}$$
$$-6 + (-2)(-6)(5)(-2) \qquad \text{Multiply left to right}$$
$$-6 + 12(5)(-2) \qquad \text{Multiply left to right}$$
$$-6 + 60(-2) \qquad \text{Multiply}$$
$$-6 - 120 \qquad \text{Subtract}$$
$$-126 \qquad \text{Our Solution}$$

It will be more common in our study of algebra that we do not know the value of the variables. In this case, we will have to simplify what we can and leave the variables in our final solution. One way we can simplify expressions is to combine like terms. **Like terms** are terms where the variables match exactly (exponents included). Examples of like terms would be $3xy$ and $-7xy$ or $3a^2b$ and $8a^2b$ or -3 and 5. If we have like terms we are allowed to add (or subtract) the numbers in front of the variables, then keep the variables the same. This is shown in the following examples

Example 32.

$$5x - 2y - 8x + 7y \quad \text{Combine like terms } 5x - 8x \text{ and } -2y + 7y$$
$$-3x + 5y \quad \text{Our Solution}$$

Example 33.

$$8x^2 - 3x + 7 - 2x^2 + 4x - 3 \quad \text{Combine like terms } 8x^2 - 2x^2 \text{ and } -3x + 4x \text{ and } 7 - 3$$
$$6x^2 + x + 4 \quad \text{Our Solution}$$

As we combine like terms we need to interpret subtraction signs as part of the following term. This means if we see a subtraction sign, we treat the following term like a negative term, the sign always stays with the term.

A final method to simplify is known as distributing. Often as we work with problems there will be a set of parenthesis that make solving a problem difficult, if not impossible. To get rid of these unwanted parenthesis we have the distributive property. Using this property we multiply the number in front of the parenthesis by each term inside of the parenthesis.

$$\textbf{Distributive Property: } \boldsymbol{a(b+c)} = \textbf{ab} + \textbf{ac}$$

Several examples of using the distributive property are given below.

Example 34.

$$4(2x - 7) \quad \text{Multiply each term by } 4$$
$$8x - 28 \quad \text{Our Solution}$$

Example 35.

$$-7(5x - 6) \quad \text{Multiply each term by } -7$$
$$-35 + 42 \quad \text{Our Solution}$$

In the previous example we again use the fact that the sign goes with the number, this means we treat the -6 as a negative number, this gives $(-7)(-6) = 42$, a positive number. The most common error in distributing is a sign error, be very careful with your signs!

It is possible to distribute just a negative through parenthesis. If we have a negative in front of parenthesis we can think of it like a -1 in front and distribute the -1 through. This is shown in the following example.

Example 36.

$$-(4x - 5y + 6) \quad \text{Negative can be thought of as} -1$$
$$-1(4x - 5y + 6) \quad \text{Multiply each term by} -1$$
$$-4x + 5y - 6 \quad \text{Our Solution}$$

Distributing through parenthesis and combining like terms can be combined into one problem. Order of operations tells us to multiply (distribute) first then add or subtract last (combine like terms). Thus we do each problem in two steps, distribute then combine.

Example 37.

$$5 + 3(2x - 4) \quad \text{Distribute 3, multipling each term}$$
$$5 + 6x - 12 \quad \text{Combine like terms } 5 - 12$$
$$-7 + 6x \quad \text{Our Solution}$$

Example 38.

$$3x - 2(4x - 5) \quad \text{Distribute} -2, \text{multilpying each term}$$
$$3x - 8x + 10 \quad \text{Combine like terms } 3x - 8x$$
$$-5x + 10 \quad \text{Our Solution}$$

In the previous example we distributed -2, not just 2. This is because we will always treat subtraction like a negative sign that goes with the number after it. This makes a big difference when we multiply by the -5 inside the parenthesis, we now have a positive answer. Following are more involved examples of distributing and combining like terms.

Example 39.

$$2(5x - 8) - 6(4x + 3) \quad \text{Distribute 2 into first parenthesis and} -6 \text{ into second}$$
$$10x - 16 - 24x - 18 \quad \text{Combine like terms } 10x - 24x \text{ and} -16 - 18$$
$$-14x - 34 \quad \text{Our Solution}$$

Example 40.

$$4(3x - 8) - (2x - 7) \quad \text{Negative (subtract) in middle can be thought of as} -1$$
$$4(3x - 8) - 1(2x - 7) \quad \text{Distribute 4 into first parenthesis,} -1 \text{ into second}$$
$$12x - 32 - 2x + 7 \quad \text{Combine like terms } 12x - 2x \text{ and} -32 + 7$$
$$10x - 25 \quad \text{Our Solution}$$

0.4 Practice - Properties of Algebra

Evaluate each using the values given.

1) $p + 1 + q - m$; use $m = 1, p = 3, q = 4$

2) $y^2 + y - z$; use $y = 5, z = 1$

3) $p - \frac{pq}{6}$; use $p = 6$ and $q = 5$

4) $\frac{6 + z - y}{3}$; use $y = 1, z = 4$

5) $c^2 - (a - 1)$; use $a = 3$ and $c = 5$

6) $x + 6z - 4y$; use $x = 6, y = 4, z = 4$

7) $5j + \frac{kh}{2}$; use $h = 5, j = 4, k = 2$

8) $5(b + a) + 1 + c$; use $a = 2, b = 6, c = 5$

9) $\frac{4 - (p - m)}{2} + q$; use $m = 4, p = 6, q = 6$

10) $z + x - (1^2)^3$; use $x = 5, z = 4$

11) $m + n + m + \frac{n}{2}$; use $m = 1$ and $n = 2$

12) $3 + z - 1 + y - 1$; use $y = 5, z = 4$

13) $q - p - (q - 1 - 3)$; use $p = 3, q = 6$

14) $p + (q - r)(6 - p)$; use $p = 6, q = 5, r = 5$

15) $y - [4 - y - (z - x)]$; use $x = 3, y = 1, z = 6$

16) $4z - (x + x - (z - z))$; use $x = 3, z = 2$

17) $k \times 3^2 - (j + k) - 5$; use $j = 4, k = 5$

18) $a^3(c^2 - c)$; use $a = 3, c = 2$

19) $zx - (z - \frac{4 + x}{6})$; use $x = 2, z = 6$

20) $5 + qp + pq - q$; use $p = 6, q = 3$

Combine Like Terms

21) $r - 9 + 10$

22) $-4x + 2 - 4$

23) $n + n$

24) $4b + 6 + 1 + 7b$

25) $8v + 7v$

26) $-x + 8x$

27) $-7x - 2x$

28) $-7a - 6 + 5$

29) $k - 2 + 7$

30) $-8p + 5p$

31) $x - 10 - 6x + 1$

32) $1 - 10n - 10$

33) $m - 2m$

34) $1 - r - 6$

35) $9n - 1 + n + 4$

36) $-4b + 9b$

Distribute

37) $-8(x-4)$

38) $3(8v+9)$

39) $8n(n+9)$

40) $-(-5+9a)$

41) $7k(-k+6)$

42) $10x(1+2x)$

43) $-6(1+6x)$

44) $-2(n+1)$

45) $8m(5-m)$

46) $-2p(9p-1)$

47) $-9x(4-x)$

48) $4(8n-2)$

49) $-9b(b-10)$

50) $-4(1+7r)$

51) $-8n(5+10n)$

52) $2x(8x-10)$

Simplify.

53) $9(b+10)+5b$

54) $4v-7(1-8v)$

55) $-3x(1-4x)-4x^2$

56) $-8x+9(-9x+9)$

57) $-4k^2-8k(8k+1)$

58) $-9-10(1+9a)$

59) $1-7(5+7p)$

60) $-10(x-2)-3$

61) $-10-4(n-5)$

62) $-6(5-m)+3m$

63) $4(x+7)+8(x+4)$

64) $-2r(1+4r)+8r(-r+4)$

65) $-8(n+6)-8n(n+8)$

66) $9(6b+5)-4b(b+3)$

67) $7(7+3v)+10(3-10v)$

68) $-7(4x-6)+2(10x-10)$

69) $2n(-10n+5)-7(6-10n)$

70) $-3(4+a)+6a(9a+10)$

71) $5(1-6k)+10(k-8)$

72) $-7(4x+3)-10(10x+10)$

73) $(8n^2-3n)-(5+4n^2)$

74) $(7x^2-3)-(5x^2+6x)$

75) $(5p-6)+(1-p)$

76) $(3x^2-x)-(7-8x)$

77) $(2-4v^2)+(3v^2+2v)$

78) $(2b-8)+(b-7b^2)$

79) $(4-2k^2)+(8-2k^2)$

80) $(7a^2+7a)-(6a^2+4a)$

81) $(x^2-8)+(2x^2-7)$

82) $(3-7n^2)+(6n^2+3)$

Chapter 1 : Solving Linear Equations

Solving Linear Equations - One Step Equations

Objective: Solve one step linear equations by balancing using inverse operations

Solving linear equations is an important and fundamental skill in algebra. In algebra, we are often presented with a problem where the answer is known, but part of the problem is missing. The missing part of the problem is what we seek to find. An example of such a problem is shown below.

Example 41.

$$4x + 16 = -4$$

Notice the above problem has a missing part, or unknown, that is marked by x. If we are given that the solution to this equation is -5, it could be plugged into the equation, replacing the x with -5. This is shown in Example 2.

Example 42.

$$
\begin{aligned}
4(-5) + 16 &= -4 \qquad &&\text{Multiply } 4(-5) \\
-20 + 16 &= -4 \qquad &&\text{Add} -20 + 16 \\
-4 &= -4 \qquad &&\text{True!}
\end{aligned}
$$

Now the equation comes out to a true statement! Notice also that if another number, for example, 3, was plugged in, we would not get a true statement as seen in Example 3.

Example 43.

$$
\begin{aligned}
4(3) + 16 &= -4 \qquad &&\text{Multiply } 4(3) \\
12 + 16 &= -4 \qquad &&\text{Add} 12 + 16 \\
28 &\neq -4 \qquad &&\text{False!}
\end{aligned}
$$

Due to the fact that this is not a true statement, this demonstates that 3 is not the solution. However, depending on the complexity of the problem, this "guess and check" method is not very efficient. Thus, we take a more algebraic approach to solving equations. Here we will focus on what are called "one-step equations" or equations that only require one step to solve. While these equations often seem very fundamental, it is important to master the pattern for solving these problems so we can solve more complex problems.

Addition Problems

To solve equations, the general rule is to do the opposite. For example, consider the following example.

Example 44.

$$\begin{array}{ll} x + 7 = -5 & \text{The 7 is added to the } x \\ \underline{-7 \quad -7} & \text{Subtract 7 from both sides to get rid of it} \\ x = -12 & \text{Our solution!} \end{array}$$

Then we get our solution, $x = -12$. The same process is used in each of the following examples.

Example 45.

$$\begin{array}{ccc} 4 + x = 8 & 7 = x + 9 & 5 = 8 + x \\ \underline{-4 \quad -4} & \underline{-9 \quad -9} & \underline{-8 -8} \\ x = 4 & -2 = x & -3 = x \end{array}$$

Table 1. Addition Examples

Subtraction Problems

In a subtraction problem, we get rid of negative numbers by adding them to both sides of the equation. For example, consider the following example.

Example 46.

$$\begin{array}{ll} x - 5 = 4 & \text{The 5 is negative, or subtracted from } x \\ \underline{+5 +5} & \text{Add 5 to both sides} \\ x = 9 & \text{Our Solution!} \end{array}$$

Then we get our solution $x = 9$. The same process is used in each of the following examples. Notice that each time we are getting rid of a negative number by adding.

Example 47.

$$-6 + x = -2 \qquad\qquad -10 = x - 7 \qquad\qquad 5 = -8 + x$$
$$\underline{+6 \quad\quad +6} \qquad\qquad \underline{+7 \quad +7} \qquad\qquad \underline{+8 \ +8}$$
$$x = 4 \qquad\qquad\qquad -3 = x \qquad\qquad\qquad 13 = x$$

Table 2. Subtraction Examples

Multiplication Problems

With a multiplication problem, we get rid of the number by dividing on both sides. For example consider the following example.

Example 48.

$$4x = 20 \qquad \text{Variable is multiplied by 4}$$
$$\overline{4 \quad 4} \qquad \text{Divide both sides by 4}$$
$$x = 5 \qquad \text{Our solution!}$$

Then we get our solution $x = 5$

With multiplication problems it is very important that care is taken with signs. If x is multiplied by a negative then we will divide by a negative. This is shown in example 9.

Example 49.

$$-5x = 30 \qquad \text{Variable is multiplied by} -5$$
$$\overline{-5 \quad -5} \qquad \text{Divide both sides by} -5$$
$$x = -6 \qquad \text{Our Solution!}$$

The same process is used in each of the following examples. Notice how negative and positive numbers are handled as each problem is solved.

Example 50.

$$\frac{8x}{8} = \frac{-24}{8} \qquad\qquad \frac{-4x}{-4} = \frac{-20}{-4} \qquad\qquad \frac{42}{7} = \frac{7x}{7}$$

$$x = -3 \qquad\qquad\qquad x = 5 \qquad\qquad\qquad 6 = x$$

Table 3. Multiplication Examples

Division Problems:

In division problems, we get rid of the denominator by multiplying on both sides. For example consider our next example.

Example 51.

$$\frac{x}{5} = -3 \qquad \text{Variable is divided by 5}$$

$$(5)\frac{x}{5} = -3(5) \qquad \text{Multiply both sides by 5}$$

$$x = -15 \qquad \text{Our Solution!}$$

Then we get our solution $x = -15$. The same process is used in each of the following examples.

Example 52.

$$\frac{x}{7} = -2 \qquad\qquad \frac{x}{8} = 5 \qquad\qquad \frac{x}{4} = 9$$

$$(-7)\frac{x}{-7} = -2(-7) \qquad (8)\frac{x}{8} = 5(8) \qquad (-4)\frac{x}{-4} = 9(-4)$$

$$x = 14 \qquad\qquad\qquad x = 40 \qquad\qquad\qquad x = -36$$

Table 4. Division Examples

The process described above is fundamental to solving equations. once this process is mastered, the problems we will see have several more steps. These problems may seem more complex, but the process and patterns used will remain the same.

World View Note: The study of algebra originally was called the "Cossic Art" from the Latin, the study of "things" (which we now call variables).

1.1 Practice - One Step Equations

Solve each equation.

1) $v + 9 = 16$

2) $14 = b + 3$

3) $x - 11 = -16$

4) $-14 = x - 18$

5) $30 = a + 20$

6) $-1 + k = 5$

7) $x - 7 = -26$

8) $-13 + p = -19$

9) $13 = n - 5$

10) $22 = 16 + m$

11) $340 = -17x$

12) $4r = -28$

13) $-9 = \frac{n}{12}$

14) $\frac{5}{9} = \frac{b}{9}$

15) $20v = -160$

16) $-20x = -80$

17) $340 = 20n$

18) $\frac{1}{2} = \frac{a}{8}$

19) $16x = 320$

20) $\frac{k}{13} = -16$

21) $-16 + n = -13$

22) $21 = x + 5$

23) $p - 8 = -21$

24) $m - 4 = -13$

25) $180 = 12x$

26) $3n = 24$

27) $20b = -200$

28) $-17 = \frac{x}{12}$

29) $\frac{r}{14} = \frac{5}{14}$

30) $n + 8 = 10$

31) $-7 = a + 4$

32) $v - 16 = -30$

33) $10 = x - 4$

34) $-15 = x - 16$

35) $13a = -143$

36) $-8k = 120$

37) $\frac{p}{20} = -12$

38) $-15 = \frac{x}{9}$

39) $9 + m = -7$

40) $-19 = \frac{n}{20}$

Linear Equations - Two-Step Equations

Objective: Solve two-step equations by balancing and using inverse opperations.

After mastering the technique for solving equations that are simple one-step equations, we are ready to consider two-step equations. As we solve two-step equations, the important thing to remember is that everything works backwards! When working with one-step equations, we learned that in order to clear a "plus five" in the equation, we would subtract five from both sides. We learned that to clear "divided by seven" we multiply by seven on both sides. The same pattern applies to the order of operations. When solving for our variable x, we use order of operations backwards as well. This means we will add or subtract first, then multiply or divide second (then exponents, and finally any parentheses or grouping symbols, but that's another lesson). So to solve the equation in the first example,

Example 53.

$$4x - 20 = -8$$

We have two numbers on the same side as the x. We need to move the 4 and the 20 to the other side. We know to move the four we need to divide, and to move the twenty we will add twenty to both sides. If order of operations is done backwards, we will add or subtract first. Therefore we will add 20 to both sides first. Once we are done with that, we will divide both sides by 4. The steps are shown below.

$$
\begin{array}{ll}
4x - 20 = -8 & \text{Start by focusing on the subtract 20} \\
\underline{+\,20 \quad +\,20} & \text{Add 20 to both sides} \\
4x \qquad = 12 & \text{Now we focus on the 4 multiplied by } x \\
\dfrac{4x}{4} \qquad \dfrac{}{4} & \text{Divide both sides by 4} \\
x = 3 & \text{Our Solution!}
\end{array}
$$

Notice in our next example when we replace the x with 3 we get a true statement.

$$
\begin{array}{ll}
4(3) - 20 = -8 & \text{Multiply 4(3)} \\
12 - 20 = -8 & \text{Subtract } 12 - 20 \\
-8 = -8 & \text{True!}
\end{array}
$$

The same process is used to solve any two-step equations. Add or subtract first, then multiply or divide. Consider our next example and notice how the same process is applied.

Example 54.

$$
\begin{array}{ll}
5x + 7 = 7 & \text{Start by focusing on the plus 7} \\
\underline{-7 \quad -7} & \text{Subtract 7 from both sides} \\
5x \quad = 0 & \text{Now focus on the multiplication by 5} \\
\overline{5} \qquad \overline{5} & \text{Divide both sides by 5} \\
x = 0 & \text{Our Solution!}
\end{array}
$$

Notice the seven subtracted out completely! Many students get stuck on this point, do not forget that we have a number for "nothing left" and that number is zero. With this in mind the process is almost identical to our first example.

A common error students make with two-step equations is with negative signs. Remember the sign always stays with the number. Consider the following example.

Example 55.

$$
\begin{array}{ll}
4 - 2x = 10 & \text{Start by focusing on the positive 4} \\
\underline{-4 \qquad -4} & \text{Subtract 4 from both sides} \\
-2x = 6 & \text{Negative (subtraction) stays on the } 2x \\
\overline{-2} \ \ \overline{-2} & \text{Divide by } -2 \\
x = -3 & \text{Our Solution!}
\end{array}
$$

The same is true even if there is no coefficient in front of the variable. Consider the next example.

Example 56.

$$
\begin{array}{ll}
8 - x = 2 & \text{Start by focusing on the positive 8} \\
\underline{-8 \qquad -8} & \text{Subtract 8 from both sides} \\
-x = -6 & \text{Negative (subtraction) stays on the } x \\
-1x = -6 & \text{Remember, no number in front of variable means 1}
\end{array}
$$

$$\underline{-1 \quad -1} \qquad \text{Divide both sides by} -1$$
$$x = 6 \qquad \text{Our Solution!}$$

Solving two-step equations is a very important skill to master, as we study algebra. The first step is to add or subtract, the second is to multiply or divide. This pattern is seen in each of the following examples.

Example 57.

$$
\begin{array}{ccc}
\begin{array}{c}
-3x + 7 = -8 \\
\underline{-7 \quad -7} \\
-3x = -15 \\
\underline{-3 \quad -3} \\
x = 5
\end{array}
&
\begin{array}{c}
-2 + 9x = 7 \\
\underline{+2 \quad\quad +2} \\
9x = 9 \\
\underline{9 \quad 9} \\
x = 1
\end{array}
&
\begin{array}{c}
8 = 2x + 10 \\
\underline{-10 \quad -10} \\
-2 = 2x \\
\underline{2 \quad 2} \\
-1 = x
\end{array}
\end{array}
$$

$$
\begin{array}{ccc}
\begin{array}{c}
7 - 5x = 17 \\
\underline{-7 \quad\quad -7} \\
-5x = 10 \\
\underline{-5 \quad -5} \\
x = -2
\end{array}
&
\begin{array}{c}
-5 - 3x = -5 \\
\underline{+5 \quad\quad +5} \\
-3x = 0 \\
\underline{-3 \quad -3} \\
x = 0
\end{array}
&
\begin{array}{c}
-3 = \frac{x}{5} - 4 \\
\underline{+4 \quad\quad +4} \\
(5)(1) = \frac{x}{5}(5) \\
5 = x
\end{array}
\end{array}
$$

Table 5. Two-Step Equation Examples

As problems in algebra become more complex the process covered here will remain the same. In fact, as we solve problems like those in the next example, each one of them will have several steps to solve, but the last two steps are a two-step equation like we are solving here. This is why it is very important to master two-step equations now!

Example 58.

$$3x^2 + 4 - x + 6 \qquad \frac{1}{x-8} + \frac{1}{x} = \frac{1}{3} \qquad \sqrt{5x - 5} + 1 = x \qquad \log_5(2x - 4) = 1$$

World View Note: Persian mathematician Omar Khayyam would solve algebraic problems geometrically by intersecting graphs rather than solving them algebraically.

1.2 Practice - Two-Step Problems

Solve each equation.

1) $5 + \frac{n}{4} = 4$

2) $-2 = -2m + 12$

3) $102 = -7r + 4$

4) $27 = 21 - 3x$

5) $-8n + 3 = -77$

6) $-4 - b = 8$

7) $0 = -6v$

8) $-2 + \frac{x}{2} = 4$

9) $-8 = \frac{x}{5} - 6$

10) $-5 = \frac{a}{4} - 1$

11) $0 = -7 + \frac{k}{2}$

12) $-6 = 15 + 3p$

13) $-12 + 3x = 0$

14) $-5m + 2 = 27$

15) $24 = 2n - 8$

16) $-37 = 8 + 3x$

17) $2 = -12 + 2r$

18) $-8 + \frac{n}{12} = -7$

19) $\frac{b}{3} + 7 = 10$

20) $\frac{x}{1} - 8 = -8$

21) $152 = 8n + 64$

22) $-11 = -8 + \frac{v}{2}$

23) $-16 = 8a + 64$

24) $-2x - 3 = -29$

25) $56 + 8k = 64$

26) $-4 - 3n = -16$

27) $-2x + 4 = 22$

28) $67 = 5m - 8$

29) $-20 = 4p + 4$

30) $9 = 8 + \frac{x}{6}$

31) $-5 = 3 + \frac{n}{2}$

32) $\frac{m}{4} - 1 = -2$

33) $\frac{r}{8} - 6 = -5$

34) $-80 = 4x - 28$

35) $-40 = 4n - 32$

36) $33 = 3b + 3$

37) $87 = 3 - 7v$

38) $3x - 3 = -3$

39) $-x + 1 = -11$

40) $4 + \frac{a}{3} = 1$

Solving Linear Equations - General Equations

Objective: Solve general linear equations with variables on both sides.

Often as we are solving linear equations we will need to do some work to set them up into a form we are familiar with solving. This section will focus on manipulating an equation we are asked to solve in such a way that we can use our pattern for solving two-step equations to ultimately arrive at the solution.

One such issue that needs to be addressed is parenthesis. Often the parenthesis can get in the way of solving an otherwise easy problem. As you might expect we can get rid of the unwanted parenthesis by using the distributive property. This is shown in the following example. Notice the first step is distributing, then it is solved like any other two-step equation.

Example 59.

$$
\begin{aligned}
4(2x-6) &= 16 && \text{Distribute 4 through parenthesis} \\
8x-24 &= 16 && \text{Focus on the subtraction first} \\
\underline{+24} &\underline{+24} && \text{Add 24 to both sides} \\
8x &= 40 && \text{Now focus on the multiply by 8} \\
\overline{8} \quad &\overline{8} && \text{Divide both sides by 8} \\
x &= 5 && \text{Our Solution!}
\end{aligned}
$$

Often after we distribute there will be some like terms on one side of the equation. Example 2 shows distributing to clear the parenthesis and then combining like terms next. Notice we only combine like terms on the same side of the equation. Once we have done this, our next example solves just like any other two-step equation.

Example 60.

$$
\begin{aligned}
3(2x-4)+9 &= 15 && \text{Distribute the 3 through the parenthesis} \\
6x-12+9 &= 15 && \text{Combine like terms, } -12+9 \\
6x-3 &= 15 && \text{Focus on the subtraction first} \\
\underline{+3} \quad &\underline{+3} && \text{Add 3 to both sides} \\
6x &= 18 && \text{Now focus on multiply by 6}
\end{aligned}
$$

$$\overline{6} \quad \overline{6} \qquad \text{Divide both sides by 6}$$
$$x = 3 \qquad \text{Our Solution}$$

A second type of problem that becomes a two-step equation after a bit of work is one where we see the variable on both sides. This is shown in the following example.

Example 61.

$$4x - 6 = 2x + 10$$

Notice here the x is on both the left and right sides of the equation. This can make it difficult to decide which side to work with. We fix this by moving one of the terms with x to the other side, much like we moved a constant term. It doesn't matter which term gets moved, $4x$ or $2x$, however, it would be the author's suggestion to move the smaller term (to avoid negative coefficients). For this reason we begin this problem by clearing the positive $2x$ by subtracting $2x$ from both sides.

$$4x - 6 = 2x + 10 \qquad \text{Notice the variable on both sides}$$
$$\underline{-2x \qquad -2x} \qquad \text{Subtract } 2x \text{ from both sides}$$
$$2x - 6 = 10 \qquad \text{Focus on the subtraction first}$$
$$\underline{+6 +6} \qquad \text{Add 6 to both sides}$$
$$2x = 16 \qquad \text{Focus on the multiplication by 2}$$
$$\overline{2} \quad \overline{2} \qquad \text{Divide both sides by 2}$$
$$x = 8 \qquad \text{Our Solution!}$$

The previous example shows the check on this solution. Here the solution is plugged into the x on both the left and right sides before simplifying.

Example 62.

$$4(8) - 6 = 2(8) + 10 \qquad \text{Multiply } 4(8) \text{ and } 2(8) \text{ first}$$
$$32 - 6 = 16 + 10 \qquad \text{Add and Subtract}$$
$$26 = 26 \qquad \text{True!}$$

The next example illustrates the same process with negative coefficients. Notice first the smaller term with the variable is moved to the other side, this time by adding because the coefficient is negative.

Example 63.

$$-3x + 9 = 6x - 27 \qquad \text{Notice the variable on both sides, } -3x \text{ is smaller}$$
$$\underline{+\,3x \qquad +\,3x} \qquad \text{Add } 3x \text{ to both sides}$$
$$9 = 9x - 27 \qquad \text{Focus on the subtraction by 27}$$
$$\underline{+\,27 \qquad +\,27} \qquad \text{Add 27 to both sides}$$
$$36 = 9x \qquad \text{Focus on the mutiplication by 9}$$
$$\overline{\;9\;}\;\;\overline{\;9\;} \qquad \text{Divide both sides by 9}$$
$$4 = x \qquad \text{Our Solution}$$

Linear equations can become particularly intersting when the two processes are combined. In the following problems we have parenthesis and the variable on both sides. Notice in each of the following examples we distribute, then combine like terms, then move the variable to one side of the equation.

Example 64.

$$2(x - 5) + 3x = x + 18 \qquad \text{Distribute the 2 through parenthesis}$$
$$2x - 10 + 3x = x + 18 \qquad \text{Combine like terms } 2x + 3x$$
$$5x - 10 = x + 18 \qquad \text{Notice the variable is on both sides}$$
$$\underline{-\,x \qquad\quad -\,x} \qquad \text{Subtract } x \text{ from both sides}$$
$$4x - 10 = 18 \qquad \text{Focus on the subtraction of 10}$$
$$\underline{+\,10 + 10} \qquad \text{Add 10 to both sides}$$
$$4x = 28 \qquad \text{Focus on multiplication by 4}$$
$$\overline{\;4\;}\;\;\overline{\;4\;} \qquad \text{Divide both sides by 4}$$
$$x = 7 \qquad \text{Our Solution}$$

Sometimes we may have to distribute more than once to clear several parenthesis. Remember to combine like terms after you distribute!

Example 65.

$$3(4x - 5) - 4(2x + 1) = 5 \qquad \text{Distribute 3 and } -4 \text{ through parenthesis}$$
$$12x - 15 - 8x - 4 = 5 \qquad \text{Combine like terms } 12x - 8x \text{ and } -15 - 4$$
$$4x - 19 = 5 \qquad \text{Focus on subtraction of 19}$$
$$\underline{+\,19 + 19} \qquad \text{Add 19 to both sides}$$
$$4x = 24 \qquad \text{Focus on multiplication by 4}$$

$$\overline{4 \quad 4} \quad \text{Divide both sides by 4}$$
$$x = 6 \quad \text{Our Solution}$$

This leads to a 5-step process to solve any linear equation. While all five steps aren't always needed, this can serve as a guide to solving equations.

1. Distribute through any parentheses.

2. Combine like terms on each side of the equation.

3. Get the variables on one side by adding or subtracting

4. Solve the remaining 2-step equation (add or subtract then multiply or divide)

5. Check your answer by plugging it back in for x to find a true statement.

The order of these steps is very important.

World View Note: The Chinese developed a method for solving equations that involved finding each digit one at a time about 2000 years ago!

We can see each of the above five steps worked through our next example.

Example 66.

$$
\begin{array}{ll}
4(2x - 6) + 9 = 3(x - 7) + 8x & \text{Distribute 4 and 3 through parenthesis} \\
8x - 24 + 9 = 3x - 21 + 8x & \text{Combine like terms} -24 + 9 \text{ and } 3x + 8x \\
8x - 15 = 11x - 21 & \text{Notice the variable is on both sides} \\
\underline{-8x \qquad -8x} & \text{Subtract } 8x \text{ from both sides} \\
-15 = 3x - 21 & \text{Focus on subtraction of 21} \\
\underline{+21 \qquad +21} & \text{Add 21 to both sides} \\
6 = 3x & \text{Focus on multiplication by 3} \\
\overline{3 \quad 3} & \text{Divide both sides by 3} \\
2 = x & \text{Our Solution}
\end{array}
$$

Check:

$$
\begin{array}{ll}
4[2(2) - 6] + 9 = 3[(2) - 7] + 8(2) & \text{Plug 2 in for each } x. \text{ Multiply inside parenthesis} \\
4[4 - 6] + 9 = 3[-5] + 8(2) & \text{Finish parentesis on left, multiply on right}
\end{array}
$$

$$4[-2]+9=-15+8(2) \qquad \text{Finish multiplication on both sides}$$
$$-8+9=-15+16 \qquad \text{Add}$$
$$1=1 \qquad \text{True!}$$

When we check our solution of $x=2$ we found a true statement, $1=1$. Therefore, we know our solution $x=2$ is the correct solution for the problem.

There are two special cases that can come up as we are solving these linear equations. The first is illustrated in the next two examples. Notice we start by distributing and moving the variables all to the same side.

Example 67.

$$3(2x-5)=6x-15 \qquad \text{Distribute 3 through parenthesis}$$
$$6x-15=6x-15 \qquad \text{Notice the variable on both sides}$$
$$\underline{-6x \qquad -6x} \qquad \text{Subtract } 6x \text{ from both sides}$$
$$-15=-15 \qquad \text{Variable is gone! True!}$$

Here the variable subtracted out completely! We are left with a true statement, $-15=-15$. If the variables subtract out completely and we are left with a true statement, this indicates that the equation is always true, no matter what x is. Thus, for our solution we say **all real numbers** or $\mathbb{R}$.

Example 68.

$$2(3x-5)-4x=2x+7 \qquad \text{Distribute 2 through parenthesis}$$
$$6x-10-4x=2x+7 \qquad \text{Combine like terms } 6x-4x$$
$$2x-10=2x+7 \qquad \text{Notice the variable is on both sides}$$
$$\underline{-2x \qquad -2x} \qquad \text{Subtract } 2x \text{ from both sides}$$
$$-10 \neq 7 \qquad \text{Variable is gone! False!}$$

Again, the variable subtracted out completely! However, this time we are left with a false statement, this indicates that the equation is never true, no matter what x is. Thus, for our solution we say **no solution** or $\varnothing$.

1.3 Practice - General Linear Equations

Solve each equation.

1) $2 - (-3a - 8) = 1$

2) $2(-3n + 8) = -20$

3) $-5(-4 + 2v) = -50$

4) $2 - 8(-4 + 3x) = 34$

5) $66 = 6(6 + 5x)$

6) $32 = 2 - 5(-4n + 6)$

7) $0 = -8(p - 5)$

8) $-55 = 8 + 7(k - 5)$

9) $-2 + 2(8x - 7) = -16$

10) $-(3 - 5n) = 12$

11) $-21x + 12 = -6 - 3x$

12) $-3n - 27 = -27 - 3n$

13) $-1 - 7m = -8m + 7$

14) $56p - 48 = 6p + 2$

15) $1 - 12r = 29 - 8r$

16) $4 + 3x = -12x + 4$

17) $20 - 7b = -12b + 30$

18) $-16n + 12 = 39 - 7n$

19) $-32 - 24v = 34 - 2v$

20) $17 - 2x = 35 - 8x$

21) $-2 - 5(2 - 4m) = 33 + 5m$

22) $-25 - 7x = 6(2x - 1)$

23) $-4n + 11 = 2(1 - 8n) + 3n$

24) $-7(1 + b) = -5 - 5b$

25) $-6v - 29 = -4v - 5(v + 1)$

26) $-8(8r - 2) = 3r + 16$

27) $2(4x - 4) = -20 - 4x$

28) $-8n - 19 = -2(8n - 3) + 3n$

29) $-a - 5(8a - 1) = 39 - 7a$

30) $-4 + 4k = 4(8k - 8)$

31) $-57 = -(-p + 1) + 2(6 + 8p)$

32) $16 = -5(1 - 6x) + 3(6x + 7)$

33) $-2(m - 2) + 7(m - 8) = -67$

34) $7 = 4(n - 7) + 5(7n + 7)$

35) $50 = 8(7 + 7r) - (4r + 6)$

36) $-8(6 + 6x) + 4(-3 + 6x) = -12$

37) $-8(n - 7) + 3(3n - 3) = 41$

38) $-76 = 5(1 + 3b) + 3(3b - 3)$

39) $-61 = -5(5r - 4) + 4(3r - 4)$

40) $-6(x - 8) - 4(x - 2) = -4$

41) $-2(8n - 4) = 8(1 - n)$

42) $-4(1 + a) = 2a - 8(5 + 3a)$

43) $-3(-7v + 3) + 8v = 5v - 4(1 - 6v)$

44) $-6(x - 3) + 5 = -2 - 5(x - 5)$

45) $-7(x - 2) = -4 - 6(x - 1)$

46) $-(n + 8) + n = -8n + 2(4n - 4)$

47) $-6(8k + 4) = -8(6k + 3) - 2$

48) $-5(x + 7) = 4(-8x - 2)$

49) $-2(1 - 7p) = 8(p - 7)$

50) $8(-8n + 4) = 4(-7n + 8)$

Solving Linear Equations - Fractions

Objective: Solve linear equations with rational coefficients by multiplying by the least common denominator to clear the fractions.

Often when solving linear equations we will need to work with an equation with fraction coefficients. We can solve these problems as we have in the past. This is demonstrated in our next example.

Example 69.

$$\frac{3}{4}x - \frac{7}{2} = \frac{5}{6} \qquad \text{Focus on subtraction}$$

$$\underline{+\frac{7}{2} \quad +\frac{7}{2}} \qquad \text{Add } \frac{7}{2} \text{ to both sides}$$

Notice we will need to get a common denominator to add $\frac{5}{6} + \frac{7}{2}$. Notice we have a common denominator of 6. So we build up the denominator, $\frac{7}{2}\left(\frac{3}{3}\right) = \frac{21}{6}$, and we can now add the fractions:

$$\frac{3}{4}x - \frac{21}{6} = \frac{5}{6} \qquad \text{Same problem, with common denominator 6}$$

$$\underline{+\frac{21}{6} \quad +\frac{21}{6}} \qquad \text{Add } \frac{21}{6} \text{ to both sides}$$

$$\frac{3}{4}x = \frac{26}{6} \qquad \text{Reduce } \frac{26}{6} \text{ to } \frac{13}{3}$$

$$\frac{3}{4}x = \frac{13}{3} \qquad \text{Focus on multiplication by } \frac{3}{4}$$

We can get rid of $\frac{3}{4}$ by dividing both sides by $\frac{3}{4}$. Dividing by a fraction is the same as multiplying by the reciprocal, so we will multiply both sides by $\frac{4}{3}$.

$$\left(\frac{4}{3}\right)\frac{3}{4}x = \frac{13}{3}\left(\frac{4}{3}\right) \qquad \text{Multiply by reciprocal}$$

$$x = \frac{52}{9} \qquad \text{Our solution!}$$

While this process does help us arrive at the correct solution, the fractions can make the process quite difficult. This is why we have an alternate method for dealing with fractions - clearing fractions. Clearing fractions is nice as it gets rid of the fractions for the majority of the problem. We can easily clear the fractions by finding the LCD and multiplying each term by the LCD. This is shown in the next example, the same problem as our first example, but this time we will solve by clearing fractions.

Example 70.

$$\frac{3}{4}x - \frac{7}{2} = \frac{5}{6} \qquad \text{LCD} = 12, \text{multiply each term by 12}$$

$$\frac{(12)3}{4}x - \frac{(12)7}{2} = \frac{(12)5}{6} \qquad \text{Reduce each 12 with denominators}$$

$$
\begin{aligned}
(3)3x - (6)7 &= (2)5 && \text{Multiply out each term} \\
9x - 42 &= 10 && \text{Focus on subtraction by 42} \\
\underline{+\,42 \qquad +\,42} && \text{Add 42 to both sides} \\
9x &= 52 && \text{Focus on multiplication by 9} \\
\overline{9} \quad \overline{9} && \text{Divide both sides by 9} \\
x &= \frac{52}{9} && \text{Our Solution}
\end{aligned}
$$

The next example illustrates this as well. Notice the 2 isn't a fraction in the origional equation, but to solve it we put the 2 over 1 to make it a fraction.

Example 71.

$$\frac{2}{3}x - 2 = \frac{3}{2}x + \frac{1}{6} \qquad \text{LCD} = 6, \text{multiply each term by 6}$$

$$\frac{(6)2}{3}x - \frac{(6)2}{1} = \frac{(6)3}{2}x + \frac{(6)1}{6} \qquad \text{Reduce 6 with each denominator}$$

$$
\begin{aligned}
(2)2x - (6)2 &= (3)3x + (1)1 && \text{Multiply out each term} \\
4x - 12 &= 9x + 1 && \text{Notice variable on both sides} \\
\underline{-\,4x \qquad\quad -\,4x} && \text{Subtract } 4x \text{ from both sides} \\
-12 &= 5x + 1 && \text{Focus on addition of 1} \\
\underline{-\,1 \qquad -\,1} && \text{Subtract 1 from both sides} \\
-13 &= 5x && \text{Focus on multiplication of 5} \\
\overline{5} \quad \overline{5} && \text{Divide both sides by 5} \\
-\frac{13}{5} &= x && \text{Our Solution}
\end{aligned}
$$

We can use this same process if there are parenthesis in the problem. We will first distribute the coefficient in front of the parenthesis, then clear the fractions. This is seen in the following example.

Example 72.

$$\frac{3}{2}\left(\frac{5}{9}x + \frac{4}{27}\right) = 3 \qquad \text{Distribute } \frac{3}{2} \text{ through parenthesis, reducing if possible}$$

$$\frac{5}{6}x + \frac{2}{9} = 3 \qquad \text{LCD} = 18, \text{ multiply each term by } 18$$

$$\frac{(18)5}{6}x + \frac{(18)2}{9} = \frac{(18)3}{9} \qquad \text{Reduce } 18 \text{ with each denominator}$$

$$(3)5x + (2)2 = (18)3 \qquad \text{Multiply out each term}$$

$$15x + 4 = 54 \qquad \text{Focus on addition of } 4$$

$$\underline{-4 \ -4} \qquad \text{Subtract } 4 \text{ from both sides}$$

$$15x = 50 \qquad \text{Focus on multiplication by } 15$$

$$.\ \overline{15} \ \ \overline{15} \qquad \text{Divide both sides by } 15. \text{ Reduce on right side.}$$

$$x = \frac{10}{3} \qquad \text{Our Solution}$$

While the problem can take many different forms, the pattern to clear the fraction is the same, after distributing through any parentheses we multiply each term by the LCD and reduce. This will give us a problem with no fractions that is much easier to solve. The following example again illustrates this process.

Example 73.

$$\frac{3}{4}x - \frac{1}{2} = \frac{1}{3}\left(\frac{3}{4}x + 6\right) - \frac{7}{2} \qquad \text{Distribute } \frac{1}{3}, \text{ reduce if possible}$$

$$\frac{3}{4}x - \frac{1}{2} = \frac{1}{4}x + 2 - \frac{7}{2} \qquad \text{LCD} = 4, \text{ multiply each term by } 4.$$

$$\frac{(4)3}{4}x - \frac{(4)1}{2} = \frac{(4)1}{4}x + \frac{(4)2}{1} - \frac{(4)7}{2} \qquad \text{Reduce } 4 \text{ with each denominator}$$

$$(1)3x - (2)1 = (1)1x + (4)2 - (2)7 \qquad \text{Multiply out each term}$$

$$3x - 2 = x + 8 - 14 \qquad \text{Combine like terms } 8 - 14$$

$$3x - 2 = x - 6 \qquad \text{Notice variable on both sides}$$

$$\underline{-x \qquad -x} \qquad \text{Subtract } x \text{ from both sides}$$

$$2x - 2 = -6 \qquad \text{Focus on subtraction by } 2$$

$$\underline{+2 \ +2} \qquad \text{Add } 2 \text{ to both sides}$$

$$2x = -4 \qquad \text{Focus on multiplication by } 2$$

$$\overline{2} \qquad \overline{2} \qquad \text{Divide both sides by } 2$$

$$x = -2 \qquad \text{Our Solution}$$

World View Note: The Egyptians were among the first to study fractions and linear equations. The most famous mathematical document from Ancient Egypt is the Rhind Papyrus where the unknown variable was called "heap"

1.4 Practice - Fractions

Solve each equation.

1) $\frac{3}{5}(1+p) = \frac{21}{20}$

2) $-\frac{1}{2} = \frac{3}{2}k + \frac{3}{2}$

3) $0 = -\frac{5}{4}(x - \frac{6}{5})$

4) $\frac{3}{2}n - \frac{8}{3} = -\frac{29}{12}$

5) $\frac{3}{4} - \frac{5}{4}m = \frac{113}{24}$

6) $\frac{11}{4} + \frac{3}{4}r = \frac{163}{32}$

7) $\frac{635}{72} = -\frac{5}{2}(-\frac{11}{4} + x)$

8) $-\frac{16}{9} = -\frac{4}{3}(\frac{5}{3} + n)$

9) $2b + \frac{9}{5} = -\frac{11}{5}$

10) $\frac{3}{2} - \frac{7}{4}v = -\frac{9}{8}$

11) $\frac{3}{2}(\frac{7}{3}n + 1) = \frac{3}{2}$

12) $\frac{41}{9} = \frac{5}{2}(x + \frac{2}{3}) - \frac{1}{3}x$

13) $-a - \frac{5}{4}(-\frac{8}{3}a + 1) = -\frac{19}{4}$

14) $\frac{1}{3}(-\frac{7}{4}k + 1) - \frac{10}{3}k = -\frac{13}{8}$

15) $\frac{55}{6} = -\frac{5}{2}(\frac{3}{2}p - \frac{5}{3})$

16) $-\frac{1}{2}(\frac{2}{3}x - \frac{3}{4}) - \frac{7}{2}x = -\frac{83}{24}$

17) $\frac{16}{9} = -\frac{4}{3}(-\frac{4}{3}n - \frac{4}{3})$

18) $\frac{2}{3}(m + \frac{9}{4}) - \frac{10}{3} = -\frac{53}{18}$

19) $-\frac{5}{8} = \frac{5}{4}(r - \frac{3}{2})$

20) $\frac{1}{12} = \frac{4}{3}x + \frac{5}{3}(x - \frac{7}{4})$

21) $-\frac{11}{3} + \frac{3}{2}b = \frac{5}{2}(b - \frac{5}{3})$

22) $\frac{7}{6} - \frac{4}{3}n = -\frac{3}{2}n + 2(n + \frac{3}{2})$

23) $-(-\frac{5}{2}x - \frac{3}{2}) = -\frac{3}{2} + x$

24) $-\frac{149}{16} - \frac{11}{3}r = -\frac{7}{4}r - \frac{5}{4}(-\frac{4}{3}r + 1)$

25) $\frac{45}{16} + \frac{3}{2}n = \frac{7}{4}n - \frac{19}{16}$

26) $-\frac{7}{2}(\frac{5}{3}a + \frac{1}{3}) = \frac{11}{4}a + \frac{25}{8}$

27) $\frac{3}{2}(v + \frac{3}{2}) = -\frac{7}{4}v - \frac{19}{6}$

28) $-\frac{8}{3} - \frac{1}{2}x = -\frac{4}{3}x - \frac{2}{3}(-\frac{13}{4}x + 1)$

29) $\frac{47}{9} + \frac{3}{2}x = \frac{5}{3}(\frac{5}{2}x + 1)$

30) $\frac{1}{3}n + \frac{29}{6} = 2(\frac{4}{3}n + \frac{2}{3})$

Solving Linear Equations - Formulas

Objective: Solve linear formulas for a given variable.

Solving formulas is much like solving general linear equations. The only difference is we will have several varaibles in the problem and we will be attempting to solve for one specific variable. For example, we may have a formula such as $A = \pi r^2 + \pi rs$ (formula for surface area of a right circular cone) and we may be interested in solving for the varaible s. This means we want to isolate the s so the equation has s on one side, and everything else on the other. So a solution might look like $s = \frac{A - \pi r^2}{\pi s}$. This second equation gives the same information as the first, they are algebraically equivalent, however, one is solved for the area, while the other is solved for s (slant height of the cone). In this section we will discuss how we can move from the first equation to the second.

When solving formulas for a variable we need to focus on the one varaible we are trying to solve for, all the others are treated just like numbers. This is shown in the following example. Two parallel problems are shown, the first is a normal one-step equation, the second is a formula that we are solving for x

Example 74.

$$\begin{array}{ll} 3x = 12 & wx = z \quad \text{In both problems, } x \text{ is multiplied by something} \\ \dfrac{3}{3} \quad \dfrac{3}{3} & \dfrac{w}{w} \quad \dfrac{w}{w} \quad \text{To isolate the } x \text{ we divide by 3 or } w. \\ x = 4 & x = \dfrac{z}{w} \quad \text{Our Solution} \end{array}$$

We use the same process to solve $3x = 12$ for x as we use to solve $wx = z$ for x. Because we are solving for x we treat all the other variables the same way we would treat numbers. Thus, to get rid of the multiplication we divided by w. This same idea is seen in the following example.

Example 75.

$$\begin{array}{ll} m + n = p \quad \text{for } n & \text{Solving for } n, \text{ treat all other variables like numbers} \\ \underline{-m \quad -m} & \text{Subtract } m \text{ from both sides} \\ n = p - m & \text{Our Solution} \end{array}$$

As p and m are not like terms, they cannot be combined. For this reason we leave the expression as $p - m$. This same one-step process can be used with grouping symbols.

Example 76.

$$\frac{a(x-y)}{(x-y)} = \frac{b}{(x-y)} \quad \text{for } a \quad \text{Solving for } a, \text{ treat } (x-y) \text{ like a number}$$

Divide both sides by $(x-y)$

$$a = \frac{b}{x-y} \qquad \text{Our Solution}$$

Because $(x-y)$ is in parenthesis, if we are not searching for what is inside the parenthesis, we can keep them together as a group and divide by that group. However, if we are searching for what is inside the parenthesis, we will have to break up the parenthesis by distributing. The following example is the same formula, but this time we will solve for x.

Example 77.

$$a(x-y) = b \quad \text{for } x \quad \text{Solving for } x, \text{ we need to distribute to clear parenthesis}$$

$ax - ay = b \qquad$ This is a two $-$ step equation, ay is subtracted from our x term

$\underline{+ ay + ay} \qquad$ Add ay to both sides

$ax = b + ay \qquad$ The x is multipied by a

$\overline{a} \qquad \overline{a} \qquad$ Divide both sides by a

$$x = \frac{b+ay}{a} \qquad \text{Our Solution}$$

Be very careful as we isolate x that we do not try and cancel the a on top and bottom of the fraction. This is not allowed if there is any adding or subtracting in the fraction. There is no reducing possible in this problem, so our final reduced answer remains $x = \frac{b+ay}{a}$. The next example is another two-step problem

Example 78.

$$y = mx + b \quad \text{for } m \quad \text{Solving for } m, \text{ focus on addition first}$$

$\underline{-b \qquad -b} \qquad$ Subtract b from both sides

$y - b = mx \qquad$ m is multipied by x.

$\overline{x} \qquad \overline{x} \qquad$ Divide both sides by x

$$\frac{y-b}{x} = m \qquad \text{Our Solution}$$

It is important to note that we know we are done with the problem when the variable we are solving for is isolated or alone on one side of the equation and it does not appear anywhere on the other side of the equation.

The next example is also a two-step equation, it is the problem we started with at the beginning of the lesson.

Example 79.

$$A = \pi r^2 + \pi r s \quad \text{for } s \quad \text{Solving for } s, \text{focus on what is added to the term with } s$$
$$\underline{-\pi r^2 \quad -\pi r^2} \qquad\qquad \text{Subtract } \pi r^2 \text{ from both sides}$$
$$A - \pi r^2 = \pi r s \qquad\quad s \text{ is multipied by } \pi r$$
$$\overline{\quad \pi r \qquad \pi r \quad} \qquad\quad \text{Divide both sides by } \pi r$$
$$\frac{A - \pi r^2}{\pi r} = s \qquad\qquad \text{Our Solution}$$

Again, we cannot reduce the πr in the numerator and denominator because of the subtraction in the problem.

Formulas often have fractions in them and can be solved in much the same way we solved with fractions before. First identify the LCD and then multiply each term by the LCD. After we reduce there will be no more fractions in the problem so we can solve like any general equation from there.

Example 80.

$$h = \frac{2m}{n} \quad \text{for } m \quad \text{To clear the fraction we use LCD} = n$$

$$(n)h = \frac{(n)2m}{n} \qquad\qquad \text{Multiply each term by } n$$

$$nh = 2m \qquad\qquad \text{Reduce } n \text{ with denominators}$$
$$\overline{\quad 2 \qquad 2 \quad} \qquad\qquad \text{Divide both sides by 2}$$
$$\frac{nh}{2} = m \qquad\qquad \text{Our Solution}$$

The same pattern can be seen when we have several fractions in our problem.

Example 81.

$$\frac{a}{b} + \frac{c}{b} = e \quad \text{for } a \quad \text{To clear the fraction we use LCD} = b$$
$$\frac{(b)a}{b} + \frac{(b)c}{b} = e\,(b) \qquad\qquad \text{Multiply each term by } b$$
$$a + c = eb \qquad\qquad \text{Reduce } b \text{ with denominators}$$
$$\underline{\quad -c \quad -c} \qquad\qquad \text{Subtract } c \text{ from both sides}$$
$$a = eb - c \qquad\qquad \text{Our Solution}$$

Depending on the context of the problem we may find a formula that uses the same letter, one capital, one lowercase. These represent different values and we must be careful not to combine a capital variable with a lower case variable.

Example 82.

$$a = \frac{A}{2-b} \quad \text{for } b \quad \text{Use LCD } (2-b) \text{ as a group}$$

$$(2 - b)a = \frac{(2-b)A}{2-b} \qquad \text{Multiply each term by } (2-b)$$

$$(2-b)a = A \qquad \text{reduce } (2-b) \text{ with denominator}$$
$$2a - ab = A \qquad \text{Distribute through parenthesis}$$
$$\underline{-2a \qquad -2a} \qquad \text{Subtract } 2a \text{ from both sides}$$
$$-ab = A - 2a \qquad \text{The } b \text{ is multipied by } -a$$
$$\underline{-a \qquad -a} \qquad \text{Divide both sides by } -a$$
$$b = \frac{A - 2a}{-a} \qquad \text{Our Solution}$$

Notice the A and a were not combined as like terms. This is because a formula will often use a capital letter and lower case letter to represent different variables. Often with formulas there is more than one way to solve for a variable. The next example solves the same problem in a slightly different manner. After clearing the denominator, we divide by a to move it to the other side, rather than distributing.

Example 83.

$$a = \frac{A}{2-b} \quad \text{for } b \quad \text{Use LCD} = (2-b) \text{ as a group}$$

$$(2-b)a = \frac{(2-b)A}{2-b} \qquad \text{Multiply each term by } (2-b)$$
$$(2-b)a = A \qquad \text{Reduce } (2-b) \text{ with denominator}$$
$$\underline{a \qquad a} \qquad \text{Divide both sides by } a$$
$$2 - b = \frac{A}{a} \qquad \text{Focus on the positive 2}$$
$$\underline{-2 \qquad -2} \qquad \text{Subtract 2 from both sides}$$
$$-b = \frac{A}{a} - 2 \qquad \text{Still need to clear the negative}$$
$$(-1)(-b) = (-1)\frac{A}{a} - 2(-1) \qquad \text{Multiply (or divide) each term by } -1$$
$$b = -\frac{A}{a} + 2 \qquad \text{Our Solution}$$

Both answers to the last two examples are correct, they are just written in a different form because we solved them in different ways. This is very common with formulas, there may be more than one way to solve for a varaible, yet both are equivalent and correct.

World View Note: The father of algebra, Persian mathematician Muhammad ibn Musa Khwarizmi, introduced the fundamental idea of blancing by subtracting the same term to the other side of the equation. He called this process al-jabr which later became the world algebra.

1.5 Practice - Formulas

Solve each of the following equations for the indicated variable.

1) $ab = c$ for b

2) $g = \frac{h}{i}$ for h

3) $\frac{f}{g}x = b$ for x

4) $p = \frac{3y}{q}$ for y

5) $3x = \frac{a}{b}$ for x

6) $\frac{ym}{b} = \frac{c}{d}$ for y

7) $E = mc^2$ for m

8) $DS = ds$ for D

9) $V = \frac{4}{3}\pi r^3$ for π

10) $E = \frac{mv^2}{2}$ for m

11) $a + c = b$ for c

12) $x - f = g$ for x

13) $c = \frac{4y}{m+n}$ for y

14) $\frac{rs}{a} \ 3 = k$ for r

15) $V = \frac{\pi Dn}{12}$ for D

16) $F = k(R - L)$ for k

17) $P = n(p - c)$ for n

18) $S = L + 2B$ for L

19) $T = \frac{D-d}{L}$ for D

20) $I = \frac{E_a \ E_q}{R}$ for E_a

21) $L = L_o(1 + at)$ for L_o

22) $ax + b = c$ for x

23) $2m + p = 4m + q$ for m

24) $q = 6(L - p)$ for L

25) $\frac{k-m}{r} = q$ for k

26) $R = aT + b$ for T

27) $h = vt - 16t^2$ for v

28) $S = \pi rh + \pi r^2$ for h

29) $Q_1 = P(Q_2 - Q_1)$ for Q_2

30) $L = \pi(r_1 + r_2) + 2d$ for r_1

31) $R = \frac{kA(T_1 + T_2)}{d}$ for T_1

32) $P = \frac{V_1(V_2 - V_1)}{g}$ for V_2

33) $ax + b = c$ for a

34) $rt = d$ for r

35) $lwh = V$ for w

36) $V = \frac{\pi r^2 h}{3}$ for h

37) $\frac{1}{a} + b = \frac{c}{a}$ for a

38) $\frac{1}{a} + b = \frac{c}{a}$ for b

39) $at - bw = s$ for t

40) $at - bw = s$ for w

41) $ax + bx = c$ for a

42) $x + 5y = 3$ for x

43) $x + 5y = 3$ for y

44) $3x + 2y = 7$ for x

45) $3x + 2y = 7$ for y

46) $5a - 7b = 4$ for a

47) $5a - 7b = 4$ for b

48) $4x - 5y = 8$ for x

49) $4x - 5y = 8$ for y

50) $C = \frac{5}{9}(F - 32)$ for F

Solving Linear Equations - Absolute Value

Objective: Solve linear absolute value equations.

When solving equations with absolute value we can end up with more than one possible answer. This is because what is in the absolute value can be either negative or positive and we must account for both possibilities when solving equations. This is illustrated in the following example.

Example 84.

$$|x| = 7 \quad \text{Absolute value can be positive or negative}$$
$$x = 7 \text{ or } x = -7 \quad \text{Our Solution}$$

Notice that we have considered two possibilities, both the positive and negative. Either way, the absolute value of our number will be positive 7.

World View Note: The first set of rules for working with negatives came from 7th century India. However, in 1758, almost a thousand years later, British mathematician Francis Maseres claimed that negatives "Darken the very whole doctrines of the equations and make dark of the things which are in their nature excessively obvious and simple."

When we have absolute values in our problem it is important to first isolate the absolute value, then remove the absolute value by considering both the positive and negative solutions. Notice in the next two examples, all the numbers outside of the absolute value are moved to the other side first before we remove the absolute value bars and consider both positive and negative solutions.

Example 85.

$$5 + |x| = 8 \quad \text{Notice absolute value is not alone}$$
$$\underline{-5 \qquad -5} \quad \text{Subtract 5 from both sides}$$
$$|x| = 3 \quad \text{Absolute value can be positive or negative}$$
$$x = 3 \text{ or } x = -3 \quad \text{Our Solution}$$

Example 86.

$$-4|x| = -20 \quad \text{Notice absolute value is not alone}$$
$$\overline{-4} \quad \overline{-4} \quad \text{Divide both sides by} -4$$

$$|x| = 5 \qquad \text{Absolute value can be positive or negative}$$
$$x = 5 \text{ or } x = -5 \qquad \text{Our Solution}$$

Notice we never combine what is inside the absolute value with what is outside the absolute value. This is very important as it will often change the final result to an incorrect solution. The next example requires two steps to isolate the absolute value. The idea is the same as a two-step equation, add or subtract, then multiply or divide.

Example 87.

$$5|x| - 4 = 26 \qquad \text{Notice the absolute value is not alone}$$
$$\underline{+4 + 4} \qquad \text{Add 4 to both sides}$$
$$5|x| = 30 \qquad \text{Absolute value still not alone}$$
$$\overline{5 \quad 5} \qquad \text{Divide both sides by 5}$$
$$|x| = 6 \qquad \text{Absolute value can be positive or negative}$$
$$x = 6 \text{ or } x = -6 \qquad \text{Our Solution}$$

Again we see the same process, get the absolute value alone first, then consider the positive and negative solutions. Often the absolute value will have more than just a variable in it. In this case we will have to solve the resulting equations when we consider the positive and negative possibilities. This is shown in the next example.

Example 88.

$$|2x - 1| = 7 \qquad \text{Absolute value can be positive or negative}$$
$$2x - 1 = 7 \text{ or } 2x - 1 = -7 \qquad \text{Two equations to solve}$$

Now notice we have two equations to solve, each equation will give us a different solution. Both equations solve like any other two-step equation.

$$
\begin{array}{ccc}
2x - 1 = 7 & & 2x - 1 = -7 \\
\underline{+1 + 1} & & \underline{+1 +1} \\
2x = 8 & \text{or} & 2x = -6 \\
\overline{2 \quad 2} & & \overline{2 \quad 2} \\
x = 4 & & x = -3
\end{array}
$$

Thus, from our previous example we have two solutions, $x = 4$ or $x = -3$.

Again, it is important to remember that the absolute value must be alone first before we consider the positive and negative possibilities. This is illustrated in below.

Example 89.

$$2 - 4|2x + 3| = -18$$

To get the absolute value alone we first need to get rid of the 2 by subtracting, then divide by -4. Notice we cannot combine the 2 and -4 becuase they are not like terms, the -4 has the absolute value connected to it. Also notice we do not distribute the -4 into the absolute value. This is because the numbers outside cannot be combined with the numbers inside the absolute value. Thus we get the absolute value alone in the following way:

$2 - 4\|2x + 3\| = -18$	Notice absolute value is not alone
$\underline{-2 \qquad\qquad\quad -2}$	Subtract 2 from both sides
$-4\|2x + 3\| = -20$	Absolute value still not alone
$\overline{-4} \qquad\qquad \overline{-4}$	Divide both sides by -4
$\|2x + 3\| = 5$	Absoloute value can be positive or negative
$2x + 3 = 5$ or $2x + 3 = -5$	Two equations to solve

Now we just solve these two remaining equations to find our solutions.

$$
\begin{array}{ccc}
2x + 3 = 5 & & 2x + 3 = -5 \\
\underline{-3 \quad -3} & & \underline{-3 \quad -3} \\
2x = 2 & \text{or} & 2x = -8 \\
\overline{2} \quad \overline{2} & & \overline{2} \quad \overline{2} \\
x = 1 & & x = -4
\end{array}
$$

We now have our two solutions, $x = 1$ and $x = -4$.

As we are solving absolute value equations it is important to be aware of special cases. Remember the result of an absolute value must always be positive. Notice what happens in the next example.

Example 90.

$$7 + |2x - 5| = 4 \qquad \text{Notice absolute value is not alone}$$
$$\underline{-7 \qquad\qquad -7} \qquad \text{Subtract 7 from both sides}$$
$$|2x - 5| = -3 \qquad \text{Result of absolute value is negative!}$$

Notice the absolute value equals a negative number! This is impossible with absolute value. When this occurs we say there is **no solution** or $\varnothing$.

One other type of absolute value problem is when two absolute values are equal to eachother. We still will consider both the positive and negative result, the difference here will be that we will have to distribute a negative into the second absolute value for the negative possibility.

Example 91.

$$|2x - 7| = |4x + 6| \qquad\qquad \text{Absolute value can be positive or negative}$$
$$2x - 7 = 4x + 6 \;\; \text{or} \;\; 2x - 7 = -(4x + 6) \qquad \text{make second part of second equation negative}$$

Notice the first equation is the positive possibility and has no significant difference other than the missing absolute value bars. The second equation considers the negative possibility. For this reason we have a negative in front of the expression which will be distributed through the equation on the first step of solving. So we solve both these equations as follows:

$$
\begin{array}{ccc}
\begin{aligned}
2x - 7 &= 4x + 6 \\
\underline{-2x \qquad -2x} & \\
-7 &= 2x + 6 \\
\underline{-6 \qquad -6} & \\
-13 &= 2x \\
\overline{\;2\;} & \overline{\;2\;} \\
\dfrac{-13}{2} &= x
\end{aligned}
& \text{or} &
\begin{aligned}
2x - 7 &= -(4x + 6) \\
2x - 7 &= -4x - 6 \\
\underline{+4x \qquad +4x} & \\
6x - 7 &= -6 \\
\underline{+7 \qquad +7} & \\
6x &= 1 \\
\overline{\;6\;} & \overline{\;6\;} \\
x &= \dfrac{1}{6}
\end{aligned}
\end{array}
$$

This gives us our two solutions, $x = \dfrac{13}{2}$ or $x = \dfrac{1}{6}$.

1.6 Practice - Absolute Value Equations

Solve each equation.

1) $|x| = 8$

2) $|n| = 7$

3) $|b| = 1$

4) $|x| = 2$

5) $|5 + 8a| = 53$

6) $|9n + 8| = 46$

7) $|3k + 8| = 2$

8) $|3 - x| = 6$

9) $|9 + 7x| = 30$

10) $|5n + 7| = 23$

11) $|8 + 6m| = 50$

12) $|9p + 6| = 3$

13) $|6 - 2x| = 24$

14) $|3n - 2| = 7$

15) $-7|-3 - 3r| = -21$

16) $|2 + 2b| + 1 = 3$

17) $7|-7x - 3| = 21$

18) $\frac{4 \quad 3n|}{4} = 2$

19) $\frac{|-4b - 10|}{8} = 3$

20) $8|5p + 8| - 5 = 11$

21) $8|x + 7| - 3 = 5$

22) $3 - |6n + 7| = -40$

23) $5|3 + 7m| + 1 = 51$

24) $4|r + 7| + 3 = 59$

25) $3 + 5|8 - 2x| = 63$

26) $5 + 8|-10n - 2| = 101$

27) $|6b - 2| + 10 = 44$

28) $7|10v - 2| - 9 = 5$

29) $-7 + 8|-7x - 3| = 73$

30) $8|3 - 3n| - 5 = 91$

31) $|5x + 3| = |2x - 1|$

32) $|2 + 3x| = |4 - 2x|$

33) $|3x - 4| = |2x + 3|$

34) $\left|\frac{2x - 5}{3}\right| = \left|\frac{3x + 4}{2}\right|$

35) $\left|\frac{4x \quad 2}{5}\right| = \left|\frac{6x + 3}{2}\right|$

36) $\left|\frac{3x + 2}{2}\right| = \left|\frac{2x - 3}{3}\right|$

Solving Linear Equations - Variation

Objective: Solve variation problems by creating variation equations and finding the variation constant.

One application of solving linear equations is variation. Often different events are related by what is called the constant of variation. For example, the time it takes to travel a certain distance is related to how fast you are traveling. The faster you travel, the less time it take to get there. This is one type of variation problem, we will look at three types of variation here. Variation problems have two or three variables and a constant in them. The constant, usually noted with a k, describes the relationship and does not change as the other variables in the problem change. There are two ways to set up a variation problem, the first solves for one of the variables, a second method is to solve for the constant. Here we will use the second method.

The greek letter pi (π) is used to represent the ratio of the circumference of a circle to its diameter.

World View Note: In the 5th centure, Chinese mathematician Zu Chongzhi calculated the value of π to seven decimal places (3.1415926). This was the most accurate value of π for the next 1000 years!

If you take any circle and divide the circumference of the circle by the diameter you will always get the same value, about 3.14159... If you have a bigger circumference you will also have a bigger diameter. This relationship is called **direct variation** or **directly proportional**. If we see this phrase in the problem we know to divide to find the constant of variation.

Example 92.

$$m \text{ is varies directly as } n \qquad \textit{"Directly" tells us to divide}$$
$$\frac{m}{n} = k \qquad \textit{Our formula for the relationship}$$

In kickboxing, one will find that the longer the board, the easier it is to break. If you multiply the force required to break a board by the length of the board you will also get a constant. Here, we are multiplying the variables, which means as one variable increases, the other variable decreases. This relationship is called **indirect variation** or **inversly proportional**. If we see this phrase in the problem we know to multiply to find the constant of variation.

Example 93.

$$y \text{ is inversely proportional to } z \qquad \textit{"Inversely" tells us to multiply}$$
$$yz = k \qquad \textit{Our formula for the relationship}$$

The formula for the area of a triangle has three variables in it. If we divide the area by the base times the height we will also get a constant, $\frac{1}{2}$. This relationship is called **joint variation** or **jointly proportional**. If we see this phrase in the problem we know to divide the first variable by the product of the other two to find the constant of variation.

Example 94.

$$A \text{ varies jointly as } x \text{ and } y \qquad \text{"Jointly" tells us to divide by the product}$$
$$\frac{A}{xy} = k \qquad \text{Our formula for the relationship}$$

Once we have our formula for the relationship in a variation problem, we use given or known information to calculate the constant of variation. This is shown for each type of variation in the next three examples.

Example 95.

$$w \text{ is directly proportional to } y \text{ and } w = 50 \text{ when } y = 5$$

$$\frac{w}{y} = k \qquad \text{"directly" tells us to divide}$$
$$\frac{(50)}{(5)} = k \qquad \text{Substitute known values}$$
$$10 = k \qquad \text{Evaluate to find our constant}$$

Example 96.

$$c \text{ varies indirectly as } d \text{ and } c = 4.5 \text{ when } d = 6$$

$$cd = k \qquad \text{"indirectly" tells us to multiply}$$
$$(4.5)(6) = k \qquad \text{Substitute known values}$$
$$27 = k \qquad \text{Evaluate to find our constant}$$

Example 97.

$$x \text{ is jointly proportional to } y \text{ and } z \text{ and } x = 48 \text{ when } y = 2 \text{ and } z = 4$$

$$\frac{x}{yz} = k \qquad \text{"Jointly" tells us to divide by the product}$$
$$\frac{(48)}{(2)(4)} = k \qquad \text{Substitute known values}$$
$$6 = k \qquad \text{Evaluate to find our constant}$$

Once we have found the constant of variation we can use it to find other combinations in the same relationship. Each of these problems we solve will have three important steps, none of which should be skipped.

1. Find the formula for the relationship using the type of variation

2. Find the constant of variation using known values

3. Answer the question using the constant of variation

The next three examples show how this process is worked out for each type of variation.

Example 98.

The price of an item varies directly with the sales tax. If a $25 item has a sales tax of $2, what will the tax be on a $40 item?

$$\frac{p}{t} = k \qquad \text{"Directly" tells us to divide price } (p) \text{ and tax } (t)$$

$$\frac{(25)}{(2)} = k \qquad \text{Substitute known values for price and tax}$$

$$12.5 = k \qquad \text{Evaluate to find our constant}$$

$$\frac{40}{t} = 12.5 \qquad \text{Using our constant, substitute 40 for price to find the tax}$$

$$\frac{(t)40}{t} = 12.5(t) \qquad \text{Multiply by LCD} = t \text{ to clear fraction}$$

$$40 = 12.5t \qquad \text{Reduce the } t \text{ with the denominator}$$

$$\overline{12.5} \ \overline{12.5} \qquad \text{Divide by } 12.5$$

$$3.2 = t \qquad \text{Our solution: Tax is } \$3.20$$

Example 99.

The speed (or rate) Josiah travels to work is inversely proportional to time it takes to get there. If he travels 35 miles per hour it will take him 2.5 hours to get to work. How long will it take him if he travels 55 miles per hour?

$$rt = k \qquad \text{"Inversely" tells us to multiply the rate and time}$$

$$(35)(2.5) = k \qquad \text{Substitute known values for rate and time}$$

$$87.5 = k \qquad \text{Evaluate to find our constant}$$

$$55t = 87.5 \qquad \text{Using our constant, substitute 55 for rate to find the time}$$

$$\overline{55} \ \overline{55} \qquad \text{Divide both sides by 55}$$

$$t \approx 1.59 \qquad \text{Our solution: It takes him 1.59 hours to get to work}$$

Example 100.

The amount of simple interest earned on an investment varies jointly as the principle (amount invested) and the time it is invested. In an account, $150 invested for 2 years earned $12 in interest. How much interest would be earned on a $220 investment for 3 years?

$$\frac{I}{Pt} = k \qquad \text{"Jointly" divide Interest } (I) \text{ by product of Principle } (P) \text{ \& time } (t)$$

$$\frac{(12)}{(150)(2)} = k \qquad \text{Substitute known values for Interest, Principle and time}$$

$$0.04 = k \qquad \text{Evaluate to find our constant}$$

$$\frac{I}{(220)(3)} = 0.04 \qquad \text{Using constant, substitute 220 for principle and 3 for time}$$

$$\frac{I}{660} = 0.04 \qquad \text{Evaluate denominator}$$

$$\frac{(660)I}{660} = 0.04(660) \qquad \text{Multiply by 660 to isolate the variable}$$

$$I = 26.4 \qquad \text{Our Solution: The investment earned \$26.40 in interest}$$

Sometimes a variation problem will ask us to do something to a variable as we set up the formula for the relationship. For example, π can be thought of as the ratio of the area and the radius squared. This is still direct variation, we say the area varies directly as the radius square and thus our variable is squared in our formula. This is shown in the next example.

Example 101.

The area of a circle is directly proportional to the square of the radius. A circle with a radius of 10 has an area of 314. What will the area be on a circle of radius 4?

$$\frac{A}{r^2} = k \qquad \text{"Direct" tells us to divide, be sure we use } r^2 \text{ for the denominator}$$

$$\frac{(314)}{(10)^2} = k \qquad \text{Substitute known values into our formula}$$

$$\frac{(314)}{100} = k \qquad \text{Exponents first}$$

$$3.14 = k \qquad \text{Divide to find our constant}$$

$$\frac{A}{(4)^2} = 3.14 \qquad \text{Using the constant, use 4 for } r, \text{ don't forget the squared!}$$

$$\frac{A}{16} = 3.14 \qquad \text{Evaluate the exponent}$$

$$\frac{(16)A}{16} = 3.14(16) \qquad \text{Multiply both sides by 16}$$

$$A = 50.24 \qquad \text{Our Solution: Area is 50.24}$$

When solving variation problems it is important to take the time to clearly state the variation formula, find the constant, and solve the final equation.

1.7 Practice - Variation

Write the formula that expresses the relationship described

1. c varies directly as a

2. x is jointly proportional to y and z

3. w varies inversely as x

4. r varies directly as the square of s

5. f varies jointly as x and y

6. j is inversely proportional to the cube of m

7. h is directly proportional to b

8. x is jointly proportional with the square of a and the square root of b

9. a is inversely proportional to b

Find the constant of variation and write the formula to express the relationship using that constant

10. a varies directly as b and $a = 15$ when $b = 5$

11. p is jointly proportional to q and r and $p = 12$ when $q = 8$ and $r = 3$

12. c varies inversely as d and $c = 7$ when $d = 4$

13. t varies directly as the square of u and $t = 6$ when $u = 3$

14. e varies jointly as f and g and $e = 24$ when $f = 3$ and $g = 2$

15. w is inversely proportional to the cube of x and w is 54 when $x = 3$

16. h is directly proportional to j and $h = 12$ when $j = 8$

17. a is jointly proportional with the square of x and the square root of y and $a = 25$ when $x = 5$ and $y = 9$

18. m is inversely proportional to n and $m = 1.8$ when $n = 2.1$

Solve each of the following variation problems by setting up a formula to express the relationship, finding the constant, and then answering

the question.

19. The electrical current, in amperes, in a circuit varies directly as the voltage. When 15 volts are applied, the current is 5 amperes. What is the current when 18 volts are applied?

20. The current in an electrical conductor varies inversely as the resistance of the conductor. If the current is 12 ampere when the resistance is 240 ohms, what is the current when the resistance is 540 ohms?

21. Hooke's law states that the distance that a spring is stretched by hanging object varies directly as the mass of the object. If the distance is 20 cm when the mass is 3 kg, what is the distance when the mass is 5 kg?

22. The volume of a gas varies inversely as the pressure upon it. The volume of a gas is 200 cm^3 under a pressure of 32 kg/cm^2. What will be its volume under a pressure of 40 kg/cm^2?

23. The number of aluminum cans used each year varies directly as the number of people using the cans. If 250 people use 60,000 cans in one year, how many cans are used each year in Dallas, which has a population of 1,008,000?

24. The time required to do a job varies inversely as the number of peopel working. It takes 5hr for 7 bricklayers to build a park well. How long will it take 10 bricklayers to complete the job?

25. According to Fidelity Investment Vision Magazine, the average weekly allowance of children varies directly as their grade level. In a recent year, the average allowance of a 9th-grade student was 9.66 dollars per week. What was the average weekly allowance of a 4th-grade student?

26. The wavelength of a radio wave varies inversely as its frequency. A wave with a frequency of 1200 kilohertz has a length of 300 meters. What is the length of a wave with a frequency of 800 kilohertz?

27. The number of kilograms of water in a human body varies directly as the mass of the body. A 96-kg person contains 64 kg of water. How many kilo grams of water are in a 60-kg person?

28. The time required to drive a fixed distance varies inversely as the speed. It takes 5 hr at a speed of 80 km/h to drive a fixed distance. How long will it take to drive the same distance at a speed of 70 km/h?

29. The weight of an object on Mars varies directly as its weight on Earth. A person weighs 95lb on Earth weighs 38 lb on Mars. How much would a 100-lb person weigh on Mars?

30. At a constant temperature, the volume of a gas varies inversely as the pres-

sure. If the pressure of a certain gas is 40 newtons per square meter when the volume is 600 cubic meters what will the pressure be when the volume is reduced by 240 cubic meters?

31. The time required to empty a tank varies inversely as the rate of pumping. If a pump can empty a tank in 45 min at the rate of 600 kL/min, how long will it take the pump to empty the same tank at the rate of 1000 kL/min?

32. The weight of an object varies inversely as the square of the distance from the center of the earth. At sea level (6400 km from the center of the earth), an astronaut weighs 100 lb. How far above the earth must the astronaut be in order to weigh 64 lb?

33. The stopping distance of a car after the brakes have been applied varies directly as the square of the speed r. If a car, traveling 60 mph can stop in 200 ft, how fast can a car go and still stop in 72 ft?

34. The drag force on a boat varies jointly as the wetted surface area and the square of the velocity of a boat. If a boat going 6.5 mph experiences a drag force of 86 N when the wetted surface area is 41.2 ft^2, how fast must a boat with 28.5 ft^2 of wetted surface area go in order to experience a drag force of 94N?

35. The intensity of a light from a light bulb varies inversely as the square of the distance from the bulb. Suppose intensity is 90 W/m^2 (watts per square meter) when the distance is 5 m. How much further would it be to a point where the intesity is 40 W/m^2?

36. The volume of a cone varies jointly as its height, and the square of its radius. If a cone with a height of 8 centimeters and a radius of 2 centimeters has a volume of 33.5 cm^3, what is the volume of a cone with a height of 6 centimeters and a radius of 4 centimeters?

37. The intensity of a television signal varies inversely as the square of the distance from the transmitter. If the intensity is 25 W/m^2 at a distance of 2 km, how far from the trasmitter are you when the intensity is 2.56 W/m^2?

38. The intensity of illumination falling on a surface from a given source of light is inversely proportional to the square of the distance from the source of light. The unit for measuring the intesity of illumination is usually the footcandle. If a given source of light gives an illumination of 1 foot-candle at a distance of 10 feet, what would the illumination be from the same source at a distance of 20 feet?

Linear Equations - Number and Geometry

Objective: Solve number and geometry problems by creating and solving a linear equation.

Word problems can be tricky. Often it takes a bit of practice to convert the English sentence into a mathematical sentence. This is what we will focus on here with some basic number problems, geometry problems, and parts problems.

A few important phrases are described below that can give us clues for how to set up a problem.

- **A number** (or unknown, a value, etc) often becomes our variable

- **Is** (or other forms of is: was, will be, are, etc) often represents equals (=)

  x is 5 becomes $x = 5$

- **More than** often represents addition and is usually built backwards, writing the second part plus the first

  Three more than a number becomes $x + 3$

- **Less than** often represents subtraction and is usually built backwards as well, writing the second part minus the first

  Four less than a number becomes $x - 4$

Using these key phrases we can take a number problem and set up and equation and solve.

Example 102.

If 28 less than five times a certain number is 232. What is the number?

$$
\begin{array}{ll}
5x - 28 & \text{Subtraction is built backwards, multiply the unknown by 5} \\
5x - 28 = 232 & \text{Is translates to equals} \\
\underline{+\,28 \quad +\,28} & \text{Add 28 to both sides} \\
5x = 260 & \text{The variable is multiplied by 5} \\
\overline{\;5\quad\;\;5\;} & \text{Divide both sides by 5} \\
x = 52 & \text{The number is 52.}
\end{array}
$$

This same idea can be extended to a more involved problem as shown in the next example.

Example 103.

Fifteen more than three times a number is the same as ten less than six times the number. What is the number

$$3x + 15 \qquad \text{First, addition is built backwards}$$
$$6x - 10 \qquad \text{Then, subtraction is also built backwards}$$
$$3x + 15 = 6x - 10 \qquad \text{Is between the parts tells us they must be equal}$$
$$\underline{-3x \qquad\quad -3x} \qquad \text{Subtract } 3x \text{ so variable is all on one side}$$
$$15 = 3x - 10 \qquad \text{Now we have } a \text{ two} - \text{step equation}$$
$$\underline{+10 \qquad +10} \qquad \text{Add 10 to both sides}$$
$$25 = 3x \qquad \text{The variable is multiplied by 3}$$
$$\overline{3} \;\; \overline{3} \qquad \text{Divide both sides by 3}$$
$$\frac{25}{3} = x \qquad \text{Our number is } \frac{25}{3}$$

Another type of number problem involves consecutive numbers. **Consecutive numbers** are numbers that come one after the other, such as 3, 4, 5. If we are looking for several consecutive numbers it is important to first identify what they look like with variables before we set up the equation. This is shown in the following example.

Example 104.

The sum of three consecutive integers is 93. What are the integers?

$$\text{First } x \qquad \text{Make the first number } x$$
$$\text{Second } x + 1 \qquad \text{To get the next number we go up one or } +1$$
$$\text{Third } x + 2 \qquad \text{Add another 1 (2 total) to get the third}$$
$$F + S + T = 93 \qquad \text{First } (F) \text{ plus Second } (S) \text{ plus Third } (T) \text{ equals 93}$$
$$(x) + (x + 1) + (x + 2) = 93 \qquad \text{Replace } F \text{ with } x, S \text{ with } x + 1, \text{ and } T \text{ with } x + 2$$
$$x + x + 1 + x + 2 = 93 \qquad \text{Here the parenthesis aren't needed.}$$
$$3x + 3 = 93 \qquad \text{Combine like terms } x + x + x \text{ and } 2 + 1$$
$$\underline{-3 - 3} \qquad \text{Add 3 to both sides}$$
$$3x = 90 \qquad \text{The variable is multiplied by 3}$$
$$\overline{3} \;\; \overline{3} \qquad \text{Divide both sides by 3}$$
$$x = 30 \qquad \text{Our solution for } x$$
$$\text{First } 30 \qquad \text{Replace } x \text{ in our origional list with 30}$$
$$\text{Second } (30) + 1 = 31 \qquad \text{The numbers are } 30, 31, \text{ and } 32$$
$$\text{Third } (30) + 2 = 32$$

Sometimes we will work consecutive even or odd integers, rather than just consecutive integers. When we had consecutive integers, we only had to add 1 to get to the next number so we had x, $x + 1$, and $x + 2$ for our first, second, and third number respectively. With even or odd numbers they are spaced apart by two. So if we want three consecutive even numbers, if the first is x, the next number would be $x + 2$, then finally add two more to get the third, $x + 4$. The same is

65

true for consecutive odd numbers, if the first is x, the next will be $x + 2$, and the third would be $x + 4$. It is important to note that we are still adding 2 and 4 even when the numbers are odd. This is because the phrase "odd" is refering to our x, not to what is added to the numbers. Consider the next two examples.

Example 105.

The sum of three consecutive even integers is 246. What are the numbers?

First x	Make the first x
Second $x + 2$	Even numbers, so we add 2 to get the next
Third $x + 4$	Add 2 more (4 total) to get the third
$F + S + T = 246$	Sum means add First (F) plus Second (S) plus Third (T)
$(x) + (x + 2) + (x + 4) = 246$	Replace each F, S, and T with what we labeled them
$x + x + 2 + x + 4 = 246$	Here the parenthesis are not needed
$3x + 6 = 246$	Combine like terms $x + x + x$ and $2 + 4$
$\underline{-6 \quad -6}$	Subtract 6 from both sides
$3x = 240$	The variable is multiplied by 3
$\overline{3 \quad 3}$	Divide both sides by 3
$x = 80$	Our solution for x
First 80	Replace x in the origional list with 80.
Second $(80) + 2 = 82$	The numbers are 80, 82, and 84.
Third $(80) + 4 = 84$	

Example 106.

Find three consecutive odd integers so that the sum of twice the first, the second and three times the third is 152.

First x	Make the first x
Second $x + 2$	Odd numbers so we add 2 (same as even!)
Third $x + 4$	Add 2 more (4 total) to get the third
$2F + S + 3T = 152$	Twice the first gives $2F$ and three times the third gives $3T$
$2(x) + (x + 2) + 3(x + 4) = 152$	Replace F, S, and T with what we labled them
$2x + x + 2 + 3x + 12 = 152$	Distribute through parenthesis
$6x + 14 = 152$	Combine like terms $2x + x + 3x$ and $2 + 14$
$\underline{-14 \quad -14}$	Subtract 14 from both sides
$6x = 138$	Variable is multiplied by 6
$\overline{6 \quad 6}$	Divide both sides by 6
$x = 23$	Our solution for x
First 23	Replace x with 23 in the original list
Second $(23) + 2 = 25$	The numbers are 23, 25, and 27
Third $(23) + 4 = 27$	

When we started with our first, second, and third numbers for both even and odd we had x, $x + 2$, and $x + 4$. The numbers added do not change with odd or even, it is our answer for x that will be odd or even.

Another example of translating English sentences to mathematical sentences comes from geometry. A well known property of triangles is that all three angles will always add to 180. For example, the first angle may be 50 degrees, the second 30 degrees, and the third 100 degrees. If you add these together, $50 + 30 + 100 = 180$. We can use this property to find angles of triangles.

World View Note: German mathematician Bernhart Thibaut in 1809 tried to prove that the angles of a triangle add to 180 without using Euclid's parallel postulate (a point of much debate in math history). He created a proof, but it was later shown to have an error in the proof.

Example 107.

The second angle of a triangle is double the first. The third angle is 40 less than the first. Find the three angles.

First x	With nothing given about the first we make that x
Second $2x$	The second is double the first,
Third $x - 40$	The third is 40 less than the first
$F + S + T = 180$	All three angles add to 180
$(x) + (2x) + (x - 40) = 180$	Replace F, S, and T with the labeled values.
$x + 2x + x - 40 = 180$	Here the parenthesis are not needed.
$4x - 40 = 180$	Combine like terms, $x + 2x + x$
$\underline{+\,40\ +\,40}$	Add 40 to both sides
$4x = 220$	The variable is multiplied by 4
$\overline{4}\quad\overline{4}$	Divide both sides by 4
$x = 55$	Our solution for x
First 55	Replace x with 55 in the original list of angles
Second $2(55) = 110$	Our angles are 55, 110, and 15
Third $(55) - 40 = 15$	

Another geometry problem involves perimeter or the distance around an object. For example, consider a rectangle has a length of 8 and a width of 3. There are two lengths and two widths in a rectangle (opposite sides) so we add $8 + 8 + 3 + 3 = 22$. As there are two lengths and two widths in a rectangle an alternative to find the perimeter of a rectangle is to use the formula $P = 2L + 2W$. So for the rectangle of length 8 and width 3 the formula would give, $P = 2(8) + 2(3) = 16 + 6 = 22$. With problems that we will consider here the formula $P = 2L + 2W$ will be used.

Example 108.

The perimeter of a rectangle is 44. The length is 5 less than double the width. Find the dimensions.

Length x	We will make the length x
Width $2x - 5$	Width is five less than two times the length
$P = 2L + 2W$	The formula for perimeter of a rectangle
$(44) = 2(x) + 2(2x - 5)$	Replace P, L, and W with labeled values
$44 = 2x + 4x - 10$	Distribute through parenthesis
$44 = 6x - 10$	Combine like terms $2x + 4x$
$\underline{+10 \qquad +10}$	Add 10 to both sides
$54 = 6x$	The variable is multiplied by 6
$\overline{6 \quad 6}$	Divide both sides by 6
$9 = x$	Our solution for x
Length 9	Replace x with 9 in the origional list of sides
Width $2(9) - 5 = 13$	The dimensions of the rectangle are 9 by 13.

We have seen that it is imortant to start by clearly labeling the variables in a short list before we begin to solve the problem. This is important in all word problems involving variables, not just consective numbers or geometry problems. This is shown in the following example.

Example 109.

A sofa and a love seat together costs \$444. The sofa costs double the love seat. How much do they each cost?

Love Seat x	With no information about the love seat, this is our x
Sofa $2x$	Sofa is double the love seat, so we multiply by 2
$S + L = 444$	Together they cost 444, so we add.
$(x) + (2x) = 444$	Replace S and L with labeled values
$3x = 444$	Parenthesis are not needed, combine like terms $x + 2x$
$\overline{3 \quad 3}$	Divide both sides by 3
$x = 148$	Our solution for x
Love Seat 148	Replace x with 148 in the origional list
Sofa $2(148) = 296$	The love seat costs \$148 and the sofa costs \$296.

Be careful on problems such as these. Many students see the phrase "double" and believe that means we only have to divide the 444 by 2 and get \$222 for one or both of the prices. As you can see this will not work. By clearly labeling the variables in the original list we know exactly how to set up and solve these problems.

1.8 Practice - Number and Geometry Problems

Solve.

1. When five is added to three more than a certain number, the result is 19. What is the number?

2. If five is subtracted from three times a certain number, the result is 10. What is the number?

3. When 18 is subtracted from six times a certain number, the result is -42. What is the number?

4. A certain number added twice to itself equals 96. What is the number?

5. A number plus itself, plus twice itself, plus 4 times itself, is equal to -104. What is the number?

6. Sixty more than nine times a number is the same as two less than ten times the number. What is the number?

7. Eleven less than seven times a number is five more than six times the number. Find the number.

8. Fourteen less than eight times a number is three more than four times the number. What is the number?

9. The sum of three consecutive integers is 108. What are the integers?

10. The sum of three consecutive integers is -126. What are the integers?

11. Find three consecutive integers such that the sum of the first, twice the second, and three times the third is -76.

12. The sum of two consecutive even integers is 106. What are the integers?

13. The sum of three consecutive odd integers is 189. What are the integers?

14. The sum of three consecutive odd integers is 255. What are the integers?

15. Find three consecutive odd integers such that the sum of the first, two times the second, and three times the third is 70.

16. The second angle of a triangle is the same size as the first angle. The third angle is 12 degrees larger than the first angle. How large are the angles?

17. Two angles of a triangle are the same size. The third angle is 12 degrees smaller than the first angle. Find the measure the angles.

18. Two angles of a triangle are the same size. The third angle is 3 times as large as the first. How large are the angles?

19. The third angle of a triangle is the same size as the first. The second angle is 4 times the third. Find the measure of the angles.

20. The second angle of a triangle is 3 times as large as the first angle. The third angle is 30 degrees more than the first angle. Find the measure of the angles.

21. The second angle of a triangle is twice as large as the first. The measure of the third angle is 20 degrees greater than the first. How large are the angles?

22. The second angle of a triangle is three times as large as the first. The measure of the third angle is 40 degrees greater than that of the first angle. How large are the three angles?

23. The second angle of a triangle is five times as large as the first. The measure of the third angle is 12 degrees greater than that of the first angle. How large are the angles?

24. The second angle of a triangle is three times the first, and the third is 12 degrees less than twice the first. Find the measures of the angles.

25. The second angle of a triangle is four times the first and the third is 5 degrees more than twice the first. Find the measures of the angles.

26. The perimeter of a rectangle is 150 cm. The length is 15 cm greater than the width. Find the dimensions.

27. The perimeter of a rectangle is 304 cm. The length is 40 cm longer than the width. Find the length and width.

28. The perimeter of a rectangle is 152 meters. The width is 22 meters less than the length. Find the length and width.

29. The perimeter of a rectangle is 280 meters. The width is 26 meters less than the length. Find the length and width.

30. The perimeter of a college basketball court is 96 meters and the length is 14 meters more than the width. What are the dimensions?

31. A mountain cabin on 1 acre of land costs $30,000. If the land cost 4 times as much as the cabin, what was the cost of each?

32. A horse and a saddle cost $5000. If the horse cost 4 times as much as the saddle, what was the cost of each?

33. A bicycle and a bicycle helmet cost $240. How much did each cost, if the bicycle cost 5 times as much as the helmet?

34. Of 240 stamps that Harry and his sister collected, Harry collected 3 times as many as his sisters. How many did each collect?

35. If Mr. Brown and his son together had $220, and Mr. Brown had 10 times as much as his son, how much money had each?

36. In a room containing 45 students there were twice as many girls as boys. How many of each were there?

37. Aaron had 7 times as many sheep as Beth, and both together had 608. How many sheep had each?

38. A man bought a cow and a calf for $990, paying 8 times as much for the cow as for the calf. What was the cost of each?

39. Jamal and Moshe began a business with a capital of $7500. If Jamal furnished half as much capital as Moshe, how much did each furnish?

40. A lab technician cuts a 12 inch piece of tubing into two pieces in such a way that one piece is 2 times longer than the other.

41. A 6 ft board is cut into two pieces, one twice as long as the other. How long are the pieces?

42. An eight ft board is cut into two pieces. One piece is 2 ft longer than the other. How long are the pieces?

43. An electrician cuts a 30 ft piece of wire into two pieces. One piece is 2 ft longer than the other. How long are the pieces?

44. The total cost for tuition plus room and board at State University is $2,584. Tuition costs $704 more than room and board. What is the tuition fee?

45. The cost of a private pilot course is $1,275. The flight portion costs $625 more than the groung school portion. What is the cost of each?

Solving Linear Equations - Age Problems

Objective: Solve age problems by creating and solving a linear equation.

An application of linear equations is what are called age problems. When we are solving age problems we generally will be comparing the age of two people both now and in the future (or past). Using the clues given in the problem we will be working to find their current age. There can be a lot of information in these problems and we can easily get lost in all the information. To help us organize and solve our problem we will fill out a three by three table for each problem. An example of the basic structure of the table is below

	Age Now	Change
Person 1		
Person 2		

Table 6. Structure of Age Table

Normally where we see "Person 1" and "Person 2" we will use the name of the person we are talking about. We will use this table to set up the following example.

Example 110.

Adam is 20 years younger than Brian. In two years Brian will be twice as old as Adam. How old are they now?

	Age Now	$+\,2$
Adam		
Brian		

We use Adam and Brian for our persons
We use $+\,2$ for change because the second phrase is two years in the future

	Age Now	$+\,2$
Adam	$x - 20$	
Brain	x	

Consider the "Now" part, **Adam is 20 years youger than Brian**. We are given information about Adam, not Brian. So Brian is x now. To show Adam is 20 years younger we subtract 20, Adam is $x - 20$.

	Age Now	$+\,2$
Adam	$x - 20$	$x - 20 + 2$
Brian	x	$x + 2$

Now the $+\,2$ column is filled in. This is done by adding 2 to both Adam's and Brian's now column as shown in the table.

	Age Now	$+\,2$
Adam	$x - 20$	$x - 18$
Brian	x	$x + 2$

Combine like terms in Adam's future age: $-\,20 + 2$ This table is now filled out and we are ready to try and solve.

$$B = 2A$$

Our equation comes from the future statement: **Brian will be twice as old as Adam.** This means the younger, Adam, needs to be multiplied by 2.

$$(x + 2) = 2(x - 18)$$

Replace B and A with the information in their future cells, Adam (A) is replaced with $x - 18$ and Brian (B) is replaced with $(x + 2)$ This is the equation to solve!

$$x + 2 = 2x - 36$$

Distribute through parenthesis

$$\underline{-x \qquad -x}$$

Subtract x from both sides to get variable on one side

$$2 = x - 36$$

Need to clear the -36

$$\underline{+36 \quad +36}$$

Add 36 to both sides

$$38 = x$$

Our solution for x

	Age now
Adam	$38 - 20 = 18$
Brian	38

The first column will help us answer the question. Replace the $x's$ with 38 and simplify. Adam is 18 and Brian is 38

Solving age problems can be summarized in the following five steps. These five steps are guidelines to help organize the problem we are trying to solve.

1. Fill in the now column. The person we know nothing about is x.

2. Fill in the future/past column by adding/subtracting the change to the now column.

3. Make an equation for the relationship in the future. This is independent of the table.

4. Replace variables in equation with information in future cells of table

5. Solve the equation for x, use the solution to answer the question

These five steps can be seen illustrated in the following example.

Example 111.

Carmen is 12 years older than David. Five years ago the sum of their ages was 28. How old are they now?

	Age Now	-5
Carmen		
David		

Five years ago is -5 in the change column.

	Age Now	-5
Carmen	$x + 12$	
David	x	

Carmen is 12 years older than David. We don't know about David so he is x, Carmen then is $x + 12$

	Age Now	-5
Carmen	$x + 12$	$x + 12 - 5$
David	x	$x - 5$

Subtract 5 from now column to get the change

73

	Age Now	-5
Carmen	$x+12$	$x+7$
David	x	$x-5$

Simplify by combining like terms $12-5$
Our table is ready!

$$C + D = 28$$ The sum of their ages will be 29. So we add C and D

$$(x+7)+(x-5)=28$$ Replace C and D with the change cells.

$$x+7+x-5=28$$ Remove parenthesis

$$2x+2=28$$ Combine like terms $x+x$ and $7-5$

$$\underline{-2 \quad -2}$$ Subtract 2 from both sides

$$2x=26$$ Notice x is multiplied by 2

$$\overline{2 \quad 2}$$ Divide both sides by 2

$$x=13$$ Our solution for x

	Age Now
Caremen	$13+12=25$
David	13

Replace x with 13 to answer the question
Carmen is 25 and David is 13

Sometimes we are given the sum of their ages right now. These problems can be tricky. In this case we will write the sum above the now column and make the first person's age now x. The second person will then turn into the subtraction problem total $- x$. This is shown in the next example.

Example 112.

The sum of the ages of Nicole and Kristin is 32. In two years Nicole will be three times as old as Kristin. How old are they now?

32

	Age Now	$+2$
Nicole	x	
Kristen	$32-x$	

The change is $+2$ for two years in the future
The total is placed above Age Now
The first person is x. The second becomes $32-x$

	Age Now	$+2$
Nicole	x	$x+2$
Kristen	$32-x$	$32-x+2$

Add 2 to each cell fill in the change column

	Age Now	$+2$
Nicole	x	$x+2$
Kristen	$32-x$	$34-x$

Combine like terms $32+2$, our table is done!

$$N = 3K$$ Nicole is three times as old as Kristin.

$$(x+2)=3(34-x)$$ Replace variables with information in change cells

$$x+2=102-3x$$ Distribute through parenthesis

$$\underline{+3x \qquad +3x}$$ Add $3x$ to both sides so variable is only on one side

74

	Age Now	
	$4x + 2 = 102$	Solve the two $-$ step equation

$$4x + 2 = 102 \quad \text{Solve the two} - \text{step equation}$$
$$\underline{-2 \quad -2} \quad \text{Subtract 2 from both sides}$$
$$4x = 100 \quad \text{The variable is multiplied by 4}$$
$$\overline{4 \quad 4} \quad \text{Divide both sides by 4}$$
$$x = 25 \quad \text{Our solution for } x$$

	Age Now
Nicole	25
Kristen	$32 - 25 = 7$

Plug 25 in for x in the now column
Nicole is 25 and Kristin is 7

A slight variation on age problems is to ask not how old the people are, but rather ask how long until we have some relationship about their ages. In this case we alter our table slightly. In the change column because we don't know the time to add or subtract we will use a variable, t, and add or subtract this from the now column. This is shown in the next example.

Example 113.

Louis is 26 years old. Her daughter is 4 years old. In how many years will Louis be double her daughter's age?

	Age Now	$+t$
Louis	26	
Daughter	4	

As we are given their ages now, these numbers go into the table. The change is unknown, so we write $+t$ for the change

	Age Now	$+t$
Louis	26	$26 + t$
Daughter	4	$4 + t$

Fill in the change column by adding t to each person's age. Our table is now complete.

$$L = 2D \quad \text{Louis will be double her daughter}$$
$$(26 + t) = 2(4 + t) \quad \text{Replace variables with information in change cells}$$
$$26 + t = 8 + 2t \quad \text{Distribute through parenthesis}$$
$$\underline{-t \quad -t} \quad \text{Subtract } t \text{ from both sides}$$
$$26 = 8 + t \quad \text{Now we have an 8 added to the } t$$
$$\underline{-8 -8} \quad \text{Subtract 8 from both sides}$$
$$18 = t \quad \text{In 18 years she will be double her daughter's age}$$

Age problems have several steps to them. However, if we take the time to work through each of the steps carefully, keeping the information organized, the problems can be solved quite nicely.

World View Note: The oldest man in the world was Shigechiyo Izumi from Japan who lived to be 120 years, 237 days. However, his exact age has been disputed.

1.9 Practice - Age Problems

1. A boy is 10 years older than his brother. In 4 years he will be twice as old as his brother. Find the present age of each.

2. A father is 4 times as old as his son. In 20 years the father will be twice as old as his son. Find the present age of each.

3. Pat is 20 years older than his son James. In two years Pat will be twice as old as James. How old are they now?

4. Diane is 23 years older than her daughter Amy. In 6 years Diane will be twice as old as Amy. How old are they now?

5. Fred is 4 years older than Barney. Five years ago the sum of their ages was 48. How old are they now?

6. John is four times as old as Martha. Five years ago the sum of their ages was 50. How old are they now?

7. Tim is 5 years older than JoAnn. Six years from now the sum of their ages will be 79. How old are they now?

8. Jack is twice as old as Lacy. In three years the sum of their ages will be 54. How old are they now?

9. The sum of the ages of John and Mary is 32. Four years ago, John was twice as old as Mary. Find the present age of each.

10. The sum of the ages of a father and son is 56. Four years ago the father was 3 times as old as the son. Find the present age of each.

11. The sum of the ages of a china plate and a glass plate is 16 years. Four years ago the china plate was three times the age of the glass plate. Find the present age of each plate.

12. The sum of the ages of a wood plaque and a bronze plaque is 20 years. Four

years ago, the bronze plaque was one-half the age of the wood plaque. Find the present age of each plaque.

13. A is now 34 years old, and B is 4 years old. In how many years will A be twice as old as B?

14. A man's age is 36 and that of his daughter is 3 years. In how many years will the man be 4 times as old as his daughter?

15. An Oriental rug is 52 years old and a Persian rug is 16 years old. How many years ago was the Oriental rug four times as old as the Persian Rug?

16. A log cabin quilt is 24 years old and a friendship quilt is 6 years old. In how may years will the log cabin quilt be three times as old as the friendship quilt?

17. The age of the older of two boys is twice that of the younger; 5 years ago it was three times that of the younger. Find the age of each.

18. A pitcher is 30 years old, and a vase is 22 years old. How many years ago was the pitcher twice as old as the vase?

19. Marge is twice as old as Consuelo. The sum of their ages seven years ago was 13. How old are they now?

20. The sum of Jason and Mandy's age is 35. Ten years ago Jason was double Mandy's age. How old are they now?

21. A silver coin is 28 years older than a bronze coin. In 6 years, the silver coin will be twice as old as the bronze coin. Find the present age of each coin.

22. A sofa is 12 years old and a table is 36 years old. In how many years will the table be twice as old as the sofa?

23. A limestone statue is 56 years older than a marble statue. In 12 years, the limestone will be three times as old as the marble statue. Find the present age of the statues.

24. A pewter bowl is 8 years old, and a silver bowl is 22 years old. In how many years will the silver bowl be twice the age of the pewter bowl?

25. Brandon is 9 years older than Ronda. In four years the sum of their ages will be 91. How old are they now?

26. A kerosene lamp is 95 years old, and an electric lamp is 55 years old. How many years ago was the kerosene lamp twice the age of the electric lamp?

27. A father is three times as old as his son, and his daughter is 3 years younger

than the son. If the sum of their ages 3 years ago was 63 years, find the present age of the father.

28. The sum of Clyde and Wendy's age is 64. In four years, Wendy will be three times as old as Clyde. How old are they now?

29. The sum of the ages of two ships is 12 years. Two years ago, the age of the older ship was three times the age of the newer ship. Find the present age of each ship.

30. Chelsea's age is double Daniel's age. Eight years ago the sum of their ages was 32. How old are they now?

31. Ann is eighteen years older than her son. One year ago, she was three times as old as her son. How old are they now?

32. The sum of the ages of Kristen and Ben is 32. Four years ago Kristen was twice as old as Ben. How old are they both now?

33. A mosaic is 74 years older than the engraving. Thirty years ago, the mosaic was three times as old as the engraving. Find the present age of each.

34. The sum of the ages of Elli and Dan is 56. Four years ago Elli was 3 times as old as Dan. How old are they now?

35. A wool tapestry is 32 years older than a linen tapestry. Twenty years ago, the wool tapestry was twice as old as the linen tapestry. Find the present age of each.

36. Carolyn's age is triple her daughter's age. In eight years the sum of their ages will be 72. How old are they now?

37. Nicole is 26 years old. Emma is 2 years old. In how many years will Nicole be triple Emma's age?

38. The sum of the ages of two children is 16 years. Four years ago, the age of the older child was three times the age of the younger child. Find the present age of each child.

39. Mike is 4 years older than Ron. In two years, the sum of their ages will be 84. How old are they now?

40. A marble bust is 25 years old, and a terra-cotta bust is 85 years old. In how many years will the terra-cotta bust be three times as old as the marble bust?

Solving Linear Equations - Distance, Rate and Time

Objective: Solve distance problems by creating and solving a linear equation.

An application of linear equations can be found in distance problems. When solving distance problems we will use the relationship $rt = d$ or rate (speed) times time equals distance. For example, if a person were to travel 30 mph for 4 hours. To find the total distance we would multiply rate times time or $(30)(4) = 120$. This person travel a distance of 120 miles. The problems we will be solving here will be a few more steps than described above. So to keep the information in the problem organized we will use a table. An example of the basic structure of the table is blow:

	Rate	Time	Distance
Person 1			
Person 2			

Table 7. Structure of Distance Problem

The third column, distance, will always be filled in by multiplying the rate and time columns together. If we are given a total distance of both persons or trips we will put this information below the distance column. We will now use this table to set up and solve the following example

Example 114.

Two joggers start from opposite ends of an 8 mile course running towards each other. One jogger is running at a rate of 4 mph, and the other is running at a rate of 6 mph. After how long will the joggers meet?

	Rate	Time	Distance
Jogger 1			
Jogger 2			

The basic table for the joggers, one and two

	Rate	Time	Distance
Jogger 1	4		
Jogger 2	6		

We are given the rates for each jogger.
These are added to the table

	Rate	Time	Distance
Jogger 1	4	t	
Jogger 2	6	t	

We only know they both start and end at the same time. We use the variable t for both times

	Rate	Time	Distance
Jogger 1	4	t	$4t$
Jogger 2	6	t	$6t$

The distance column is filled in by multiplying rate by time

$$8$$ We have **total distance**, 8 miles, under distance

$$4t + 6t = 8$$ The distance column gives equation by adding

$$10t = 8$$ Combine like terms, $4t + 6t$

$$\overline{10}\ \overline{10}$$ Divide both sides by 10

$$t = \frac{4}{5}$$ Our solution for t, $\frac{4}{5}$ hour (48 minutes)

As the example illustrates, once the table is filled in, the equation to solve is very easy to find. This same process can be seen in the following example

Example 115.

Bob and Fred start from the same point and walk in opposite directions. Bob walks 2 miles per hour faster than Fred. After 3 hours they are 30 miles apart. How fast did each walk?

	Rate	Time	Distance
Bob		3	
Fred		3	

The basic table with given times filled in
Both traveled 3 hours

	Rate	Time	Distance
Bob	$r+2$	3	
Fred	r	3	

Bob walks 2 mph faster than Fred
We know nothing about Fred, so use r for his rate
Bob is $r+2$, showing 2 mph faster

	Rate	Time	Distance
Bob	$r+2$	3	$3r+6$
Fred	r	3	$3r$

Distance column is filled in by multiplying rate by Time. Be sure to distribute the $3(r+2)$ for Bob.

$$30$$

Total distance is put under distance

$$3r+6+3r=30$$

The distance columns is our equation, by adding

$$6r+6=30$$

Combine like terms $3r+3r$

$$-6\quad -6$$

Subtract 6 from both sides

$$6r=24$$

The variable is multiplied by 6

$$\overline{6}\quad \overline{6}$$

Divide both sides by 6

$$r=4$$

Our solution for r

	Rate
Bob	$4+2=6$
Fred	4

To answer the question completely we plug 4 in for r in the table. Bob traveled 6 miles per hour and Fred traveled 4 mph

Some problems will require us to do a bit of work before we can just fill in the cells. One example of this is if we are given a total time, rather than the individual times like we had in the previous example. If we are given total time we will write this above the time column, use t for the first person's time, and make a subtraction problem, Total $- t$, for the second person's time. This is shown in the next example

Example 116.

Two campers left their campsite by canoe and paddled downstream at an average speed of 12 mph. They turned around and paddled back upstream at an average rate of 4 mph. The total trip took 1 hour. After how much time did the campers turn around downstream?

	Rate	Time	Distance
Down	12		
Up	4		

Basic table for down and upstream
Given rates are filled in

$$1$$

Total time is put above time column

	Rate	Time	Distance
Down	12	t	
Up	4	$1-t$	

As we have the total time, in the first time we have t, the second time becomes the subtraction, total $- t$

	Rate	Time	Distance	
Down	12	t	$12t$	=
Up	4	$1-t$	$4-4t$	

Distance column is found by multiplying rate by time. Be sure to distribute $4(1-t)$ for upstream. As they cover the **same distance**, $=$ is put after the down distance

$$12t = 4 - 4t$$ With equal sign, distance colum is equation

$$\underline{+4t \qquad +4t}$$ Add $4t$ to both sides so variable is only on one side

$$16t = 4$$ Variable is multiplied by 16

$$\overline{16 \quad 16}$$ Divide both sides by 16

$$t = \frac{1}{4}$$ Our solution, turn around after $\frac{1}{4}$ hr (15 min)

Another type of a distance problem where we do some work is when one person catches up with another. Here a slower person has a head start and the faster person is trying to catch up with him or her and we want to know how long it will take the fast person to do this. Our startegy for this problem will be to use t for the faster person's time, and add amount of time the head start was to get the slower person's time. This is shown in the next example.

Example 117.

Mike leaves his house traveling 2 miles per hour. Joy leaves 6 hours later to catch up with him traveling 8 miles per hour. How long will it take her to catch up with him?

	Rate	Time	Distance
Mike	2		
Joy	8		

Basic table for Mike and Joy
The given rates are filled in

	Rate	Time	Distance
Mike	2	$t+6$	
Joy	8	t	

Joy, the faster person, we use t for time
Mike's time is $t+6$ showing his 6 hour head start

	Rate	Time	Distance	
Mike	2	$t+6$	$2t+12$	=
Joy	8	t	$8t$	

Distance column is found by multiplying the rate by time. Be sure to distribute the $2(t+6)$ for Mike As they cover the **same distance**, $=$ is put after Mike's distance

$$2t+12 = 8t$$ Now the distance column is the equation

$$\underline{-2t \qquad -2t}$$ Subtract $2t$ from both sides

$$12 = 6t$$ The variable is multiplied by 6

$$\overline{6 \quad 6}$$ Divide both sides by 6

$$2 = t \qquad \text{Our solution for } t \text{, she catches him after 2 hours}$$

World View Note: The 10,000 race is the longest standard track event. 10,000 meters is approximately 6.2 miles. The current (at the time of printing) world record for this race is held by Ethiopian Kenenisa Bekele with a time of 26 minutes, 17.53 second. That is a rate of 12.7 miles per hour!

As these example have shown, using the table can help keep all the given information organized, help fill in the cells, and help find the equation we will solve. The final example clearly illustrates this.

Example 118.

On a 130 mile trip a car travled at an average speed of 55 mph and then reduced its speed to 40 mph for the remainder of the trip. The trip took 2.5 hours. For how long did the car travel 40 mph?

	Rate	Time	Distance
Fast	55		
Slow	40		

Basic table for fast and slow speeds
The given rates are filled in

2.5

	Rate	Time	Distance
Fast	55	t	
Slow	40	$2.5 - t$	

Total time is put above the time column
As we have total time, the first time we have t
The second time is the subtraction problem
$2.5 - t$

2.5

	Rate	Time	Distance
Fast	55	t	$55t$
Slow	40	$2.5 - t$	$100 - 40t$

Distance column is found by multiplying rate by time. Be sure to distribute $40(2.5 - t)$ for slow

$$130$$

Total distance is put under distance

$$55t + 100 - 40t = 130 \qquad \text{The distance column gives our equation by adding}$$

$$15t + 100 = 130 \qquad \text{Combine like terms } 55t - 40t$$

$$\underline{-100 \quad -100} \qquad \text{Subtract 100 from both sides}$$

$$15t = 30 \qquad \text{The variable is multiplied by 30}$$

$$\overline{15 \quad 15} \qquad \text{Divide both sides by 15}$$

$$t = 2 \qquad \text{Our solution for } t.$$

	Time
Fast	2
Slow	$2.5 - 2 = 0.5$

To answer the question we plug 2 in for t
The car traveled 40 mph for 0.5 hours (30 minutes)

1.10 Practice - Distance, Rate, and Time Problems

1. A is 60 miles from B. An automobile at A starts for B at the rate of 20 miles an hour at the same time that an automobile at B starts for A at the rate of 25 miles an hour. How long will it be before the automobiles meet?

2. Two automobiles are 276 miles apart and start at the same time to travel toward each other. They travel at rates differing by 5 miles per hour. If they meet after 6 hours, find the rate of each.

3. Two trains travel toward each other from points which are 195 miles apart. They travel at rate of 25 and 40 miles an hour respectively. If they start at the same time, how soon will they meet?

4. A and B start toward each other at the same time from points 150 miles apart. If A went at the rate of 20 miles an hour, at what rate must B travel if they meet in 5 hours?

5. A passenger and a freight train start toward each other at the same time from two points 300 miles apart. If the rate of the passenger train exceeds the rate of the freight train by 15 miles per hour, and they meet after 4 hours, what must the rate of each be?

6. Two automobiles started at the same time from a point, but traveled in opposite directions. Their rates were 25 and 35 miles per hour respectively. After how many hours were they 180 miles apart?

7. A man having ten hours at his disposal made an excursion, riding out at the rate of 10 miles an hour and returning on foot, at the rate of 3 miles an hour.

Find the distance he rode.

8. A man walks at the rate of 4 miles per hour. How far can he walk into the country and ride back on a trolley that travels at the rate of 20 miles per hour, if he must be back home 3 hours from the time he started?

9. A boy rides away from home in an automobile at the rate of 28 miles an hour and walks back at the rate of 4 miles an hour. The round trip requires 2 hours. How far does he ride?

10. A motorboat leaves a harbor and travels at an average speed of 15 mph toward an island. The average speed on the return trip was 10 mph. How far was the island from the harbor if the total trip took 5 hours?

11. A family drove to a resort at an average speed of 30 mph and later returned over the same road at an average speed of 50 mph. Find the distance to the resort if the total driving time was 8 hours.

12. As part of his flight trainging, a student pilot was required to fly to an airport and then return. The average speed to the airport was 90 mph, and the average speed returning was 120 mph. Find the distance between the two airports if the total flying time was 7 hours.

13. A, who travels 4 miles an hour starts from a certain place 2 hours in advance of B, who travels 5 miles an hour in the same direction. How many hours must B travel to overtake A?

14. A man travels 5 miles an hour. After traveling for 6 hours another man starts at the same place, following at the rate of 8 miles an hour. When will the second man overtake the first?

15. A motorboat leaves a harbor and travels at an average speed of 8 mph toward a small island. Two hours later a cabin cruiser leaves the same harbor and travels at an average speed of 16 mph toward the same island. In how many hours after the cabin cruiser leaves will the cabin cuiser be alongside the motorboat?

16. A long distance runner started on a course running at an average speed of 6 mph. One hour later, a second runner began the same course at an average speed of 8 mph. How long after the second runner started will the second runner overtake the first runner?

17. A car traveling at 48 mph overtakes a cyclist who, riding at 12 mph, has had a 3 hour head start. How far from the starting point does the car overtake the cyclist?

18. A jet plane traveling at 600 mph overtakes a propeller-driven plane which has

had a 2 hour head start. The propeller-driven plane is traveling at 200 mph. How far from the starting point does the jet overtake the propeller-driven plane?

19. Two men are traveling in opposite directions at the rate of 20 and 30 miles an hour at the same time and from the same place. In how many hours will they be 300 miles apart?

20. Running at an average rate of 8 m/s, a sprinter ran to the end of a track and then jogged back to the starting point at an average rate of 3 m/s. The sprinter took 55 s to run to the end of the track and jog back. Find the length of the track.

21. A motorboat leaves a harbor and travels at an average speed of 18 mph to an island. The average speed on the return trip was 12 mph. How far was the island from the harbor if the total trip took 5 h?

22. A motorboat leaves a harbor and travels at an average speed of 9 mph toward a small island. Two hours later a cabin cruiser leaves the same harbor and travels at an average speed of 18 mph toward the same island. In how many hours after the cabin cruiser leaves will the cabin cruiser be alongside the motorboat?

23. A jet plane traveling at 570 mph overtakes a propeller-driven plane that has had a 2 h head start. The propeller-driven plane is traveling at 190 mph. How far from the starting point does the jet overtake the propeller-driven plane?

24. Two trains start at the same time from the same place and travel in opposite directions. If the rate of one is 6 miles per hour more than the rate of the other and they are 168 miles apart at the end of 4 hours, what is the rate of each?

25. As part of flight traning, a student pilot was required to fly to an airport and then return. The average speed on the way to the airport was 100 mph, and the average speed returning was 150 mph. Find the distance between the two airports if the total flight time was 5 h.

26. Two cyclists start from the same point and ride in opposite directions. One cyclist rides twice as fast as the other. In three hours they are 72 miles apart. Find the rate of each cyclist.

27. A car traveling at 56 mph overtakes a cyclist who, riding at 14 mph, has had a 3 h head start. How far from the starting point does the car overtake the cyclist?

28. Two small planes start from the same point and fly in opposite directions.

The first plan is flying 25 mph slower than the second plane. In two hours the planes are 430 miles apart. Find the rate of each plane.

29. A bus traveling at a rate of 60 mph overtakes a car traveling at a rate of 45 mph. If the car had a 1 h head start, how far from the starting point does the bus overtake the car?

30. Two small planes start from the same point and fly in opposite directions. The first plane is flying 25 mph slower than the second plane. In 2 h, the planes are 470 mi apart. Find the rate of each plane.

31. A truck leaves a depot at 11 A.M. and travels at a speed of 45 mph. At noon, a van leaves the same place and travels the same route at a speed of 65 mph. At what time does the van overtake the truck?

32. A family drove to a resort at an average speed of 25 mph and later returned over the same road at an average speed of 40 mph. Find the distance to the resort if the total driving time was 13 h.

33. Three campers left their campsite by canoe and paddled downstream at an average rate of 10 mph. They then turned around and paddled back upstream at an average rate of 5 mph to return to their campsite. How long did it take the campers to canoe downstream if the total trip took 1 hr?

34. A motorcycle breaks down and the rider has to walk the rest of the way to work. The motorcycle was being driven at 45 mph, and the rider walks at a speed of 6 mph. The distance from home to work is 25 miles, and the total time for the trip was 2 hours. How far did the motorcycle go before if broke down?

35. A student walks and jogs to college each day. The student averages 5 km/hr walking and 9 km/hr jogging. The distance from home to college is 8 km, and the student makes the trip in one hour. How far does the student jog?

36. On a 130 mi trip, a car traveled at an average speed of 55 mph and then reduced its speed to 40 mph for the remainder of the trip. The trip took a total of 2.5 h. For how long did the car travel at 40 mph?

37. On a 220 mi trip, a car traveled at an average speed of 50 mph and then reduced its average speed to 35 mph for the remainder of the trip. The trip took a total of 5 h. How long did the car travel at each speed?

38. An executive drove from home at an average speed of 40 mph to an airport where a helicopter was waiting. The executive boarded the helicopter and flew to the corporate offices at and average speed of 60 mph. The entire distance was 150 mi. The entire trip took 3 h. Find the distance from the airport to the corporate offices.

Chapter 2 : Graphing

Graphing - Points and Lines

Objective: Graph points and lines using xy coordinates.

Often, to get an idea of the behavior of an equation we will make a picture that represents the solutions to the equations. A **graph** is simply a picture of the solutions to an equation. Before we spend much time on making a visual representation of an equation, we first have to understand the basis of graphing. Following is an example of what is called the coordinate plane.

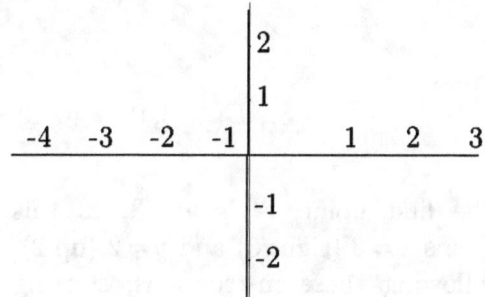

The plane is divided into four sections by a horizontal number line (x-axis) and a vertical number line (y-axis). Where the two lines meet in the center is called the origin. This center origin is where $x = 0$ and $y = 0$. As we move to the right the numbers count up from zero, representing $x = 1, 2, 3....$

To the left the numbers count down from zero, representing $x = -1, -2, -3....$ Similarly, as we move up the number count up from zero, $y = 1, 2, 3....$, and as we move down count down from zero, $y = -1, -2, -3$. We can put dots on the graph which we will call points. Each point has an "address" that defines its location. The first number will be the value on the $x-$axis or horizontal number line. This is the distance the point moves left/right from the origin. The second number will represent the value on the $y-$axis or vertical number line. This is the distance the point moves up/down from the origin. The points are given as an ordered pair (x, y).

World View Note: Locations on the globe are given in the same manner, each number is a distance from a central point, the origin which is where the prime meridian and the equator. This "origin is just off the western coast of Africa.

The following example finds the address or coordinate pair for each of several points on the coordinate plane.

Example 119.

Give the coordinates of each point.

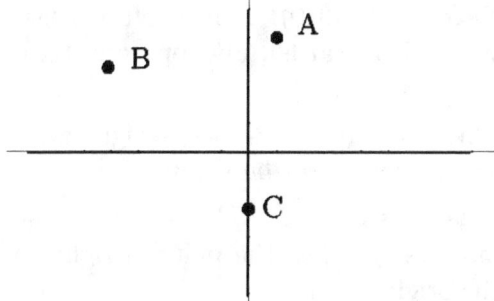

Tracing from the origin, point A is right 1, up 4. This becomes A(1, 4). Point B is left 5, up 3. Left is backwards or negative so we have B(− 5, 3). C is straight down 2 units. There is no left or right. This means we go right zero so the point is C(0, − 2).

$$A(1, 4), B(-5, 3), C(0, -2) \quad \text{Our Solution}$$

Just as we can give the coordinates for a set of points, we can take a set of points and plot them on the plane.

Example 120.

Graph the points $A(3, 2), B(-2, 1), C(3, -4), D(-2, -3), E(-3, 0), F(0, 2), G(0, 0)$

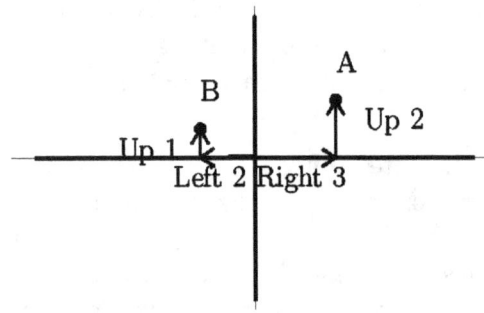

The first point, A is at (3, 2) this means $x = 3$ (right 3) and $y = 2$ (up 2). Following these instructions, starting from the origin, we get our point.

The second point, $B(-2, 1)$, is left 2 (negative moves backwards), up 1. This is also illustrated on the graph.

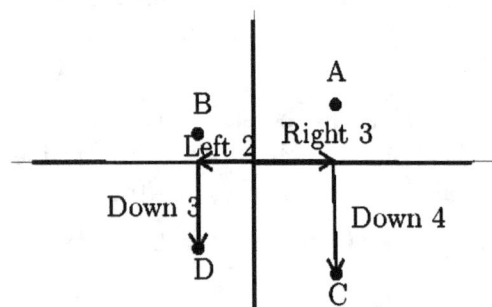

The third point, $C(3, -4)$ is right 3, down 4 (negative moves backwards).

The fourth point, D (− 2, − 3) is left 2, down 3 (both negative, both move backwards)

The last three points have zeros in them. We still treat these points just like the other points. If there is a zero there is just no movement.

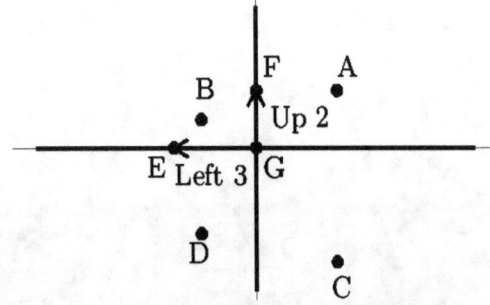

Next is $E(-3, 0)$. This is left 3 (negative is backwards), and up zero, right on the x − axis.

Then is $F(0, 2)$. This is right zero, and up two, right on the y − axis.

Finally is $G(0, 0)$. This point has no movement. Thus the point is right on the origin.

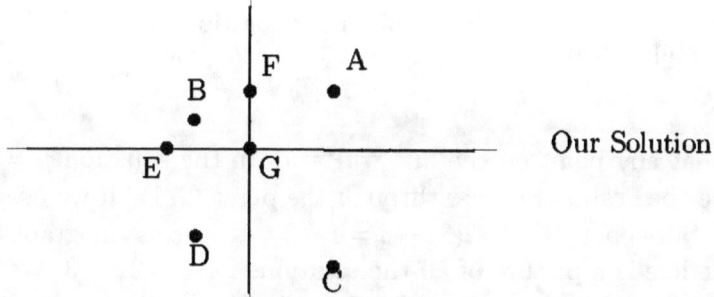

Our Solution

The main purpose of graphs is not to plot random points, but rather to give a picture of the solutions to an equation. We may have an equation such as $y = 2x - 3$. We may be interested in what type of solution are possible in this equation. We can visualize the solution by making a graph of possible x and y combinations that make this equation a true statement. We will have to start by finding possible x and y combinations. We will do this using a table of values.

Example 121.

Graph $y = 2x - 3$ We make a table of values

x	y
-1	
0	
1	

We will test three values for x. Any three can be used

x	y
-1	-5
0	-3
1	-1

Evaluate each by replacing x with the given value
$x = -1; y = 2(-1) - 3 = -2 - 3 = -5$
$x = 0; y = 2(0) - 3 = 0 - 3 = -3$
$x = 1; y = 2(1) - 3 = 2 - 3 = -1$

$(-1, -5), (0, -3), (1, -1)$ These then become the points to graph on our equation

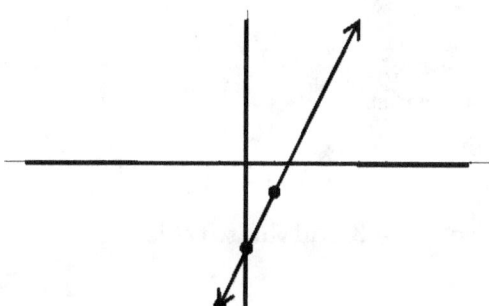

Plot each point.

Once the point are on the graph, con-

nect the dots to make a line.

The graph is our solution

What this line tells us is that any point on the line will work in the equation $y = 2x - 3$. For example, notice the graph also goes through the point $(2, 1)$. If we use $x = 2$, we should get $y = 1$. Sure enough, $y = 2(2) - 3 = 4 - 3 = 1$, just as the graph suggests. Thus we have the line is a picture of all the solutions for $y = 2x - 3$. We can use this table of values method to draw a graph of any linear equation.

Example 122.

Graph $2x - 3y = 6$ We will use a table of values

x	y
-3	
0	
3	

We will test three values for x. Any three can be used.

$$2(-3) - 3y = 6 \qquad \text{Substitute each value in for } x \text{ and solve for } y$$
$$-6 - 3y = 6 \qquad \text{Start with } x = -3, \text{ multiply first}$$
$$\underline{+6 \qquad +6} \qquad \text{Add 6 to both sides}$$
$$-3y = 12 \qquad \text{Divide both sides by } -3$$
$$\overline{-3} \quad \overline{-3}$$
$$y = -4 \qquad \text{Solution for } y \text{ when } x = -3, \text{ add this to table}$$

$$2(0) - 3y = 6 \qquad \text{Next } x = 0$$
$$-3y = 6 \qquad \text{Multiplying clears the constant term}$$
$$\overline{-3} \quad \overline{-3} \qquad \text{Divide each side by } -3$$
$$y = -2 \qquad \text{Solution for } y \text{ when } x = 0, \text{ add this to table}$$

$$2(3) - 3y = 6 \qquad \text{Next } x = 3$$
$$6 - 3y = 6 \qquad \text{Multiply}$$
$$\underline{-6 \qquad -6} \qquad \text{Subtract 9 from both sides}$$
$$-3y = 0 \qquad \text{Divide each side by } -3$$
$$\overline{-3} \quad \overline{-3}$$
$$y = 0 \qquad \text{Solution for } y \text{ when } x = -3, \text{ add this to table}$$

x	y
-3	-4
0	-2
3	0

Our completed table.

$(-3, -4), (0, 2), (3, 0)$ Table becomes points to graph

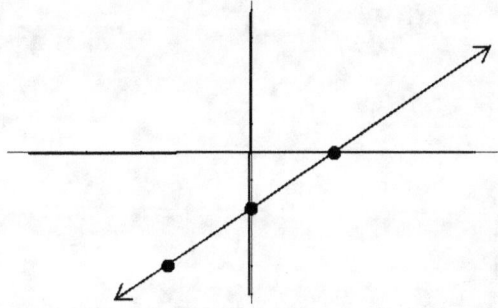

Graph points and connect dots

Our Solution

2.1 Practice - Points and Lines

State the coordinates of each point.

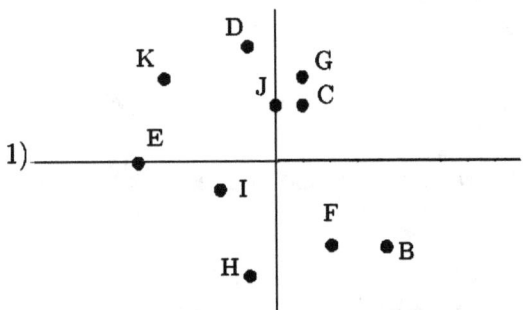

1)

Plot each point.

2) L$(-5,5)$ K$(1,0)$ J$(-3,4)$

   I$(-3,0)$ H$(-4,2)$ G$(4,-2)$

   F$(-2,-2)$ E$(3,-2)$ D$(0,3)$

   C$(0,4)$

Sketch the graph of each line.

3) $y = -\frac{1}{4}x - 3$

4) $y = x - 1$

5) $y = -\frac{5}{4}x - 4$

6) $y = -\frac{3}{5}x + 1$

7) $y = -4x + 2$

8) $y = \frac{5}{3}x + 4$

9) $y = \frac{3}{2}x - 5$

10) $y = -x - 2$

11) $y = -\frac{4}{5}x - 3$

12) $y = \frac{1}{2}x$

13) $x + 5y = -15$

14) $8x - y = 5$

15) $4x + y = 5$

16) $3x + 4y = 16$

17) $2x - y = 2$

18) $7x + 3y = -12$

19) $x + y = -1$

20) $3x + 4y = 8$

21) $x - y = -3$

22) $9x - y = -4$

Graphing - Slope

Objective: Find the slope of a line given a graph or two points.

As we graph lines, we will want to be able to identify different properties of the lines we graph. One of the most important properties of a line is its slope. **Slope** is a measure of steepness. A line with a large slope, such as 25, is very steep. A line with a small slope, such as $\frac{1}{10}$ is very flat. We will also use slope to describe the direction of the line. A line that goes up from left to right will have a positive slope and a line that goes down from left to right will have a negative slope.

As we measure steepness we are interested in how fast the line rises compared to how far the line runs. For this reason we will describe slope as the fraction $\frac{\text{rise}}{\text{run}}$. Rise would be a vertical change, or a change in the y-values. Run would be a horizontal change, or a change in the x-values. So another way to describe slope would be the fraction $\frac{\text{change in } y}{\text{change in } x}$. It turns out that if we have a graph we can draw vertical and horiztonal lines from one point to another to make what is called a slope triangle. The sides of the slope triangle give us our slope. The following examples show graphs that we find the slope of using this idea.

Example 123.

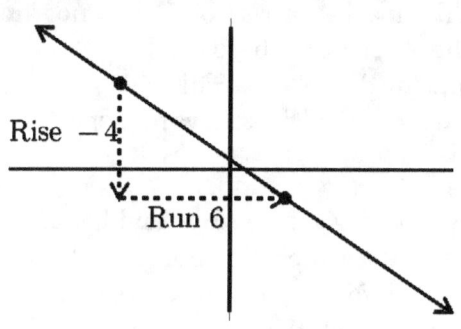

To find the slope of this line we will consider the rise, or verticle change and the run or horizontal change. Drawing these lines in makes a slope triangle that we can use to count from one point to the next the graph goes down 4, right 6. This is rise -4, run 6. As a fraction it would be, $-\frac{4}{6}$. Reduce the fraction to get $-\frac{2}{3}$.

$$-\frac{2}{3} \quad \text{Our Solution}$$

World View Note: When French mathematicians Rene Descartes and Pierre de Fermat first developed the coordinate plane and the idea of graphing lines (and other functions) the y-axis was not a verticle line!

Example 124.

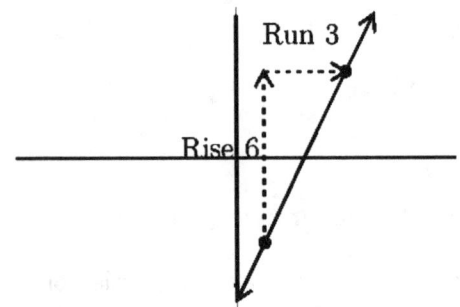

To find the slope of this line, the rise is up 6, the run is right 3. Our slope is then written as a fraction, $\frac{\text{rise}}{\text{run}}$ or $\frac{6}{3}$. This fraction reduces to 2. This will be our slope.

2   Our Solution

There are two special lines that have unique slopes that we need to be aware of. They are illustrated in the following example.

Example 125.

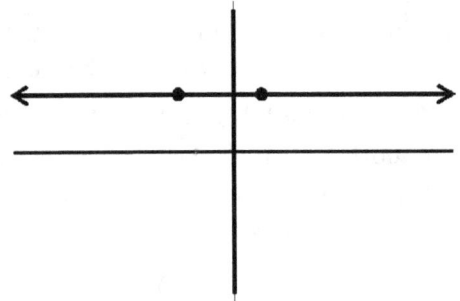

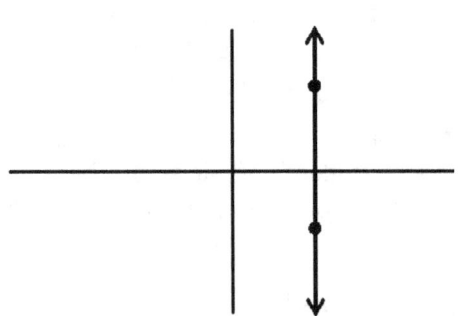

In this graph there is no rise, but the run is 3 units. This slope becomes

$\frac{0}{3}$ = 0. This line, and all horizontal lines have *a* zero slope.

This line has a rise of 5, but no run. The slope becomes $\frac{5}{0}$ = undefined. This line, and all vertical lines, have no slope.

As you can see there is a big difference between having a zero slope and having no slope or undefined slope. Remember, slope is a measure of steepness. The first slope is not steep at all, in fact it is flat. Therefore it has a zero slope. The second slope can't get any steeper. It is so steep that there is no number large enough to express how steep it is. This is an undefined slope.

We can find the slope of a line through two points without seeing the points on a graph. We can do this using a slope formula. If the rise is the change in y values, we can calculate this by subtracting the y values of a point. Similarly, if run is a change in the x values, we can calculate this by subtracting the x values of a point. In this way we get the following equation for slope.

The slope of *a* line through (x_1, y_1) and (x_2, y_2) is $\frac{y_2 - y_1}{x_2 - x_1}$

When mathematicians began working with slope, it was called the modular slope. For this reason we often represent the slope with the variable m. Now we have the following for slope.

$$\text{Slope} = m = \frac{\text{rise}}{\text{run}} = \frac{\text{change in } y}{\text{change in } x} = \frac{y_2 - y_1}{x_2 - x_1}$$

As we subtract the y values and the x values when calculating slope it is important we subtract them in the same order. This process is shown in the following examples.

Example 126.

Find the slope between $(-4, 3)$ and $(2, -9)$ Identify x_1, y_1, x_2, y_2

(x_1, y_1) and (x_2, y_2) Use slope formula, $m = \dfrac{y_2 - y_1}{x_2 - x_1}$

$m = \dfrac{-9 - 3}{2 - (-4)}$ Simplify

$m = \dfrac{-12}{6}$ Reduce

$m = -2$ Our Solution

Example 127.

Find the slope between $(4, \quad 6)$ and $(2, -1)$ Identify x_1, y_1, x_2, y_2

(x_1, y_1) and (x_2, y_2) Use slope formula, $m = \dfrac{y_2 - y_1}{x_2 - x_1}$

$m = \dfrac{-1 - 6}{2 - 4}$ Simplify

$m = \dfrac{-7}{-2}$ Reduce, dividing by -1

$m = \dfrac{7}{2}$ Our Solution

We may come up against a problem that has a zero slope (horiztonal line) or no slope (vertical line) just as with using the graphs.

Example 128.

Find the slope between $(-4, -1)$ and $(-4, -5)$ Identify x_1, y_1, x_2, y_2

$$(x_1, y_1) \text{ and } (x_2, y_2) \qquad \text{Use slope formula, } m = \frac{y_2 - y_1}{x_2 - x_1}$$

$$m = \frac{-5 - (-1)}{-4 - (-4)} \qquad \text{Simplify}$$

$$m = \frac{-4}{0} \qquad \text{Can't divide by zero, undefined}$$

$$m = \text{no slope} \qquad \text{Our Solution}$$

Example 129.

$$\text{Find the slope between } (3, 1) \text{ and } (-2, 1) \qquad \text{Identify } x_1, y_1, x_2, y_2$$

$$(x_1, y_1) \text{ and } (x_2, y_2) \qquad \text{Use slope formula, } m = \frac{y_2 - y_1}{x_2 - x_1}$$

$$m = \frac{1 - 1}{-2 - 3} \qquad \text{Simplify}$$

$$m = \frac{0}{-5} \qquad \text{Reduce}$$

$$m = 0 \qquad \text{Our Solution}$$

Again, there is a big difference between no slope and a zero slope. Zero is an integer and it has a value, the slope of a flat horizontal line. No slope has no value, it is undefined, the slope of a vertical line.

Using the slope formula we can also find missing points if we know what the slope is. This is shown in the following two examples.

Example 130.

Find the value of y between the points $(2, y)$ and $(5, -1)$ with slope -3

$$m = \frac{y_2 - y_1}{x_2 - x_1} \qquad \text{We will plug values into slope formula}$$

$$-3 = \frac{-1 - y}{5 - 2} \qquad \text{Simplify}$$

$$-3 = \frac{-1 - y}{3} \qquad \text{Multiply both sides by 3}$$

$$-3(3) = \frac{-1 - y}{3}(3) \qquad \text{Simplify}$$

$$-9 = -1 - y \qquad \text{Add 1 to both sides}$$

$$\underline{+1 \quad +1}$$

$$-8 = -y \qquad \text{Divide both sides by } -1$$

$$\overline{-1 \quad -1}$$

$$8 = y \qquad \text{Our Solution}$$

Example 131.

Find the value of x between the points $(-3, 2)$ and $(x, 6)$ with slope $\frac{2}{5}$

$$m = \frac{y_2 - y_1}{x_2 - x_1}$$ We will plug values into slope formula

$$\frac{2}{5} = \frac{6 - 2}{x - (-3)}$$ Simplify

$$\frac{2}{5} = \frac{4}{x + 3}$$ Multiply both sides by $(x + 3)$

$$\frac{2}{5}(x + 3) = 4$$ Multiply by 5 to clear fraction

$$(5)\frac{2}{5}(x + 3) = 4(5)$$ Simplify

$$2(x + 3) = 20$$ Distribute

$$2x + 6 = 20$$ Solve.

$$\underline{-6 \quad -6}$$ Subtract 6 from both sides

$$\frac{2x = 14}{2 \quad 2}$$ Divide each side by 2

$$x = 7$$ Our Solution

2.2 Practice - Slope

Find the slope of each line.

1)

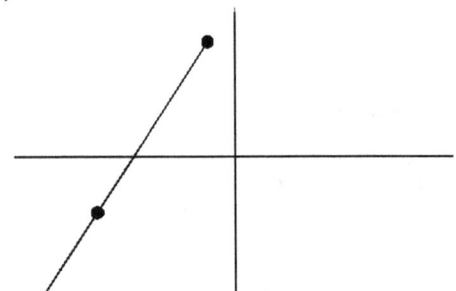

2)

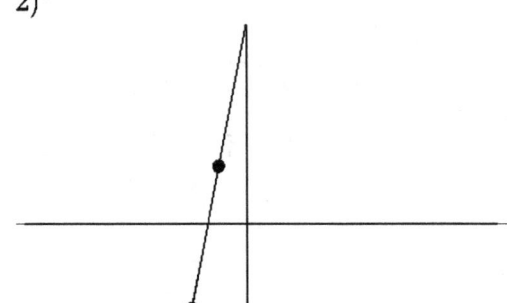

3)

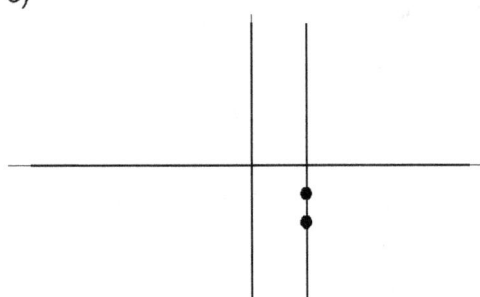

4)

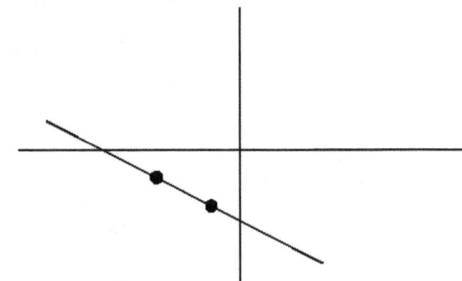

5)

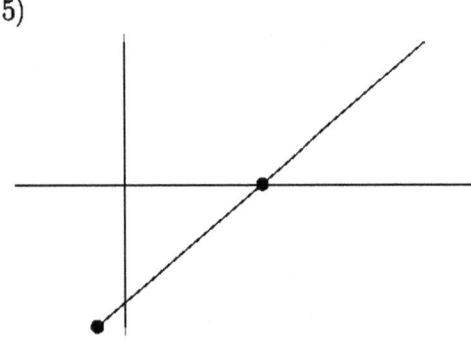

6)

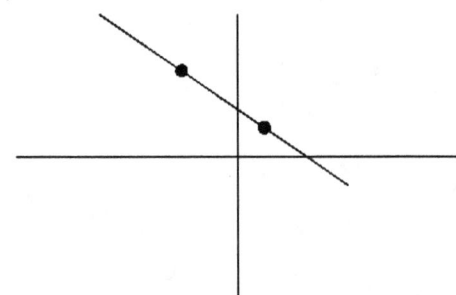

7)

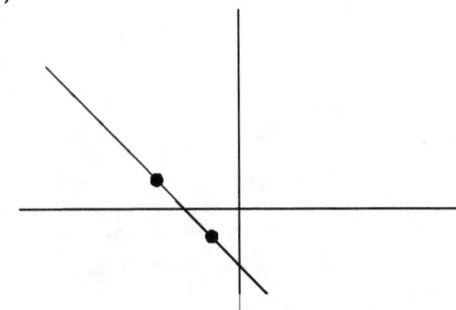

8)

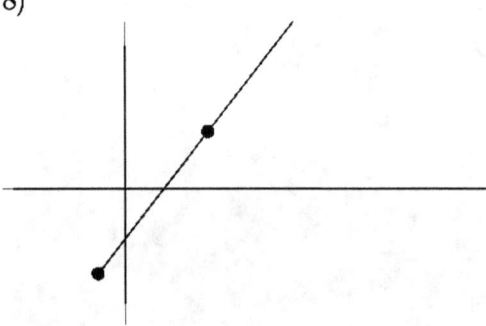

9)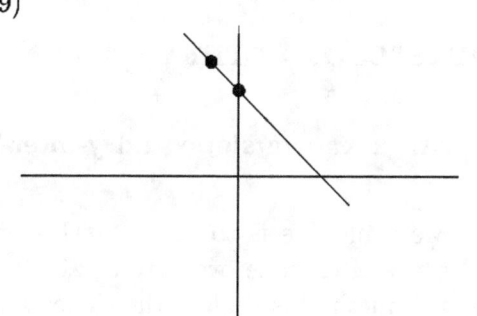

10)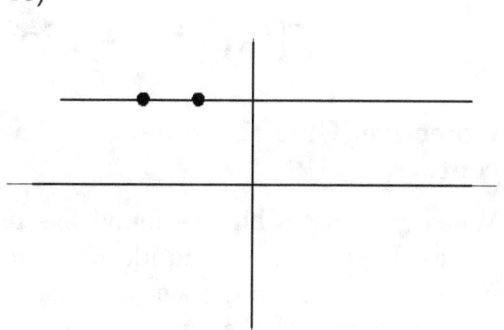

Find the slope of the line through each pair of points.

11) $(-2, 10), (-2, -15)$

12) $(1, 2), (-6, -14)$

13) $(-15, 10), (16, -7)$

14) $(13, -2), (7, 7)$

15) $(10, 18), (-11, -10)$

16) $(-3, 6), (-20, 13)$

17) $(-16, -14), (11, -14)$

18) $(13, 15), (2, 10)$

19) $(-4, 14), (-16, 8)$

20) $(9, -6), (-7, -7)$

21) $(12, -19), (6, 14)$

22) $(-16, 2), (15, -10)$

23) $(-5, -10), (-5, 20)$

24) $(8, 11), (-3, -13)$

25) $(-17, 19), (10, -7)$

26) $(11, -2), (1, 17)$

27) $(7, -14), (-8, -9)$

28) $(-18, -5), (14, -3)$

29) $(-5, 7), (-18, 14)$

30) $(19, 15), (5, 11)$

Find the value of x or y so that the line through the points has the given slope.

31) $(2, 6)$ and $(x, 2)$; slope: $\frac{4}{7}$

32) $(8, y)$ and $(-2, 4)$; slope: $-\frac{1}{5}$

33) $(-3, -2)$ and $(x, 6)$; slope: $-\frac{8}{5}$

34) $(-2, y)$ and $(2, 4)$; slope: $\frac{1}{4}$

35) $(-8, y)$ and $(-1, 1)$; slope: $\frac{6}{7}$

36) $(x, -1)$ and $(-4, 6)$; slope: $-\frac{7}{10}$

37) $(x, -7)$ and $(-9, -9)$; slope: $\frac{2}{5}$

38) $(2, -5)$ and $(3, y)$; slope: 6

39) $(x, 5)$ and $(8, 0)$; slope: $-\frac{5}{6}$

40) $(6, 2)$ and $(x, 6)$; slope: $-\frac{4}{5}$

Graphing - Slope-Intercept Form

Objective: Give the equation of a line with a known slope and y-intercept.

When graphing a line we found one method we could use is to make a table of values. However, if we can identify some properties of the line, we may be able to make a graph much quicker and easier. One such method is finding the slope and the y-intercept of the equation. The slope can be represented by m and the y-intercept, where it crosses the axis and $x = 0$, can be represented by $(0, b)$ where b is the value where the graph crosses the vertical y-axis. Any other point on the line can be represented by (x, y). Using this information we will look at the slope formula and solve the formula for y.

Example 132.

$$m, (0, b), (x, y) \qquad \text{Using the slope formula gives:}$$
$$\frac{y - b}{x - 0} = m \qquad \text{Simplify}$$
$$\frac{y - b}{x} = m \qquad \text{Multiply both sides by } x$$
$$y - b = mx \qquad \text{Add } b \text{ to both sides}$$
$$\underline{+b \quad +b}$$
$$y = mx + b \qquad \text{Our Solution}$$

This equation, $y = mx + b$ can be thought of as the equation of any line that as a slope of m and a y-intercept of b. This formula is known as the slope-intercept equation.

$$\textbf{Slope} - \textbf{Intercept Equation: } \boldsymbol{y = mx + b}$$

If we know the slope and the y-intercept we can easily find the equation that represents the line.

Example 133.

$$\text{Slope} = \frac{3}{4}, y - \text{intercept} = -3 \qquad \text{Use the slope} - \text{intercept equation}$$
$$y = mx + b \qquad m \text{ is the slope, } b \text{ is the } y - \text{intercept}$$
$$y = \frac{3}{4}x - 3 \qquad \text{Our Solution}$$

We can also find the equation by looking at a graph and finding the slope and y-intercept.

Example 134.

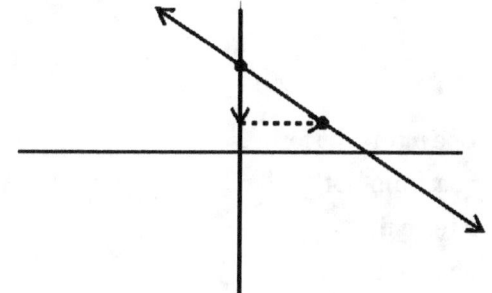

Identify the point where the graph crosses the y-axis (0,3). This means the y-intercept is 3.

Idenfity one other point and draw a slope triangle to find the slope. The slope is $-\frac{2}{3}$

$$y = mx + b \qquad \text{Slope-intercept equation}$$

$$y = -\frac{2}{3}x + 3 \qquad \text{Our Solution}$$

We can also move the opposite direction, using the equation identify the slope and y-intercept and graph the equation from this information. However, it will be important for the equation to first be in slope intercept form. If it is not, we will have to solve it for y so we can identify the slope and the y-intercept.

Example 135.

$$\begin{aligned}
\text{Write in slope} - \text{intercept form:} \; 2x - 4y &= 6 && \text{Solve for } y \\
-2x \qquad &\;\; -2x && \text{Subtract } 2x \text{ from both sides} \\
-4y &= -2x + 6 && \text{Put } x \text{ term first} \\
\overline{-4} \quad &\overline{-4} \; \overline{-4} && \text{Divide each term by } -4 \\
y &= \frac{1}{2}x - \frac{3}{2} && \text{Our Solution}
\end{aligned}$$

Once we have an equation in slope-intercept form we can graph it by first plotting the y-intercept, then using the slope, find a second point and connecting the dots.

Example 136.

$$\begin{aligned}
\text{Graph } y &= \frac{1}{2}x - 4 && \text{Recall the slope} - \text{intercept formula} \\
y &= mx + b && \text{Idenfity the slope, } m, \text{ and the } y - \text{intercept, } b \\
m &= \frac{1}{2}, b = -4 && \text{Make the graph}
\end{aligned}$$

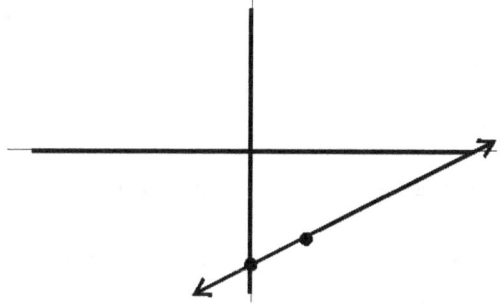

Starting with a point at the y-intercept of -4,

Then use the slope $\frac{\text{rise}}{\text{run}}$, so we will rise 1 unit and run 2 units to find the next point.

Once we have both points, connect the dots to get our graph.

World View Note: Before our current system of graphing, French Mathematician Nicole Oresme, in 1323 sugggested graphing lines that would look more like a bar graph with a constant slope!

Example 137.

$$\text{Graph } 3x + 4y = 12 \qquad \text{Not in slope intercept form}$$
$$\underline{-3x \qquad\quad -3x} \qquad \text{Subtract } 3x \text{ from both sides}$$
$$4y = -3x + 12 \qquad \text{Put the } x \text{ term first}$$
$$\overline{4 \qquad 4 \quad\; 4} \qquad \text{Divide each term by 4}$$
$$y = -\frac{3}{4}x + 3 \qquad \text{Recall slope} - \text{intercept equation}$$
$$y = mx + b \qquad \text{Idenfity } m \text{ and } b$$
$$m = -\frac{3}{4}, b = 3 \qquad \text{Make the graph}$$

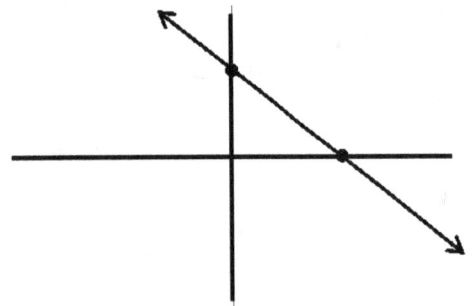

Starting with a point at the y-intercept of 3,

Then use the slope $\frac{\text{rise}}{\text{run}}$, but its negative so it will go downhill, so we will drop 3 units and run 4 units to find the next point.

Once we have both points, connect the dots to get our graph.

We want to be very careful not to confuse using slope to find the next point with use a coordinate such as $(4, -2)$ to find an individule point. Coordinates such as $(4, -2)$ start from the origin and move horizontally first, and vertically second. Slope starts from a point on the line that could be anywhere on the graph. The numerator is the vertical change and the denominator is the horizontal change.

Lines with zero slope or no slope can make a problem seem very different. Zero slope, or horiztonal line, will simply have a slope of zero which when multiplied by x gives zero. So the equation simply becomes $y = b$ or y is equal to the y-coordinate of the graph. If we have no slope, or a vertical line, the equation can't be written in slope intercept at all because the slope is undefined. There is no y in these equations. We will simply make x equal to the x-coordinate of the graph.

Example 138.

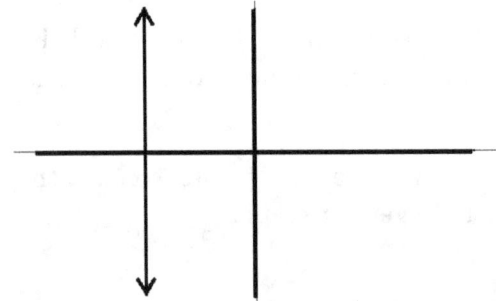

Give the equation of the line in the graph.

Because we have a vertical line and no slope there is no slope-intercept equation we can use. Rather we make x equal to the x-coordinate of -4

$$x = -4 \qquad \text{Our Solution}$$

2.3 Practice - Slope-Intercept

Write the slope-intercept form of the equation of each line given the slope and the y-intercept.

1) Slope = 2, y-intercept = 5

2) Slope = −6, y-intercept = 4

3) Slope = 1, y-intercept = −4

4) Slope = −1, y-intercept = −2

5) Slope = −$\frac{3}{4}$, y-intercept = −1

6) Slope = −$\frac{1}{4}$, y-intercept = 3

7) Slope = $\frac{1}{3}$, y-intercept = 1

8) Slope = $\frac{2}{5}$, y-intercept = 5

Write the slope-intercept form of the equation of each line.

9)

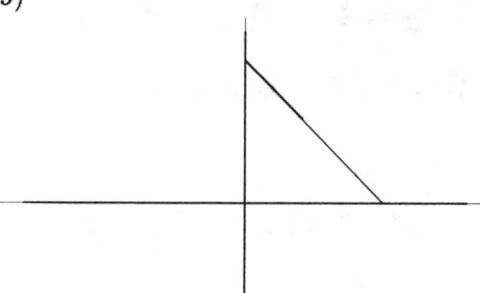

10)

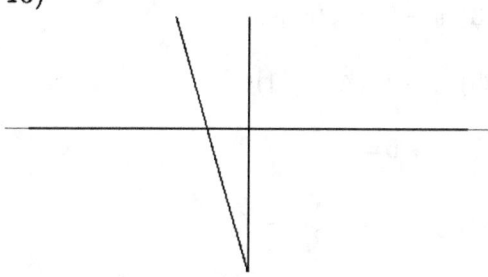

11)

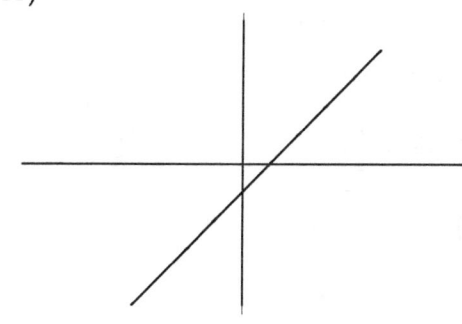

12)

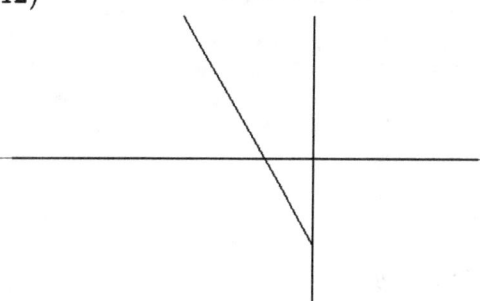

13)

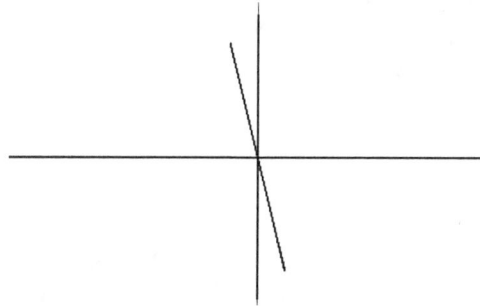

14)

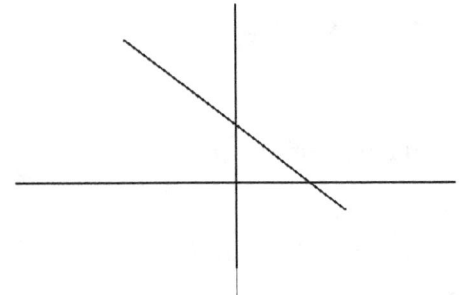

15) $x + 10y = -37$

16) $x - 10y = 3$

17) $2x + y = -1$

18) $6x - 11y = -70$

19) $7x - 3y = 24$

20) $4x + 7y = 28$

21) $x = -8$

22) $x - 7y = -42$

23) $y - 4 = -(x + 5)$

24) $y - 5 = \frac{5}{2}(x - 2)$

25) $y - 4 = 4(x - 1)$

26) $y - 3 = -\frac{2}{3}(x + 3)$

27) $y + 5 = -4(x - 2)$

28) $0 = x - 4$

29) $y + 1 = -\frac{1}{2}(x - 4)$

30) $y + 2 = \frac{6}{5}(x + 5)$

Sketch the graph of each line.

31) $y = \frac{1}{3}x + 4$

32) $y = -\frac{1}{5}x - 4$

33) $y = \frac{6}{5}x - 5$

34) $y = -\frac{3}{2}x - 1$

35) $y = \frac{3}{2}x$

36) $y = -\frac{3}{4}x + 1$

37) $x - y + 3 = 0$

38) $4x + 5 = 5y$

39) $-y - 4 + 3x = 0$

40) $-8 = 6x - 2y$

41) $-3y = -5x + 9$

42) $-3y = 3 - \frac{3}{2}x$

Graphing - Point-Slope Form

Objective: Give the equation of a line with a known slope and point.

The slope-intercept form has the advantage of being simple to remember and use, however, it has one major disadvantage: we must know the y-intercept in order to use it! Generally we do not know the y-intercept, we only know one or more points (that are not the y-intercept). In these cases we can't use the slope intercept equation, so we will use a different more flexible formula. If we let the slope of an equation be m, and a specific point on the line be (x_1, y_1), and any other point on the line be (x, y). We can use the slope formula to make a second equation.

Example 139.

$$
\begin{aligned}
&m, (x_1, y_1), (x, y) &&\text{Recall slope formula} \\
&\frac{y_2 - y_1}{x_2 - x_1} = m &&\text{Plug in values} \\
&\frac{y - y_1}{x - x_1} = m &&\text{Multiply both sides by } (x - x_1) \\
&y - y_1 = m(x - x_1) &&\text{Our Solution}
\end{aligned}
$$

If we know the slope, m of an equation and any point on the line (x_1, y_1) we can easily plug these values into the equation above which will be called the point-slope formula.

$$\textbf{Point} - \textbf{Slope Formula: } \boldsymbol{y - y_1 = m(x - x_1)}$$

Example 140.

Write the equation of the line through the point $(3, -4)$ with a slope of $\frac{3}{5}$.

$$
\begin{aligned}
&y - y_1 = m(x - x_1) &&\text{Plug values into point} - \text{slope formula} \\
&y - (-4) = \frac{3}{5}(x - 3) &&\text{Simplify signs} \\
&y + 4 = \frac{3}{5}(x - 3) &&\text{Our Solution}
\end{aligned}
$$

Often, we will prefer final answers be written in slope intercept form. If the directions ask for the answer in slope-intercept form we will simply distribute the slope, then solve for y.

Example 141.

Write the equation of the line through the point $(-6, 2)$ with a slope of $-\frac{2}{3}$ in slope-intercept form.

$$
\begin{aligned}
y - y_1 &= m(x - x_1) \qquad \text{Plug values into point} - \text{slope formula} \\
y - 2 &= -\frac{2}{3}(x - (-6)) \qquad \text{Simplify signs} \\
y - 2 &= -\frac{2}{3}(x + 6) \qquad \text{Distribute slope} \\
y - 2 &= -\frac{2}{3}x - 4 \qquad \text{Solve for } y \\
\underline{+2 \qquad\qquad +2} & \\
y &= -\frac{2}{3}x - 2 \qquad \text{Our Solution}
\end{aligned}
$$

An important thing to observe about the point slope formula is that the operation between the x's and y's is subtraction. This means when you simplify the signs you will have the opposite of the numbers in the point. We need to be very careful with signs as we use the point-slope formula.

In order to find the equation of a line we will always need to know the slope. If we don't know the slope to begin with we will have to do some work to find it first before we can get an equation.

Example 142.

Find the equation of the line through the points $(-2, 5)$ and $(4, -3)$.

$$
\begin{aligned}
m &= \frac{y_2 - y_1}{x_2 - x_1} \qquad \text{First we must find the slope} \\
m &= \frac{-3 - 5}{4 - (-2)} = \frac{-8}{6} = -\frac{4}{3} \qquad \text{Plug values in slope formula and evaluate} \\
y - y_1 &= m(x - x_1) \qquad \text{With slope and either point, use point} - \text{slope formula} \\
y - 5 &= -\frac{4}{3}(x - (-2)) \qquad \text{Simplify signs} \\
y - 5 &= -\frac{4}{3}(x + 2) \qquad \text{Our Solution}
\end{aligned}
$$

Example 143.

Find the equation of the line through the points $(-3, 4)$ and $(-1, -2)$ in slope-intercept form.

$$m = \frac{y_2 - y_1}{x_2 - x_1}$$ First we must find the slope

$$m = \frac{-2 - 4}{-1 - (-3)} = \frac{-6}{2} = -3$$ Plug values in slope formula and evaluate

$$y - y_1 = m(x - x_1)$$ With slope and either point, point − slope formula

$$y - 4 = -3(x - (-3))$$ Simplify signs

$$y - 4 = -3(x + 3)$$ Distribute slope

$$y - 4 = -3x - 9$$ Solve for y

$$\underline{+4 \qquad\qquad +4}$$ Add 4 to both sides

$$y = -3x - 5$$ Our Solution

Example 144.

Find the equation of the line through the points $(6, -2)$ and $(-4, 1)$ in slope-intercept form.

$$m = \frac{y_2 - y_1}{x_2 - x_1}$$ First we must find the slope

$$m = \frac{1 - (-2)}{-4 - 6} = \frac{3}{-10} = -\frac{3}{10}$$ Plug values into slope formula and evaluate

$$y - y_1 = m(x - x_1)$$ Use slope and either point, use point − slope formula

$$y - (-2) = -\frac{3}{10}(x - 6)$$ Simplify signs

$$y + 2 = -\frac{3}{10}(x - 6)$$ Distribute slope

$$y + 2 = -\frac{3}{10}x + \frac{9}{5}$$ Solve for y. Subtract 2 from both sides

$$\underline{-2 \qquad\qquad -\frac{10}{5}}$$ Using $\frac{10}{5}$ on right so we have a common denominator

$$y = -\frac{3}{10}x - \frac{1}{5}$$ Our Solution

World View Note: The city of Konigsberg (now Kaliningrad, Russia) had a river that flowed through the city breaking it into several parts. There were 7 bridges that connected the parts of the city. In 1735 Leonhard Euler considered the question of whether it was possible to cross each bridge exactly once and only once. It turned out that this problem was impossible, but the work laid the foundation of what would become graph theory.

2.4 Practice - Point-Slope Form

Write the point-slope form of the equation of the line through the given point with the given slope.

1) through $(2, 3)$, slope $=$ undefined

2) through $(1, 2)$, slope $=$ undefined

3) through $(2, 2)$, slope $= \frac{1}{2}$

4) through $(2, 1)$, slope $= -\frac{1}{2}$

5) through $(-1, -5)$, slope $= 9$

6) through $(2, -2)$, slope $= -2$

7) through $(-4, 1)$, slope $= \frac{3}{4}$

8) through $(4, -3)$, slope $= -2$

9) through $(0, -2)$, slope $= -3$

10) through $(-1, 1)$, slope $= 4$

11) through $(0, -5)$, slope $= -\frac{1}{4}$

12) through $(0, 2)$, slope $= -\frac{5}{4}$

13) through $(-5, -3)$, slope $= \frac{1}{5}$

14) through $(-1, -4)$, slope $= -\frac{2}{3}$

15) through $(-1, 4)$, slope $= -\frac{5}{4}$

16) through $(1, -4)$, slope $= -\frac{3}{2}$

Write the slope-intercept form of the equation of the line through the given point with the given slope.

17) through: $(-1, -5)$, slope $= 2$

18) through: $(2, -2)$, slope $= -2$

19) through: $(5, -1)$, slope $= -\frac{3}{5}$

20) through: $(-2, -2)$, slope $= -\frac{2}{3}$

21) through: $(-4, 1)$, slope $= \frac{1}{2}$

22) through: $(4, -3)$, slope $= -\frac{7}{4}$

23) through: $(4, -2)$, slope $= -\frac{3}{2}$

24) through: $(-2, 0)$, slope $= -\frac{5}{2}$

25) through: $(-5, -3)$, slope $= -\frac{2}{5}$

26) through: $(3, 3)$, slope $= \frac{7}{3}$

27) through: $(2, -2)$, slope $= 1$

28) through: $(-4, -3)$, slope $= 0$

29) through:$(-3, 4)$, slope$-$undefined

30) through: $(-2, -5)$, slope $= 2$

31) through: $(-4, 2)$, slope $= -\frac{1}{2}$

32) through: $(5, 3)$, slope $= \frac{6}{5}$

Write the point-slope form of the equation of the line through the given points.

33) through: $(-4, 3)$ and $(-3, 1)$

34) through: $(1, 3)$ and $(-3, 3)$

35) through: $(5, 1)$ and $(-3, 0)$

36) through: $(-4, 5)$ and $(4, 4)$

37) through: $(-4, -2)$ and $(0, 4)$

38) through: $(-4, 1)$ and $(4, 4)$

39) through: $(3, 5)$ and $(-5, 3)$

40) through: $(-1, -4)$ and $(-5, 0)$

41) through: $(3, -3)$ and $(-4, 5)$

42) through: $(-1, -5)$ and $(-5, -4)$

Write the slope-intercept form of the equation of the line through the given points.

43) through: $(-5, 1)$ and $(-1, -2)$

44) through: $(-5, -1)$ and $(5, -2)$

45) through: $(-5, 5)$ and $(2, -3)$

46) through: $(1, -1)$ and $(-5, -4)$

47) through: $(4, 1)$ and $(1, 4)$

48) through: $(0, 1)$ and $(-3, 0)$

49) through: $(0, 2)$ and $(5, -3)$

50) through: $(0, 2)$ and $(2, 4)$

51) through: $(0, 3)$ and $(-1, -1)$

52) through: $(-2, 0)$ and $(5, 3)$

Graphing - Parallel and Perpendicular Lines

Objective: Identify the equation of a line given a parallel or perpendicular line.

There is an interesting connection between the slope of lines that are parallel and the slope of lines that are perpendicular (meet at a right angle). This is shown in the following example.

Example 145.

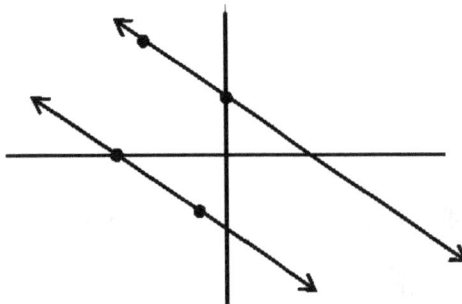 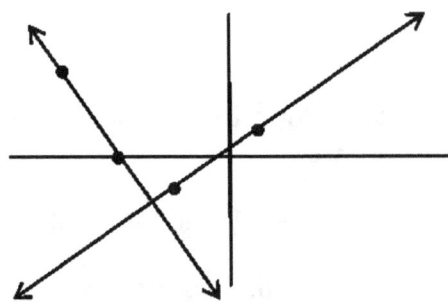

The above graph has two parallel lines. The slope of the top line is down 2, run 3, or $-\frac{2}{3}$. The slope of the bottom line is down 2, run 3 as well, or $-\frac{2}{3}$.

The above graph has two perpendicular lines. The slope of the flatter line is up 2, run 3 or $\frac{2}{3}$. The slope of the steeper line is down 3, run 2 or $-\frac{3}{2}$.

World View Note: Greek Mathematician Euclid lived around 300 BC and published a book titled, *The Elements*. In it is the famous parallel postulate which mathematicians have tried for years to drop from the list of postulates. The attempts have failed, yet all the work done has developed new types of geometries!

As the above graphs illustrate, parallel lines have the same slope and perpendicular lines have opposite (one positive, one negative) reciprocal (flipped fraction) slopes. We can use these properties to make conclusions about parallel and perpendicular lines.

Example 146.

Find the slope of a line parallel to $5y - 2x = 7$.

$$5y - 2x = 7 \qquad \text{To find the slope we will put equation in slope} - \text{intercept form}$$
$$\underline{+2x + 2x} \qquad \text{Add } 2x \text{ to both sides}$$
$$5y = 2x + 7 \qquad \text{Put } x \text{ term first}$$
$$\overline{5} \quad \overline{5} \quad \overline{5} \qquad \text{Divide each term by 5}$$
$$y = \frac{2}{5}x + \frac{7}{5} \qquad \text{The slope is the coefficient of } x$$

112

$$m = \frac{2}{5} \quad \text{Slope of first line. Parallel lines have the same slope}$$

$$m = \frac{2}{5} \quad \text{Our Solution}$$

Example 147.

Find the slope of a line perpendicular to $3x - 4y = 2$

$$\begin{aligned}
3x - 4y &= 2 & &\text{To find slope we will put equation in slope} - \text{intercept form} \\
\underline{-3x \qquad -3x} & & &\text{Subtract } 3x \text{ from both sides} \\
-4y &= -3x + 2 & &\text{Put } x \text{ term first} \\
\overline{-4 \quad -4 -4} & & &\text{Divide each term by} - 4 \\
y &= \frac{3}{4}x - \frac{1}{2} & &\text{The slope is the coefficient of } x
\end{aligned}$$

$$m = \frac{3}{4} \quad \text{Slope of first lines. Perpendicular lines have opposite reciprocal slopes}$$

$$m = -\frac{4}{3} \quad \text{Our Solution}$$

Once we have a slope, it is possible to find the complete equation of the second line if we know one point on the second line.

Example 148.

Find the equation of a line through $(4, -5)$ and parallel to $2x - 3y = 6$.

$$\begin{aligned}
2x - 3y &= 6 & &\text{We first need slope of parallel line} \\
\underline{-2x \qquad -2x} & & &\text{Subtract } 2x \text{ from each side} \\
-3y &= -2x + 6 & &\text{Put } x \text{ term first} \\
\overline{-3 \quad -3 -3} & & &\text{Divide each term by} - 3 \\
y &= \frac{2}{3}x - 2 & &\text{Identify the slope, the coefficient of } x
\end{aligned}$$

$$m = \frac{2}{3} \quad \text{Parallel lines have the same slope}$$

$$m = \frac{2}{3} \quad \text{We will use this slope and our point } (4, -5)$$

$$\begin{aligned}
y - y_1 &= m(x - x_1) & &\text{Plug this information into point slope formula} \\
y - (-5) &= \frac{2}{3}(x - 4) & &\text{Simplify signs}
\end{aligned}$$

$$y + 5 = \frac{2}{3}(x - 4) \quad \text{Our Solution}$$

Example 149.

Find the equation of the line through $(6, -9)$ perpendicular to $y = -\frac{3}{5}x + 4$ in slope-intercept form.

$$y = -\frac{3}{5}x + 4 \qquad \text{Identify the slope, coefficient of } x$$

$$m = -\frac{3}{5} \qquad \text{Perpendicular lines have opposite reciprocal slopes}$$

$$m = \frac{5}{3} \qquad \text{We will use this slope and our point } (6, -9)$$

$$y - y_1 = m(x - x_1) \qquad \text{Plug this information into point} - \text{slope formula}$$

$$y - (-9) = \frac{5}{3}(x - 6) \qquad \text{Simplify signs}$$

$$y + 9 = \frac{5}{3}(x - 6) \qquad \text{Distribute slope}$$

$$y + 9 = \frac{5}{3}x - 10 \qquad \text{Solve for } y$$
$$\underline{-9 \qquad\qquad -9} \qquad \text{Subtract 9 from both sides}$$
$$y = \frac{5}{3}x - 19 \qquad \text{Our Solution}$$

Zero slopes and no slopes may seem like opposites (one is a horizontal line, one is a vertical line). Because a horizontal line is perpendicular to a vertical line we can say that no slope and zero slope are actually perpendicular slopes!

Example 150.

Find the equation of the line through $(3, 4)$ perpendicular to $x = -2$

$$x = -2 \qquad \text{This equation has no slope, } a \text{ vertical line}$$
$$\text{no slope} \qquad \text{Perpendicular line then would have } a \text{ zero slope}$$
$$m = 0 \qquad \text{Use this and our point } (3, 4)$$
$$y - y_1 = m(x - x_1) \qquad \text{Plug this information into point} - \text{slope formula}$$
$$y - 4 = 0(x - 3) \qquad \text{Distribute slope}$$
$$y - 4 = 0 \qquad \text{Solve for } y$$
$$\underline{+4 + 4} \qquad \text{Add 4 to each side}$$
$$y = 4 \qquad \text{Our Solution}$$

Being aware that to be perpendicular to a vertical line means we have a horizontal line through a y value of 4, thus we could have jumped from this point right to the solution, $y = 4$.

114

2.5 Practice - Parallel and Perpendicular Lines

Find the slope of a line parallel to each given line.

1) $y = 2x + 4$

2) $y = -\frac{2}{3}x + 5$

3) $y = 4x - 5$

4) $y = -\frac{10}{3}x - 5$

5) $x - y = 4$

6) $6x - 5y = 20$

7) $7x + y = -2$

8) $3x + 4y = -8$

Find the slope of a line perpendicular to each given line.

9) $x = 3$

10) $y = -\frac{1}{2}x - 1$

11) $y = -\frac{1}{3}x$

12) $y = \frac{4}{5}x$

13) $x - 3y = -6$

14) $3x - y = -3$

15) $x + 2y = 8$

16) $8x - 3y = -9$

Write the point-slope form of the equation of the line described.

17) through: $(2, 5)$, parallel to $x = 0$

18) through: $(5, 2)$, parallel to $y = \frac{7}{5}x + 4$

19) through: $(3, 4)$, parallel to $y = \frac{9}{2}x - 5$

20) through: $(1, -1)$, parallel to $y = -\frac{3}{4}x + 3$

21) through: $(2, 3)$, parallel to $y = \frac{7}{5}x + 4$

22) through: $(-1, 3)$, parallel to $y = -3x - 1$

23) through: $(4, 2)$, parallel to $x = 0$

24) through: $(1, 4)$, parallel to $y = \frac{7}{5}x + 2$

25) through: $(1, -5)$, perpendicular to $-x + y = 1$

26) through: $(1, -2)$, perpendicular to $-x + 2y = 2$

27) through: $(5, 2)$, perpendicular to $5x + y = -3$

28) through: $(1, 3)$, perpendicular to $-x + y = 1$

29) through: $(4, 2)$, perpendicular to $-4x + y = 0$

30) through: $(-3, -5)$, perpendicular to $3x + 7y = 0$

31) through: $(2, -2)$ perpendicular to $3y - x = 0$

32) through: $(-2, 5)$. perpendicular to $y - 2x = 0$

Write the slope-intercept form of the equation of the line described.

33) through: $(4, -3)$, parallel to $y = -2x$

34) through: $(-5, 2)$, parallel to $y = \frac{3}{5}x$

35) through: $(-3, 1)$, parallel to $y = -\frac{4}{3}x - 1$

36) through: $(-4, 0)$, parallel to $y = -\frac{5}{4}x + 4$

37) through: $(-4, -1)$, parallel to $y = -\frac{1}{2}x + 1$

38) through: $(2, 3)$, parallel to $y = \frac{5}{2}x - 1$

39) through: $(-2, -1)$, parallel to $y = -\frac{1}{2}x - 2$

40) through: $(-5, -4)$, parallel to $y = \frac{3}{5}x - 2$

41) through: $(4, 3)$, perpendicular to $x + y = -1$

42) through: $(-3, -5)$, perpendicular to $x + 2y = -4$

43) through: $(5, 2)$, perpendicular to $x = 0$

44) through: $(5, -1)$, perpendicular to $-5x + 2y = 10$

45) through: $(-2, 5)$, perpendicular to $-x + y = -2$

46) through: $(2, -3)$, perpendicular to $-2x + 5y = -10$

47) through: $(4, -3)$, perpendicular to $-x + 2y = -6$

48) through: $(-4, 1)$, perpendicular to $4x + 3y = -9$

Chapter 3 : Inequalities

Inequalities - Solve and Graph Inequalities

Objective: Solve, graph, and give interval notation for the solution to linear inequalities.

When we have an equation such as $x = 4$ we have a specific value for our variable. With inequalities we will give a range of values for our variable. To do this we will not use equals, but one of the following symbols:

$>$	Greater than
$\geqslant$	Greater than or equal to
$<$	Less than
$\leqslant$	Less than or equal to

World View Note: English mathematician Thomas Harriot first used the above symbols in 1631. However, they were not immediately accepted as symbols such as $\sqsubset$ and $\sqsupset$ were already coined by another English mathematician, William Oughtred.

If we have an expression such as $x < 4$, this means our variable can be any number smaller than 4 such as $-2, 0, 3, 3.9$ or even 3.999999999 as long as it is smaller than 4. If we have an expression such as $x \geqslant -2$, this means our variable can be any number greater than or equal to -2, such as $5, 0, -1, -1.9999$, or even -2.

Because we don't have one set value for our variable, it is often useful to draw a picture of the solutions to the inequality on a number line. We will start from the value in the problem and bold the lower part of the number line if the variable is smaller than the number, and bold the upper part of the number line if the variable is larger. The value itself we will mark with brackets, either) or (for less than or greater than respectively, and] or [for less than or equal to or greater than or equal to respectively.

Once the graph is drawn we can quickly convert the graph into what is called interval notation. Interval notation gives two numbers, the first is the smallest value, the second is the largest value. If there is no largest value, we can use ∞ (infinity). If there is no smallest value, we can use $-\infty$ negative infinity. If we use either positive or negative infinity we will always use a curved bracket for that value.

Example 151.

Graph the inequality and give the interval notation

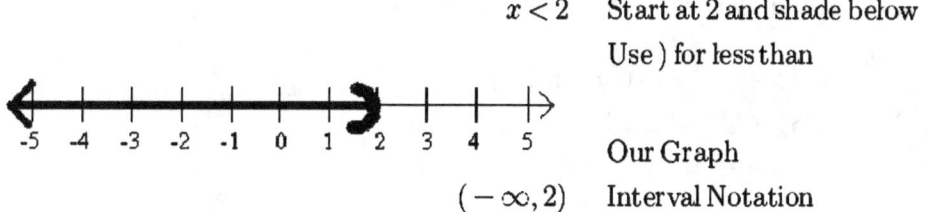

$x < 2$ Start at 2 and shade below
Use) for less than

Our Graph

$(-\infty, 2)$ Interval Notation

Example 152.

Graph the inequality and give the interval notation

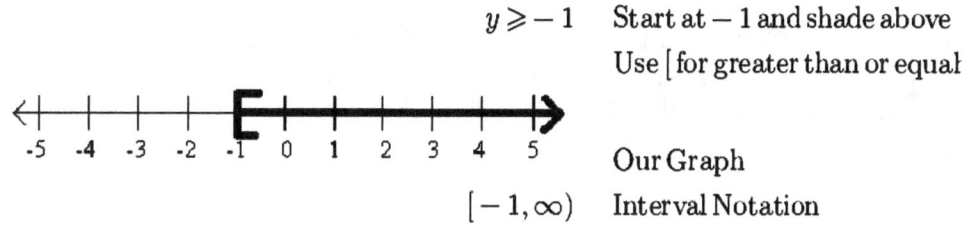

$y \geqslant -1$ Start at -1 and shade above
Use [for greater than or equal

Our Graph

$[-1, \infty)$ Interval Notation

We can also take a graph and find the inequality for it.

Example 153.

Give the inequality for the graph:

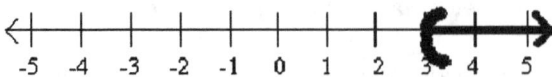

Graph starts at 3 and goes up or greater. Curved bracket means just greater than

$$x > 3 \quad \text{Our Solution}$$

Example 154.

Give the inequality for the graph:

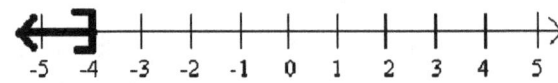

Graph starts at -4 and goes down or less. Square bracket means less than or equal to

$$x \leqslant -4 \quad \text{Our Solution}$$

Generally when we are graphing and giving interval notation for an inequality we will have to first solve the inequality for our variable. Solving inequalities is very similar to solving equations with one exception. Consider the following inequality and what happens when various operations are done to it. Notice what happens to the inequality sign as we add, subtract, multiply and divide by both positive and negative numbers to keep the statment a true statement.

$$
\begin{array}{ll}
5 > 1 & \text{Add 3 to both sides} \\
8 > 4 & \text{Subtract 2 from both sides} \\
6 > 2 & \text{Multiply both sides by 3} \\
12 > 6 & \text{Divide both sides by 2} \\
6 > 3 & \text{Add} -1 \text{ to both sides} \\
5 > 2 & \text{Subtract} -4 \text{ from both sides} \\
9 > 6 & \text{Multiply both sides by} -2 \\
-18 < -12 & \text{Divide both sides by} -6 \\
3 > 2 & \text{Symbol flipped when we multiply or divide by } a \text{ negative!}
\end{array}
$$

As the above problem illustrates, we can add, subtract, multiply, or divide on both sides of the inequality. But if we multiply or divide by a negative number, the symbol will need to flip directions. We will keep that in mind as we solve inequalities.

Example 155.

Solve and give interval notation

$$5 - 2x \geqslant 11 \quad \text{Subtract 5 from both sides}$$

$$\underline{-5 \qquad -5}$$

$-2x \geqslant 6$ Divide both sides by -2

$\overline{-2\ -2}$ Divide by a negative – flip symbol!

$x \leqslant -3$ Graph, starting at -3, going down with] for less than or equal to

$(-\infty, -3]$ Interval Notation

The inequality we solve can get as complex as the linear equations we solved. We will use all the same patterns to solve these inequalities as we did for solving equations. Just remember that any time we multiply or divide by a negative the symbol switches directions (multiplying or dividing by a positive does not change the symbol!)

Example 156.

Solve and give interval notation

$3(2x - 4) + 4x < 4(3x - 7) + 8$ Distribute

$6x - 12 + 4x < 12x - 28 + 8$ Combine like terms

$10x - 12 < 12x - 20$ Move variable to one side

$\underline{-10x \qquad\ -10x}$ Subtract $10x$ from both sides

$-12 < 2x - 20$ Add 20 to both sides

$\underline{+20 \qquad +20}$

$8 < 2x$ Divide both sides by 2

$\overline{2\ \ 2}$

$4 < x$ Be careful with graph, x is larger!

$(4, \infty)$ Interval Notation

It is important to be careful when the inequality is written backwards as in the previous example ($4 < x$ rather than $x > 4$). Often students draw their graphs the wrong way when this is the case. The inequality symbol opens to the variable, this means the variable is greater than 4. So we must shade above the 4.

3.1 Practice - Solve and Graph Inequalities

Draw a graph for each inequality and give interval notation.

1) $n > -5$

2) $n > 4$

3) $-2 \geqslant k$

4) $1 \geqslant k$

5) $5 \geqslant x$

6) $-5 < x$

Write an inequality for each graph.

7)

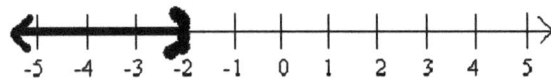

8)

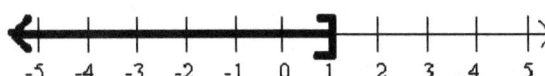

9)

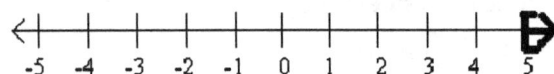

10)

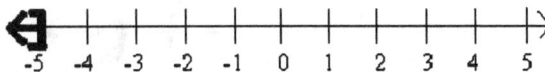

11)

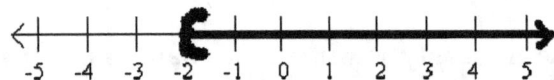

12)

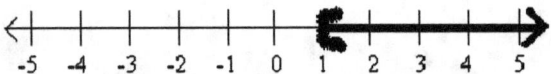

Solve each inequality, graph each solution, and give interval notation.

13) $\frac{x}{11} \geqslant 10$

14) $-2 \leqslant \frac{n}{13}$

15) $2 + r < 3$

16) $\frac{m}{5} \leqslant -\frac{6}{5}$

17) $8 + \frac{n}{3} \geqslant 6$

18) $11 > 8 + \frac{x}{2}$

19) $2 > \frac{a-2}{5}$

20) $\frac{v}{4} \frac{9}{} \leqslant 2$

21) $-47 \geqslant 8 - 5x$

22) $\frac{6+x}{12} \leqslant -1$

23) $-2(3 + k) < -44$

24) $-7n - 10 \geqslant 60$

25) $18 < -2(-8 + p)$

26) $5 \geqslant \frac{x}{5} + 1$

27) $24 \geqslant -6(m - 6)$

28) $-8(n - 5) \geqslant 0$

29) $-r - 5(r - 6) < -18$

30) $-60 \geqslant -4(-6x - 3)$

31) $24 + 4b < 4(1 + 6b)$

32) $-8(2 - 2n) \geqslant -16 + n$

33) $-5v - 5 < -5(4v + 1)$

34) $-36 + 6x > -8(x + 2) + 4x$

35) $4 + 2(a + 5) < -2(-a - 4)$

36) $3(n + 3) + 7(8 - 8n) < 5n + 5 + 2$

37) $-(k - 2) > -k - 20$

38) $-(4 - 5p) + 3 \geqslant -2(8 - 5p)$

Inequalities - Compound Inequalities

Objective: Solve, graph and give interval notation to the solution of compound inequalities.

Several inequalities can be combined together to form what are called compound inequalities. There are three types of compound inequalities which we will investigate in this lesson.

The first type of a compound inequality is an OR inequality. For this type of inequality we want a true statment from either one inequality OR the other inequality OR both. When we are graphing these type of inequalities we will graph each individual inequality above the number line, then move them both down together onto the actual number line for our graph that combines them together.

When we give interval notation for our solution, if there are two different parts to the graph we will put a $\cup$ (union) symbol between two sets of interval notation, one for each part.

Example 157.

Solve each inequality, graph the solution, and give interval notation of solution

$$2x - 5 > 3 \quad \text{or} \quad 4 - x \geqslant 6 \qquad \text{Solve each inequality}$$
$$\underline{+5 + 5 \quad\quad -4 \quad\quad -4} \qquad \text{Add or subtract first}$$
$$2x > 8 \quad \text{or} \quad -x \geqslant 2 \qquad \text{Divide}$$
$$\overline{2 \quad 2} \quad\quad \overline{-1 \, -1} \qquad \text{Dividing by negative flips sign}$$
$$x > 4 \quad \text{or} \quad x \leqslant -2 \qquad \text{Graph the inequalities separatly above number line}$$

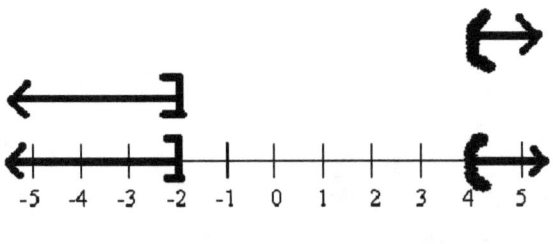

$$(-\infty, -2] \cup (4, \infty) \quad \text{Interval Notation}$$

World View Note: The symbol for infinity was first used by the Romans, although at the time the number was used for 1000. The greeks also used the symbol for 10,000.

There are several different results that could result from an OR statement. The graphs could be pointing different directions, as in the graph above, or pointing in the same direction as in the graph below on the left, or pointing opposite directions, but overlapping as in the graph below on the right. Notice how interval notation works for each of these cases.

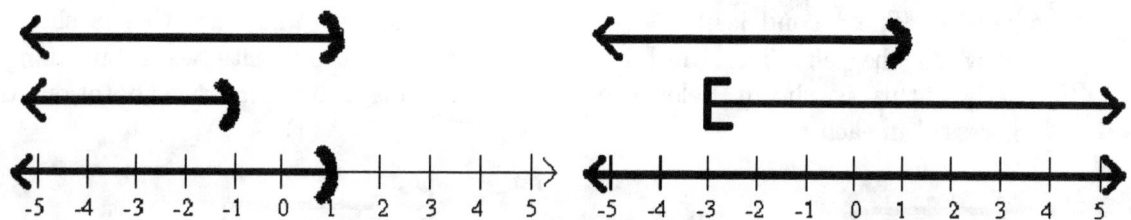

As the graphs overlap, we take the largest graph for our solution.	When the graphs are combined they cover the entire number line.
Interval Notation: $(-\infty, 1)$	Interval Notation: $(-\infty, \infty)$ or $\mathbb{R}$

The second type of compound inequality is an AND inequality. AND inequalities require both statements to be true. If one is false, they both are false. When we graph these inequalities we can follow a similar process, first graph both inequalities above the number line, but this time only where they overlap will be drawn onto the number line for our final graph. When our solution is given in interval notation it will be expressed in a manner very similar to single inequalities (there is a symbol that can be used for AND, the intersection - $\cap$, but we will not use it here).

Example 158.

Solve each inequality, graph the solution, and express it interval notation.

$$2x + 8 \geqslant 5x - 7 \text{ and } 5x - 3 > 3x + 1 \quad \text{Move variables to one side}$$
$$\underline{-2x \qquad -2x \qquad \quad -3x \quad -3x}$$
$$8 \geqslant 3x - 7 \text{ and } 2x - 3 > 1 \quad \text{Add 7 or 3 to both sides}$$
$$\underline{+7 \qquad +7 \qquad \qquad +3+3}$$
$$\frac{15}{3} \geqslant \frac{3x}{3} \text{ and } \frac{2x}{2} > \frac{4}{2} \quad \text{Divide}$$
$$5 \geqslant x \text{ and } x > 2 \quad \text{Graph, } x \text{ is smaller (or equal) than 5, greater than 2}$$

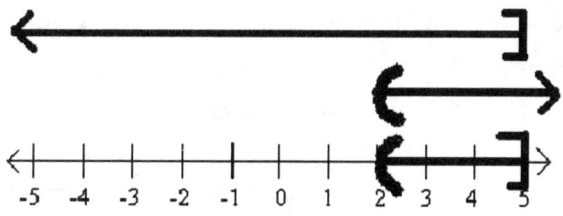

$(2, 5]$ Interval Notation

Again, as we graph AND inequalities, only the overlapping parts of the individual graphs makes it to the final number line. As we graph AND inequalities there are also three different types of results we could get. The first is shown in the above

125

example. The second is if the arrows both point the same way, this is shown below on the left. The third is if the arrows point opposite ways but don't overlap, this is shown below on the right. Notice how interval notation is expressed in each case.

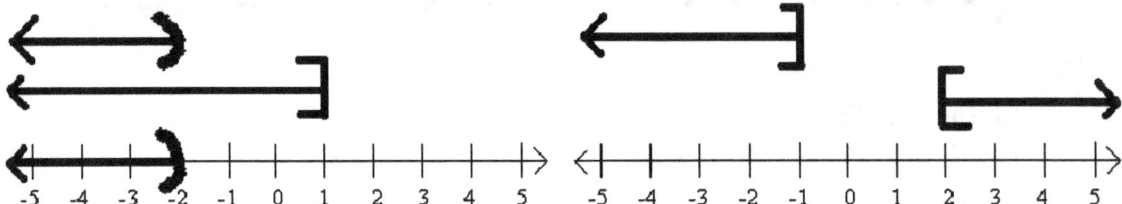

In this graph, the overlap is only the smaller graph, so this is what makes it to the final number line.

Interval Notation: $(-\infty, -2)$

In this graph there is no overlap of the parts. Because their is no overlap, no values make it to the final number line.

Interval Notation: No Solution or $\varnothing$

The third type of compound inequality is a special type of AND inequality. When our variable (or expression containing the variable) is between two numbers, we can write it as a single math sentence with three parts, such as $5 < x \leqslant 8$, to show x is between 5 and 8 (or equal to 8). When solving these type of inequalities, because there are three parts to work with, to stay balanced we will do the same thing to all three parts (rather than just both sides) to isolate the variable in the middle. The graph then is simply the values between the numbers with appropriate brackets on the ends.

Example 159.

Solve the inequality, graph the solution, and give interval notation.

$$-6 \leqslant -4x + 2 < 2 \qquad \text{Subtract 2 from all three parts}$$
$$\underline{-2 \qquad\quad -2-2}$$
$$-8 \leqslant -4x < 0 \qquad \text{Divide all three parts by} -4$$
$$\overline{-4} \quad \overline{-4} \ \overline{-4} \qquad \text{Dividing by } a \text{ negative flips the symbols}$$
$$2 \geqslant x > 0 \qquad \text{Flip entire statement so values get larger left to right}$$
$$0 < x \leqslant 2 \qquad \text{Graph } x \text{ between 0 and 2}$$

$(0, 2]$ Interval Notation

3.2 Practice - Compound Inequalities

Solve each compound inequality, graph its solution, and give interval notation.

1) $\frac{n}{3} \leqslant -3$ or $-5n \leqslant -10$

2) $6m \geqslant -24$ or $m - 7 < -12$

3) $x + 7 \geqslant 12$ or $9x < -45$

4) $10r > 0$ or $r - 5 < -12$

5) $x - 6 < -13$ or $6x \leqslant -60$

6) $9 + n < 2$ or $5n > 40$

7) $\frac{v}{8} > -1$ and $v - 2 < 1$

8) $-9x < 63$ and $\frac{x}{4} < 1$

9) $-8 + b < -3$ and $4b < 20$

10) $-6n \leqslant 12$ and $\frac{n}{3} \leqslant 2$

11) $a + 10 \geqslant 3$ and $8a \leqslant 48$

12) $-6 + v \geqslant 0$ and $2v > 4$

13) $3 \leqslant 9 + x \leqslant 7$

14) $0 \geqslant \frac{x}{9} \geqslant -1$

15) $11 < 8 + k \leqslant 12$

16) $-11 \leqslant n - 9 \leqslant -5$

17) $-3 < x - 1 < 1$

18) $1 \leqslant \frac{p}{8} \leqslant 0$

19) $-4 < 8 - 3m \leqslant 11$

20) $3 + 7r > 59$ or $-6r - 3 > 33$

21) $-16 \leqslant 2n - 10 \leqslant -22$

22) $-6 - 8x \geqslant -6$ or $2 + 10x > 82$

23) $-5b + 10 \leqslant 30$ and $7b + 2 \leqslant -40$

24) $n + 10 \geqslant 15$ or $4n - 5 < -1$

25) $3x - 9 < 2x + 10$ and $5 + 7x \leqslant 10x - 10$

26) $4n + 8 < 3n - 6$ or $10n - 8 \geqslant 9 + 9n$

27) $-8 - 6v \leqslant 8 - 8v$ and $7v + 9 \leqslant 6 + 10v$

28) $5 - 2a \geqslant 2a + 1$ or $10a - 10 \geqslant 9a + 9$

29) $1 + 5k \leqslant 7k - 3$ or $k - 10 > 2k + 10$

30) $8 - 10r \leqslant 8 + 4r$ or $-6 + 8r < 2 + 8r$

31) $2x + 9 \geqslant 10x + 1$ and $3x - 2 < 7x + 2$

32) $-9m + 2 < -10 - 6m$ or $-m + 5 \geqslant 10 + 4m$

Inequalities - Absolute Value Inequalities

Objective: Solve, graph and give interval notation for the solution to inequalities with absolute values.

When an inequality has an absolute value we will have to remove the absolute value in order to graph the solution or give interval notation. The way we remove the absolute value depends on the direction of the inequality symbol.

Consider $|x| < 2$.

Absolute value is defined as distance from zero. Another way to read this inequality would be the distance from zero is less than 2. So on a number line we will shade all points that are less than 2 units away from zero.

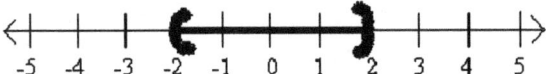

This graph looks just like the graphs of the three part compound inequalities! When the absolute value is **less than** a number we will remove the absolute value by changing the problem to a three part inequality, with the negative value on the left and the positive value on the right. So $|x| < 2$ becomes $-2 < x < 2$, as the graph above illustrates.

Consider $|x| > 2$.

Absolute value is defined as distance from zero. Another way to read this inequality would be the distance from zero is greater than 2. So on the number line we shade all points that are more than 2 units away from zero.

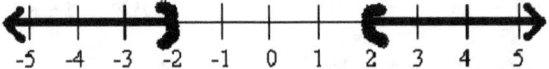

This graph looks just like the graphs of the OR compound inequalities! When the absolute value is **greater than** a number we will remove the absolute value by changing the problem to an OR inequality, the first inequality looking just like the problem with no absolute value, the second flipping the inequality symbol and changing the value to a negative. So $|x| > 2$ becomes $x > 2$ or $x < -2$, as the graph above illustrates.

World View Note: The phrase "absolute value" comes from German mathematician Karl Weierstrass in 1876, though he used the absolute value symbol for complex numbers. The first known use of the symbol for integers comes from a 1939 edition of a college algebra text!

For all absolute value inequalities we can also express our answers in interval notation which is done the same way it is done for standard compound inequalities.

We can solve absolute value inequalities much like we solved absolute value equations. Our first step will be to isolate the absolute value. Next we will remove the absolute value by making a three part inequality if the absolute value is less than a number, or making an OR inequality if the absolute value is greater than a number. Then we will solve these inequalites. Remember, if we multiply or divide by a negative the inequality symbol will switch directions!

Example 160.

Solve, graph, and give interval notation for the solution

$$|4x - 5| \geqslant 6 \qquad \text{Absolute value is greater, use OR}$$

$$4x - 5 \geqslant 6 \ \text{ OR } \ 4x - 5 \leqslant -6 \qquad \text{Solve}$$

$$\underline{+5 + 5} \qquad\qquad \underline{+5 \ +5} \qquad \text{Add 5 to both sides}$$

$$\frac{4x}{4} \geqslant \frac{11}{4} \ \text{ OR } \ \frac{4x}{4} \leqslant \frac{-1}{4} \qquad \text{Divide both sides by 4}$$

$$x \geqslant \frac{11}{4} \ \text{ OR } \ x \leqslant -\frac{1}{4} \qquad \text{Graph}$$

$$\left(-\infty, -\frac{1}{4}\right] \cup \left[\frac{11}{4}, \infty\right) \qquad \text{Interval notation}$$

Example 161.

Solve, graph, and give interval notation for the solution

$$-4 - 3|x| \leqslant -16 \qquad \text{Add 4 to both sides}$$

$$\underline{+4 \qquad\qquad +4}$$

$$-3|x| \leqslant -12 \qquad \text{Divide both sides by } -3$$

$$\overline{-3} \qquad \overline{-3} \qquad \text{Dividing by } a \text{ negative switches the symbol}$$

$$|x| \geqslant 4 \qquad \text{Absolute value is greater, use OR}$$

129

$x \geqslant 4$ OR $x \leqslant -4$ Graph

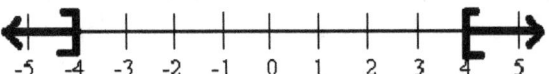

$(-\infty, -4] \cup [4, \infty)$ Interval Notation

In the previous example, we cannot combine -4 and -3 because they are not like terms, the -3 has an absolute value attached. So we must first clear the -4 by adding 4, then divide by -3. The next example is similar.

Example 162.

Solve, graph, and give interval notation for the solution

$$9 - 2|4x+1| > 3 \quad \text{Subtract 9 from both sides}$$
$$\underline{-9 \qquad\qquad -9}$$
$$-2|4x+1| > -6 \quad \text{Divide both sides by } -2$$
$$\overline{\quad -2 \qquad -2} \quad \text{Dividing by negative switches the symbol}$$
$$|4x+1| < 3 \quad \text{Absolute value is less, use three part}$$
$$-3 < 4x+1 < 3 \quad \text{Solve}$$
$$\underline{-1 \qquad -1-1} \quad \text{Subtract 1 from all three parts}$$
$$-4 < 4x < 2 \quad \text{Divide all three parts by 4}$$
$$\overline{\quad 4 \quad 4 \quad 4}$$
$$-1 < x < \frac{1}{2} \quad \text{Graph}$$

$\left(-1, \dfrac{1}{2}\right)$ Interval Notation

In the previous example, we cannot distribute the -2 into the absolute value. We can never distribute or combine things outside the absolute value with what is inside the absolute value. Our only way to solve is to first isolate the absolute value by clearing the values around it, then either make a compound inequality (and OR or a three part) to solve.

It is important to remember as we are solving these equations, the absolute value is always positive. If we end up with an absolute value is less than a negative number, then we will have no solution because absolute value will always be positive, greater than a negative. Similarly, if absolute value is greater than a negative, this will always happen. Here the answer will be all real numbers.

Example 163.

Solve, graph, and give interval notation for the solution

$$12 + 4|6x - 1| < 4 \qquad \text{Subtract 12 from both sides}$$
$$\underline{-12 \qquad\qquad -12}$$
$$\frac{4|6x - 1|}{4} < \frac{-8}{4} \qquad \text{Divide both sides by 4}$$
$$|6x - 1| < -2 \qquad \text{Absolute value can't be less than } a \text{ negative}$$

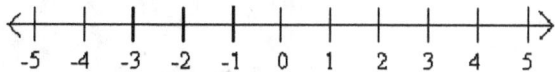

$$\text{No Solution or } \varnothing$$

Example 164.

Solve, graph, and give interval notation for the solution

$$5 - 6|x + 7| \leqslant 17 \qquad \text{Subtract 5 from both sides}$$
$$\underline{-5 \qquad\qquad -5}$$
$$-6|x + 7| \leqslant 12 \qquad \text{Divide both sides by } -6$$
$$\frac{}{-6} \quad \frac{}{-6} \qquad \text{Dividing by } a \text{ negative flips the symbol}$$
$$|x + 7| \geqslant -2 \qquad \text{Absolute value always greater than negative}$$

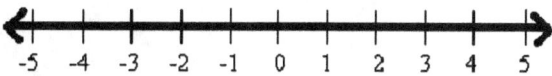

$$\text{All Real Numbers or } \mathbb{R}$$

3.3 Practice - Absolute Value Inequalities

Solve each inequality, graph its solution, and give interval notation.

1) $|x| < 3$

2) $|x| \leqslant 8$

3) $|2x| < 6$

4) $|x + 3| < 4$

5) $|x - 2| < 6$

6) $x - 8| < 12$

7) $|x - 7| < 3$

8) $|x + 3| \leqslant 4$

9) $|3x - 2| < 9$

10) $|2x + 5| < 9$

11) $1 + 2|x - 1| \leqslant 9$

12) $10 - 3|x - 2| \geqslant 4$

13) $6 - |2x - 5| >= 3$

14) $|x| > 5$

15) $|3x| > 5$

16) $|x - 4| > 5$

17) $|x = 3| >= 3$

18) $|2x - 4| > 6$

19) $|3x - 5| > \geqslant 3$

20) $3 - |2 - x| < 1$

21) $4 + 3|x - 1| >= 10$

22) $3 - 2|3x - 1| \geqslant -7$

23) $3 - 2|x - 5| \leqslant -15$

24) $4 - 6|-6 - 3x| \leqslant -5$

25) $-2 - 3|4 - 2x| \geqslant -8$

26) $-3 - 2|4x - 5| \geqslant 1$

27) $4 - 5|-2x - 7| < -1$

28) $-2 + 3|5 - x| \leqslant 4$

29) $3 - 2|4x - 5| \geqslant 1$

30) $-2 - 3|-3x - 5 \geqslant -5$

31) $-5 - 2|3x - 6| < -8$

32) $6 - 3|1 - 4x| < -3$

33) $4 - 4|-2x + 6| > -4$

34) $-3 - 4|-2x - 5| \geqslant -7$

35) $|-10 + x| \geqslant 8$

Chapter 4 : Systems of Equations

Systems of Equations - Graphing

Objective: Solve systems of equations by graphing and identifying the point of intersection.

We have solved problems like $3x - 4 = 11$ by adding 4 to both sides and then dividing by 3 (solution is $x = 5$). We also have methods to solve equations with more than one variable in them. It turns out that to solve for more than one variable we will need the same number of equations as variables. For example, to solve for two variables such as x and y we will need two equations. When we have several equations we are using to solve, we call the equations a **system of equations**. When solving a system of equations we are looking for a solution that works in both equations. This solution is usually given as an ordered pair (x, y). The following example illustrates a solution working in both equations

Example 165.

Show $(2,1)$ is the solution to the system $\begin{matrix} 3x - y = 5 \\ x + y = 3 \end{matrix}$

$$(2, 1) \quad \text{Identify } x \text{ and } y \text{ from the orderd pair}$$
$$x = 2, y = 1 \quad \text{Plug these values into each equation}$$

$$3(2) - (1) = 5 \quad \text{First equation}$$
$$6 - 1 = 5 \quad \text{Evaluate}$$
$$5 = 5 \quad \text{True}$$

$$(2) + (1) = 3 \quad \text{Second equation, evaluate}$$
$$3 = 3 \quad \text{True}$$

As we found a true statement for both equations we know $(2,1)$ is the solution to the system. It is in fact the only combination of numbers that works in both equations. In this lesson we will be working to find this point given the equations. It seems to follow that if we use points to describe the solution, we can use graphs to find the solutions.

If the graph of a line is a picture of all the solutions, we can graph two lines on the same coordinate plane to see the solutions of both equations. We are interested in the point that is a solution for both lines, this would be where the lines intersect! If we can find the intersection of the lines we have found the solution that works in both equations.

Example 166.

$$y = -\frac{1}{2}x + 3$$
$$y = \frac{3}{4}x - 2$$

To graph we identify slopes and $y-$intercepts

First: $m = -\frac{1}{2}, b = 3$

Second: $m = \frac{3}{4}, b = -2$

Now we can graph both lines on the same plane.

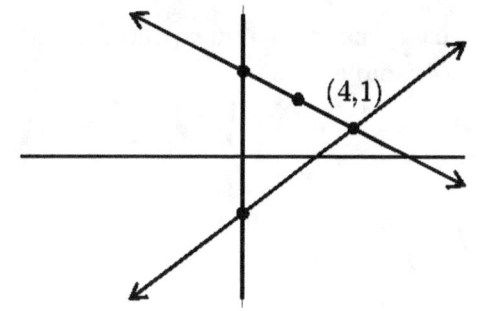

(4,1)

To graph each equation, we start at the y-intercept and use the slope $\frac{\text{rise}}{\text{run}}$ to get the next point and connect the dots.

Remember a negative slope is downhill!

Find the intersection point, (4,1)

(4,1)   Our Solution

Often our equations won't be in slope-intercept form and we will have to solve both equations for y first so we can idenfity the slope and y-intercept.

Example 167.

$$6x - 3y = -9$$
$$2x + 2y = -6$$

Solve each equation for y

$6x - 3y = -9$	$2x + 2y = -6$
$\underline{-6x \qquad -6x}$	$\underline{-2x \qquad -2x}$
$-3y = -6x - 9$	$2y = -2x - 6$
$\overline{-3 \quad -3\,-3}$	$\overline{2 \quad 2\ 2}$
$y = 2x + 3$	$y = -x - 3$

Subtract x terms

Put x terms first

Divide by coefficient of y

Identify slope and $y-$intercepts

135

First: $m = \dfrac{2}{1}, b = 3$

Second: $m = -\dfrac{1}{1}, b = -3$

Now we can graph both lines on the same plane

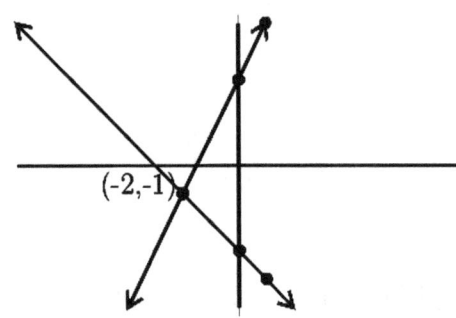

To graph each equation, we start at the y-intercept and use the slope $\frac{\text{rise}}{\text{run}}$ to get the next point and connect the dots.

Remember a negative slope is down-hill!

Find the intersection point, $(-2, -1)$

$(-2, -1)$ Our Solution

As we are graphing our lines, it is possible to have one of two unexpected results. These are shown and discussed in the next two example.

Example 168.

$$y = \frac{3}{2}x - 4$$
$$y = \frac{3}{2}x + 1$$

Identify slope and $y-$ intercept of each equation

First: $m = \dfrac{3}{2}, b = -4$

Second: $m = \dfrac{3}{2}, b = 1$

Now we can graph both equations on the same plane

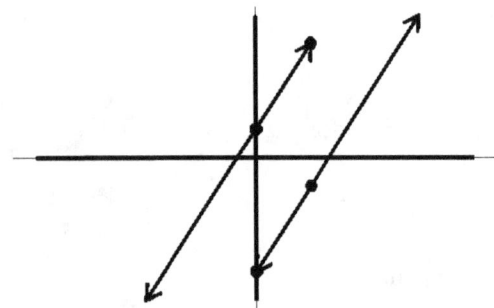

To graph each equation, we start at the y-intercept and use the slope $\frac{\text{rise}}{\text{run}}$ to get the next point and connect the dots.

The two lines do not intersect! They are parallel! If the lines do not intersect we know that there is no point that works in both equations, there is no solution

$\varnothing$ No Solution

We also could have noticed that both lines had the same slope. Remembering that parallel lines have the same slope we would have known there was no solution even without having to graph the lines.

Example 169.

$$2x - 6y = 12$$
$$3x - 9y = 18$$

Solve each equation for y

$$2x - 6y = 12 \qquad 3x - 9y = 18$$
$$\underline{-2x \qquad -2x} \quad \underline{-3x \qquad -3x}$$

Subtract x terms

$$-6y = -2x + 12 \qquad -9y = -3x + 18$$

Put x terms first

$$\overline{-6} \quad \overline{-6} \ \overline{-6} \qquad \overline{-9} \quad \overline{-9} \ \overline{-9}$$

Divide by coefficient of y

$$y = \frac{1}{3}x - 2 \qquad\qquad y = \frac{1}{3}x - 2$$

Identify the slopes and y − intercepts

$$\text{First: } m = \frac{1}{3}, b = -2$$
$$\text{Second: } m = \frac{1}{3}, b = -2$$

Now we can graph both equations together

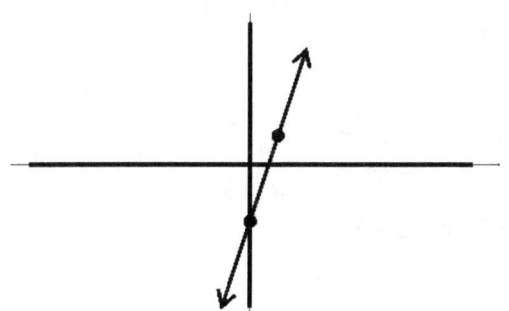

To graph each equation, we start at the y-intercept and use the slope $\frac{\text{rise}}{\text{run}}$ to get the next point and connect the dots.

Both equations are the same line! As one line is directly on top of the other line, we can say that the lines "intersect" at all the points! Here we say we have infinite solutions

Once we had both equations in slope-intercept form we could have noticed that the equations were the same. At this point we could have stated that there are infinite solutions without having to go through the work of graphing the equations.

World View Note: The Babylonians were the first to work with systems of equations with two variables. However, their work with systems was quickly passed by the Greeks who would solve systems of equations with three or four variables and around 300 AD, developed methods for solving systems with any number of unknowns!

4.1 Practice - Graphing

Solve each equation by graphing.

1) $y = -x + 1$
   $y = -5x - 3$

2) $y = -\frac{5}{4}x - 2$
   $y = -\frac{1}{4}x + 2$

3) $y = -3$
   $y = -x - 4$

4) $y = -x - 2$
   $y = \frac{2}{3}x + 3$

5) $y = -\frac{3}{4}x + 1$
   $y = -\frac{3}{4}x + 2$

6) $y = 2x + 2$
   $y = -x - 4$

7) $y = \frac{1}{3}x + 2$
   $y = -\frac{5}{3}x - 4$

8) $y = 2x - 4$
   $y = \frac{1}{2}x + 2$

9) $y = \frac{5}{3}x + 4$
   $y = -\frac{2}{3}x - 3$

10) $y = \frac{1}{2}x + 4$
    $y = \frac{1}{2}x + 1$

11) $x + 3y = -9$
    $5x + 3y = 3$

12) $x + 4y = -12$
    $2x + y = 4$

13) $x - y = 4$
    $2x + y = -1$

14) $6x + y = -3$
    $x + y = 2$

15) $2x + 3y = -6$
    $2x + y = 2$

16) $3x + 2y = 2$
    $3x + 2y = -6$

17) $2x + y = 2$
    $x - y = 4$

18) $x + 2y = 6$
    $5x - 4y = 16$

19) $2x + y = -2$
    $x + 3y = 9$

20) $x - y = 3$
    $5x + 2y = 8$

21) $0 = -6x - 9y + 36$
    $12 = 6x - 3y$

22) $-2y + x = 4$
    $2 = -x + \frac{1}{2}y$

23) $2x - y = -1$
    $0 = -2x - y - 3$

24) $-2y = -4 - x$
    $-2y = -5x + 4$

25) $3 + y = -x$
    $-4 - 6x = -y$

26) $16 = -x - 4y$
    $-2x = -4 - 4y$

27) $-y + 7x = 4$
    $-y - 3 + 7x = 0$

28) $-4 + y = x$
    $x + 2 = -y$

29) $-12 + x = 4y$
    $12 - 5x = 4y$

30) $-5x + 1 = -y$
    $-y + x = -3$

Systems of Equations - Substitution

Objective: Solve systems of equations using substitution.

When solving a system by graphing has several limitations. First, it requires the graph to be perfectly drawn, if the lines are not straight we may arrive at the wrong answer. Second, graphing is not a great method to use if the answer is really large, over 100 for example, or if the answer is a decimal the that graph will not help us find, 3.2134 for example. For these reasons we will rarely use graphing to solve our systems. Instead, an algebraic approach will be used.

The first algebraic approach is called substitution. We will build the concepts of substitution through several example, then end with a five-step process to solve problems using this method.

Example 170.

$$x = 5$$
$$y = 2x - 3 \qquad \text{We already know } x = 5, \text{ substitute this into the other equation}$$
$$y = 2(5) - 3 \qquad \text{Evaluate, multiply first}$$
$$y = 10 - 3 \qquad \text{Subtract}$$
$$y = 7 \qquad \text{We now also have } y$$
$$(5, 7) \qquad \text{Our Solution}$$

When we know what one variable equals we can plug that value (or expression) in for the variable in the other equation. It is very important that when we substitute, the substituted value goes in parenthesis. The reason for this is shown in the next example.

Example 171.

$$2x - 3y = 7$$
$$y = 3x - 7 \qquad \text{We know } y = 3x - 7, \text{ substitute this into the other equation}$$
$$2x - 3(3x - 7) = 7 \qquad \text{Solve this equation, distributing } -3 \text{ first}$$

$$2x - 9x + 21 = 7 \quad \text{Combine like terms } 2x - 9x$$
$$-7x + 21 = 7 \quad \text{Subtract } 21$$
$$\underline{-21 \quad -21}$$
$$-7x = -14 \quad \text{Divide by } -7$$
$$\underline{-7 \quad -7}$$
$$x = 2 \quad \text{We now have our } x, \text{ plug into the } y = \text{ equation to find } y$$
$$y = 3(2) - 7 \quad \text{Evaluate, multiply first}$$
$$y = 6 - 7 \quad \text{Subtract}$$
$$y = -1 \quad \text{We now also have } y$$
$$(2, -1) \quad \text{Our Solution}$$

By using the entire expression $3x - 7$ to replace y in the other equation we were able to reduce the system to a single linear equation which we can easily solve for our first variable. However, the lone variable (a variable without a coefficient) is not always alone on one side of the equation. If this happens we can isolate it by solving for the lone variable.

Example 172.

$$3x + 2y = 1$$
$$\boldsymbol{x - 5y = 6} \quad \text{Lone variable is } x, \text{ isolate by adding } 5y \text{ to both sides.}$$
$$\underline{+5y + 5y}$$
$$x = 6 + 5y \quad \text{Substitute this into the untouched equation}$$
$$3(\boldsymbol{6 + 5y}) + 2y = 1 \quad \text{Solve this equation, distributing 3 first}$$
$$18 + 15y + 2y = 1 \quad \text{Combine like terms } 15y + 2y$$
$$18 + 17y = 1 \quad \text{Subtract 18 from both sides}$$
$$\underline{-18 \quad -18}$$
$$17y = -17 \quad \text{Divide both sides by 17}$$
$$\underline{17 \quad 17}$$
$$y = -1 \quad \text{We have our } y, \text{ plug this into the } x = \text{ equation to find } x$$
$$x = 6 + 5(\boldsymbol{-1}) \quad \text{Evaluate, multiply first}$$
$$x = 6 - 5 \quad \text{Subtract}$$
$$x = 1 \quad \text{We now also have } x$$
$$(1, -1) \quad \text{Our Solution}$$

The process in the previous example is how we will solve problems using substitution. This process is described and illustrated in the following table which lists the five steps to solving by substitution.

Problem	$4x - 2y = 2$ $2x + y = -5$
1. Find the lone variable	Second Equation, y $2x + \boldsymbol{y} = -5$
2. Solve for the lone variable	$\underline{-2x \qquad -2x}$ $\boldsymbol{y = -5 - 2x}$
3. Substitute into the untouched equation	$4x - 2(-5 - 2x) = 2$
4. Solve	$4x + 10 + 4x = 2$ $8x + 10 = 2$ $\underline{-10 - 10}$ $8x = -8$ $\overline{8} \quad \overline{8}$ $\boldsymbol{x = -1}$
5. Plug into lone variable equation and evaluate	$y = -5 - 2(-1)$ $y = -5 + 2$ $\boldsymbol{y = -3}$
Solution	$(-1, -3)$

Sometimes we have several lone variables in a problem. In this case we will have the choice on which lone variable we wish to solve for, either will give the same final result.

Example 173.

$\begin{aligned} \boldsymbol{x + y} &= 5 \end{aligned}$ Find the lone variable: x or y in first, or x in second.

$x - y = -1$ We will chose x in the first

$\boldsymbol{x + y} = 5$ Solve for the lone variable, subtract y from both sides

$\underline{\boldsymbol{-y - y}}$

$x = 5 - y$ Plug into the untouched equation, the second equation

$(\boldsymbol{5 - y}) - y = -1$ Solve, parenthesis are not needed here, combine like terms

$5 - 2y = -1$ Subtract 5 from both sides

$\underline{-5 \qquad -5}$

$-2y = -6$ Divide both sides by -2

$\overline{-2} \quad \overline{-2}$

$y = 3$ We have our y!

$x = 5 - (3)$ Plug into lone variable equation, evaluate

$x = 2$ Now we have our x

$(2, 3)$ Our Solution

Just as with graphing it is possible to have no solution $\varnothing$ (parallel lines) or infinite solutions (same line) with the substitution method. While we won't have a parallel line or the same line to look at and conclude if it is one or the other, the process takes an interesting turn as shown in the following example.

Example 174.

$$\begin{array}{ll}
\mathbf{y} + 4 = 3x & \text{Find the lone variable, } y \text{ in the first equation} \\
2y - 6x = -8 & \\
y + 4 = 3x & \text{Solve for the lone variable, subtract 4 from both sides} \\
\underline{-4 \quad -4} & \\
y = 3x - 4 & \text{Plug into untouched equation} \\
2(\mathbf{3x - 4}) - 6x = -8 & \text{Solve, distribute through parenthesis} \\
6x - 8 - 6x = -8 & \text{Combine like terms } 6x - 6x \\
-8 = -8 & \text{Variables are gone! } A \text{ true statement.} \\
\text{Infinite solutions} & \text{Our Solution}
\end{array}$$

Because we had a true statement, and no variables, we know that anything that works in the first equation, will also work in the second equation. However, we do not always end up with a true statement.

Example 175.

$$\begin{array}{ll}
6x - 3y = -9 & \text{Find the lone variable, } y \text{ in the second equation} \\
-2x + \mathbf{y} = 5 & \\
-2x + y = 5 & \text{Solve for the lone variable, add } 2x \text{ to both sides} \\
\underline{+2x \quad +2x} & \\
y = 5 + 2x & \text{Plug into untouched equation} \\
6x - 3(\mathbf{5 + 2x}) = -9 & \text{Solve, distribute through parenthesis} \\
6x - 15 - 6x = -9 & \text{Combine like terms } 6x - 6x \\
-15 \neq -9 & \text{Variables are gone! } A \text{ false statement.} \\
\text{No Solution } \varnothing & \text{Our Solution}
\end{array}$$

Because we had a false statement, and no variables, we know that nothing will work in both equations.

142

World View Note: French mathematician Rene Descartes wrote a book which included an appendix on geometry. It was in this book that he suggested using letters from the end of the alphabet for unknown values. This is why often we are solving for the variables $x, y,$ and z.

One more question needs to be considered, what if there is no lone variable? If there is no lone variable substitution can still work to solve, we will just have to select one variable to solve for and use fractions as we solve.

Example 176.

$$5x - 6y = -14 \qquad \text{No lone variable,}$$
$$-2x + 4y = 12 \qquad \text{we will solve for } x \text{ in the first equation}$$
$$5x - 6y = -14 \qquad \text{Solve for our variable, add } 6y \text{ to both sides}$$
$$\underline{+6y \quad +6y}$$
$$\frac{5x}{5} = \frac{-14}{5} + \frac{6y}{5} \qquad \text{Divide each term by } 5$$
$$x = \frac{-14}{5} + \frac{6y}{5} \qquad \text{Plug into untouched equation}$$
$$-2\left(\frac{-14}{5} + \frac{6y}{5}\right) + 4y = 12 \qquad \text{Solve, distribute through parenthesis}$$
$$\frac{28}{5} - \frac{12y}{5} + 4y = 12 \qquad \text{Clear fractions by multiplying by } 5$$
$$\frac{28(5)}{5} - \frac{12y(5)}{5} + 4y(5) = 12(5) \qquad \text{Reduce fractions and multiply}$$
$$28 - 12y + 20y = 60 \qquad \text{Combine like terms } -12y + 20y$$
$$28 + 8y = 60 \qquad \text{Subtract } 28 \text{ from both sides}$$
$$\underline{-28 \qquad -28}$$
$$\frac{8y}{8} = \frac{32}{8} \qquad \text{Divide both sides by } 8$$
$$y = 4 \qquad \text{We have our } y$$
$$x = \frac{-14}{5} + \frac{6(4)}{5} \qquad \text{Plug into lone variable equation, multiply}$$
$$x = \frac{-14}{5} + \frac{24}{5} \qquad \text{Add fractions}$$
$$x = \frac{10}{5} \qquad \text{Reduce fraction}$$
$$x = 2 \qquad \text{Now we have our } x$$
$$(2, 4) \qquad \text{Our Solution}$$

Using the fractions does make the problem a bit more tricky. This is why we have another method for solving systems of equations that will be discussed in another lesson.

4.2 Practice - Substitution

Solve each system by substitution.

1) $y = -3x$
   $y = 6x - 9$

2) $y = x + 5$
   $y = -2x - 4$

3) $y = -2x - 9$
   $y = 2x - 1$

4) $y = -6x + 3$
   $y = 6x + 3$

5) $y = 6x + 4$
   $y = -3x - 5$

6) $y = 3x + 13$
   $y = -2x - 22$

7) $y = 3x + 2$
   $y = -3x + 8$

8) $y = -2x - 9$
   $y = -5x - 21$

9) $y = 2x - 3$
   $y = -2x + 9$

10) $y = 7x - 24$
    $y = -3x + 16$

11) $y = 6x - 6$
    $-3x - 3y = -24$

12) $-x + 3y = 12$
    $y = 6x + 21$

13) $y = -6$
    $3x - 6y = 30$

14) $6x - 4y = -8$
    $y = -6x + 2$

15) $y = -5$
    $3x + 4y = -17$

16) $7x + 2y = -7$
    $y = 5x + 5$

17) $-2x + 2y = 18$
    $y = 7x + 15$

18) $y = x + 4$
    $3x - 4y = -19$

19) $y = -8x + 19$
    $-x + 6y = 16$

20) $y = -2x + 8$
    $-7x - 6y = -8$

21) $7x - 2y = -7$
    $y = 7$

22) $x - 2y = -13$
    $4x + 2y = 18$

23) $x - 5y = 7$
    $2x + 7y = -20$

24) $3x - 4y = 15$
    $7x + y = 4$

25) $-2x - y = -5$
    $x - 8y = -23$

26) $6x + 4y = 16$
    $-2x + y = -3$

27) $-6x + y = 20$
    $-3x - 3y = -18$

28) $7x + 5y = -13$
    $x - 4y = -16$

29) $3x + y = 9$
    $2x + 8y = -16$

30) $-5x - 5y = -20$
    $-2x + y = 7$

31) $2x + y = 2$
$3x + 7y = 14$

32) $2x + y = -7$
$5x + 3y = -21$

33) $x + 5y = 15$
$-3x + 2y = 6$

34) $2x + 3y = -10$
$7x + y = 3$

35) $-2x + 4y = -16$
$y = -2$

36) $-2x + 2y = -22$
$-5x - 7y = -19$

37) $-6x + 6y = -12$
$8x - 3y = 16$

38) $-8x + 2y = -6$
$-2x + 3y = 11$

39) $2x + 3y = 16$
$-7x - y = 20$

40) $-x - 4y = -14$
$-6x + 8y = 12$

Systems of Equations - Addition/Elimination

Objective: Solve systems of equations using the addition/elimination method.

When solving systems we have found that graphing is very limited when solving equations. We then considered a second method known as substituion. This is probably the most used idea in solving systems in various areas of algebra. However, substitution can get ugly if we don't have a lone variable. This leads us to our second method for solving systems of equations. This method is known as either Elimination or Addition. We will set up the process in the following examples, then define the five step process we can use to solve by elimination.

Example 177.

$$3x - 4y = 8$$
$$\underline{5x + 4y = -24}$$
    Notice opposites in front of $y's$. Add columns.

$$\underline{8x \quad\quad = -16}$$ Solve for x, divide by 8
$$8 \qquad\quad 8$$

$$x = -2$$ We have our x!

$$5(-2) + 4y = -24$$ Plug into either original equation, simplify

$$-10 + 4y = -24$$ Add 10 to both sides
$$\underline{+10 \qquad\quad +10}$$

$$\underline{4y = -14}$$ Divide by 4
$$4 \quad\quad 4$$

$$y = \frac{-7}{2}$$ Now we have our y!

$$\left(-2, \frac{-7}{2}\right)$$ Our Solution

In the previous example one variable had opposites in front of it, $-4y$ and $4y$. Adding these together eliminated the y completely. This allowed us to solve for the x. This is the idea behind the addition method. However, generally we won't have opposites in front of one of the variables. In this case we will manipulate the equations to get the opposites we want by multiplying one or both equations (on both sides!). This is shown in the next example.

Example 178.

$$-6x + 5y = 22$$ We can get opposites in front of x, by multiplying the
$$2x + 3y = 2$$ second equation by 3, to get $-6x$ and $+6x$

$$3(2x + 3y) = (2)3$$ Distribute to get new second equation.

146

$$6x + 9y = 6 \quad \text{New second equation}$$
$$\underline{-6x + 5y = 22} \quad \text{First equation still the same, add}$$
$$14y = 28 \quad \text{Divide both sides by 14}$$
$$\overline{14} \quad \overline{14}$$
$$y = 2 \quad \text{We have our } y!$$
$$2x + 3(2) = 2 \quad \text{Plug into one of the original equations, simplify}$$
$$2x + 6 = 2 \quad \text{Subtract 6 from both sides}$$
$$\underline{-6 - 6}$$
$$2x = -4 \quad \text{Divide both sides by 2}$$
$$\overline{2} \quad \overline{2}$$
$$x = -2 \quad \text{We also have our } x!$$
$$(-2, 2) \quad \text{Our Solution}$$

When we looked at the x terms, $-6x$ and $2x$ we decided to multiply the $2x$ by 3 to get the opposites we were looking for. What we are looking for with our opposites is the least common multiple (LCM) of the coefficients. We also could have solved the above problem by looking at the terms with y, $5y$ and $3y$. The LCM of 3 and 5 is 15. So we would want to multiply both equations, the $5y$ by 3, and the $3y$ by -5 to get opposites, $15y$ and $-15y$. This illustrates an important point, some problems we will have to multiply both equations by a constant (on both sides) to get the opposites we want.

Example 179.

$$3x + 6y = -9 \quad \text{We can get opposites in front of } x, \text{ find LCM of 6 and 9,}$$
$$2x + 9y = -26 \quad \text{The LCM is 18. We will multiply to get } 18y \text{ and } -18y$$

$$3(3x + 6y) = (-9)3 \quad \text{Multiply the first equation by 3, both sides!}$$
$$9x + 18y = -27$$

$$-2(2x + 9y) = (-26)(-2) \quad \text{Multiply the second equation by } -2, \text{ both sides!}$$
$$-4x - 18y = 52$$

$$9x + 18y = -27 \quad \text{Add two new equations together}$$
$$\underline{-4x - 18y = 52}$$
$$5x \quad\quad = 25 \quad \text{Divide both sides by 5}$$
$$\overline{5} \quad\quad\quad \overline{5}$$
$$x = 5 \quad \text{We have our solution for } x$$
$$3(5) + 6y = -9 \quad \text{Plug into either original equation, simplify}$$
$$15 + 6y = -9 \quad \text{Subtract 15 from both sides}$$
$$\underline{-15 \quad\quad -15}$$

$$6y = -24 \quad \text{Divide both sides by 6}$$
$$\frac{6}{6} \quad \frac{6}{6}$$
$$y = -4 \quad \text{Now we have our solution for } y$$
$$(5, -4) \quad \text{Our Solution}$$

It is important for each problem as we get started that all variables and constants are lined up before we start multiplying and adding equations. This is illustrated in the next example which includes the five steps we will go through to solve a problem using elimination.

Problem	$2x - 5y = -13$ $-3y + 4 = -5x$
1. Line up the variables and constants	Second Equation: $-3y + 4 = -5x$ $+5x - 4 \quad +5x - 4$ $5x - 3y = -4$
2. Multiply to get opposites (use LCD)	$2x - 5y = -13$ $5x - 3y = -4$ First Equation: multiply by -5 $-5(2x - 5y) = (-13)(-5)$ $-10x + 25y = 65$ Second Equation: multiply by 2 $2(5x - 3y) = (-4)2$ $10x - 6y = -8$ $-10x + 25y = 65$ $10x - 6y = -8$
3. Add	$19y = 57$
4. Solve	$\dfrac{19y}{19} = \dfrac{57}{19}$ $y = 3$
5. Plug into either original and solve	$2x - 5(3) = -13$ $2x - 15 = -13$ $\dfrac{+15 \quad +15}{2x \quad\quad = 2}$ $\dfrac{2x}{2} = \dfrac{2}{2}$ $x = 1$
Solution	$(1, 3)$

World View Note: The famous mathematical text, *The Nine Chapters on the Mathematical Art*, which was printed around 179 AD in China describes a formula very similar to Gaussian elimination which is very similar to the addition method.

Just as with graphing and substution, it is possible to have no solution or infinite solutions with elimination. Just as with substitution, if the variables all disappear from our problem, a true statment will indicate infinite solutions and a false statment will indicate no solution.

Example 180.

$$2x - 5y = 3$$
$$-6x + 15y = -9$$

To get opposites in front of x, multiply first equation by 3

$$3(2x - 5y) = (3)3$$

Distribute

$$6x - 15y = 9$$

$$6x - 15y = 9$$

Add equations together

$$\underline{-6x + 15y = -9}$$

$$0 = 0$$

True statement

Infinite solutions

Our Solution

Example 181.

$$4x - 6y = 8$$
$$6x - 9y = 15$$

LCM for $x's$ is 12.

$$3(4x - 6y) = (8)3$$
$$12x - 18y = 24$$

Multiply first equation by 3

$$-2(6x - 9y) = (15)(-2)$$
$$-12x + 18y = -30$$

Multiply second equation by -2

$$12x - 18y = 24$$

Add both new equations together

$$\underline{-12x + 18y = -30}$$

$$0 = -6$$

False statement

No Solution

Our Solution

We have covered three different methods that can be used to solve a system of two equations with two variables. While all three can be used to solve any system, graphing works great for small integer solutions. Substitution works great when we have a lone variable, and addition works great when the other two methods fail. As each method has its own strengths, it is important you are familiar with all three methods.

4.3 Practice - Addition/Elimination

Solve each system by elimination.

1) $4x + 2y = 0$
   $-4x - 9y = -28$

2) $-7x + y = -10$
   $-9x - y = -22$

3) $-9x + 5y = -22$
   $9x - 5y = 13$

4) $-x - 2y = -7$
   $x + 2y = 7$

5) $-6x + 9y = 3$
   $6x - 9y = -9$

6) $5x - 5y = -15$
   $5x - 5y = -15$

7) $4x - 6y = -10$
   $4x - 6y = -14$

8) $-3x + 3y = -12$
   $-3x + 9y = -24$

9) $-x - 5y = 28$
   $-x + 4y = -17$

10) $-10x - 5y = 0$
    $-10x - 10y = -30$

11) $2x - y = 5$
    $5x + 2y = -28$

12) $-5x + 6y = -17$
    $x - 2y = 5$

13) $10x + 6y = 24$
    $-6x + y = 4$

14) $x + 3y = -1$
    $10x + 6y = -10$

15) $2x + 4y = 24$
    $4x - 12y = 8$

16) $-6x + 4y = 12$
    $12x + 6y = 18$

17) $-7x + 4y = -4$
    $10x - 8y = -8$

18) $-6x + 4y = 4$
    $-3x - y = 26$

19) $5x + 10y = 20$
    $-6x - 5y = -3$

20) $-9x - 5y = -19$
    $3x - 7y = -11$

21) $-7x - 3y = 12$
    $-6x - 5y = 20$

22) $-5x + 4y = 4$
    $-7x - 10y = -10$

23) $9x - 2y = -18$
    $5x - 7y = -10$

24) $3x + 7y = -8$
    $4x + 6y = -4$

25) $9x + 6y = -21$
    $-10x - 9y = 28$

26) $-4x - 5y = 12$
    $-10x + 6y = 30$

27) $-7x + 5y = -8$
    $-3x - 3y = 12$

28) $8x + 7y = -24$
    $6x + 3y = -18$

29) $-8x - 8y = -8$
    $10x + 9y = 1$

30) $-7x + 10y = 13$
    $4x + 9y = 22$

31) $9y = 7 - x$
    $-18y + 4x = -26$

32) $0 = -9x - 21 + 12y$
    $1 + \frac{4}{3}y + \frac{7}{3}x = 0$

33) $0 = 9x + 5y$
    $y = \frac{2}{7}x$

34) $-6 - 42y = -12x$
    $x - \frac{1}{2} - \frac{7}{2}y = 0$

150

Systems of Equations - Three Variables

Objective: Solve systems of equations with three variables using addition/elimination.

Solving systems of equations with 3 variables is very similar to how we solve systems with two variables. When we had two variables we reduced the system down to one with only one variable (by substitution or addition). With three variables we will reduce the system down to one with two variables (usually by addition), which we can then solve by either addition or substitution.

To reduce from three variables down to two it is very important to keep the work organized. We will use addition with two equations to eliminate one variable. This new equation we will call (A). Then we will use a different pair of equations and use addition to eliminate the **same** variable. This second new equation we will call (B). Once we have done this we will have two equations (A) and (B) with the same two variables that we can solve using either method. This is shown in the following examples.

Example 182.

$$3x + 2y - z = -1$$
$$-2x - 2y + 3z = 5$$
$$5x + 2y - z = 3$$

We will eliminate y using two different pairs of equations

$3x + 2y - z = -1$	Using the first two equations,
$-2x - 2y + 3z = 5$	Add the first two equations
$(A) \quad x \qquad + 2z = 4$	This is equation (A), our first equation

$-2x - 2y + 3z = 5$	Using the second two equations
$5x + 2y - z = 3$	Add the second two equations
$(B) \quad 3x \qquad + 2z = 8$	This is equation (B), our second equation

$(A) \quad x + 2z = 4$	Using (A) and (B) we will solve this system.
$(B) \quad 3x + 2z = 8$	We will solve by addition

$-1(x + 2z) = (4)(-1)$	Multiply (A) by -1
$-x - 2z = -4$	

$-x - 2z = -4$	Add to the second equation, unchanged
$3x + 2z = 8$	
$2x = 4$	Solve, divide by 2
$\overline{2 \quad 2}$	
$x = 2$	We now have x! Plug this into either (A) or (B)

$(2) + 2z = 4$	We plug it into (A), solve this equation, subtract 2
$-2 \qquad -2$	
$2z = 2$	Divide by 2
$\overline{2 \quad 2}$	
$z = 1$	We now have z! Plug this and x into any original equation

$3(2) + 2y - (1) = -1$	We use the first, multiply $3(2) = 6$ and combine with -1
$2y + 5 = -1$	Solve, subtract 5
$-5 \quad -5$	
$2y = -6$	Divide by 2
$\overline{2 \quad 2}$	
$y = -3$	We now have y!

$(2, -3, 1)$	Our Solution

As we are solving for $x, y,$ and z we will have an ordered triplet (x, y, z) instead of just the ordered pair (x, y). In this above problem, y was easily eliminated using the addition method. However, sometimes we may have to do a bit of work to get a variable to eliminate. Just as with addition of two equations, we may have to multiply equations by something on both sides to get the opposites we want so a variable eliminates. As we do this remmeber it is improtant to eliminate the **same** variable both times using two **different** pairs of equations.

Example 183.

$$4x - 3y + 2z = -29 \qquad \text{No variable will easily eliminate.}$$
$$6x + 2y - z = -16 \qquad \text{We could choose any variable, so we chose } x$$
$$-8x - y + 3z = 23 \qquad \text{We will eliminate } x \text{ twice.}$$

$$4x - 3y + 2z = -29 \qquad \text{Start with first two equations. LCM of 4 and 6 is 12.}$$
$$6x + 2y - z = -16 \qquad \text{Make the first equation have } 12x, \text{ the second } -12x$$

$$3(4x - 3y + 2z) = (-29)3 \qquad \text{Multiply the first equation by 3}$$
$$12x - 9y + 6z = -87$$

$$-2(6x + 2y - z) = (-16)(-2) \qquad \text{Multiply the second equation by } -2$$
$$-12x - 4y + 2z = 32$$

$$12x - 9y + 6z = -87 \qquad \text{Add these two equations together}$$
$$\underline{-12x - 4y + 2z = 32}$$
$$(A) \qquad -13y + 8z = -55 \qquad \text{This is our } (A) \text{ equation}$$

$$6x + 2y - z = -16 \qquad \text{Now use the second two equations } (a \text{ different pair})$$
$$-8x - y + 3z = 23 \qquad \text{The LCM of 6 and } -8 \text{ is 24.}$$

$$4(6x + 2y - z) = (-16)4 \qquad \text{Multiply the first equation by 4}$$
$$24x + 8y - 4 = -64$$

$$3(-8x - y + 3z) = (23)3 \qquad \text{Multiply the second equation by 3}$$
$$-24x - 3y + 9z = 69$$

$$24x + 8y - 4 = -64 \qquad \text{Add these two equations together}$$
$$\underline{-24x - 3y + 9z = 69}$$
$$(B) \qquad 5y + 5z = 5 \qquad \text{This is our } (B) \text{ equation}$$

153

$$(A) \quad -13y + 8z = -55 \qquad \text{Using } (A) \text{ and } (B) \text{ we will solve this system}$$
$$(B) \quad 5y + 5z = 5 \qquad \text{The second equation is solved for } z \text{ to use substitution}$$

$$5y + 5z = 5 \qquad \text{Solving for } z, \text{ subtract } 5y$$
$$\underline{-5y \quad\quad -5y}$$
$$\frac{5z}{5} = \frac{5}{5} - \frac{5y}{5} \qquad \text{Divide each term by } 5$$

$$z = 1 - y \qquad \text{Plug into untouched equation}$$
$$-13y + 8(1 - y) = -55 \qquad \text{Distribute}$$
$$-13y + 8 - 8y = -55 \qquad \text{Combine like terms } -13y - 8y$$
$$-21y + 8 = -55 \qquad \text{Subtract } 8$$
$$\underline{-8 \quad\quad -8}$$
$$\frac{-21y}{-21} = \frac{-63}{-21} \qquad \text{Divide by } -21$$

$$y = 3 \qquad \text{We have our } y! \text{ Plug this into } z = \text{equations}$$
$$z = 1 - (3) \qquad \text{Evaluate}$$
$$z = -2 \qquad \text{We have } z, \text{ now find } x \text{ from original equation.}$$

$$4x - 3(3) + 2(-2) = -29 \qquad \text{Multiply and combine like terms}$$
$$4x - 13 = -29 \qquad \text{Add } 13$$
$$\underline{+13 \quad\quad +13}$$
$$\frac{4x}{4} = \frac{-16}{4} \qquad \text{Divide by } 4$$

$$x = -4 \qquad \text{We have our } x!$$

$$(-4, 3, -2) \qquad \text{Our Solution!}$$

World View Note: Around 250, *The Nine Chapters on the Mathematical Art* were published in China. This book had 246 problems, and chapter 8 was about solving systems of equations. One problem had four equations with five variables!

Just as with two variables and two equations, we can have special cases come up with three variables and three equations. The way we interpret the result is identical.

Example 184.

$$5x - 4y + 3z = -4$$
$$-10x + 8y - 6z = 8 \qquad \text{We will eliminate } x, \text{ start with first two equations}$$

$$15x - 12y + 9z = -12$$

$$5x - 4y + 3z = -4 \qquad \text{LCM of 5 and } -10 \text{ is 10.}$$
$$-10x + 8y - 6z = 8$$

$$2(5x - 4y + 3z) = -4(2) \qquad \text{Multiply the first equation by 2}$$
$$10x - 8y + 6z = -8$$

$$10x - 8y + 6z = -8 \qquad \text{Add this to the second equation, unchanged}$$
$$\underline{-10x + 8y - 6 = 8}$$
$$0 = 0 \qquad \textit{A} \text{ true statment}$$
$$\text{Infinite Solutions} \qquad \text{Our Solution}$$

Example 185.

$$3x - 4y + z = 2$$
$$-9x + 12y - 3z = -5 \qquad \text{We will eliminate } z, \text{ starting with the first two equations}$$
$$4x - 2y - z = 3$$

$$3x - 4y + z = 2 \qquad \text{The LCM of 1 and } -3 \text{ is 3}$$
$$-9x + 12y - 3z = -5$$

$$3(3x - 4y + z) = (2)3 \qquad \text{Multiply the first equation by 3}$$
$$9x - 12y + 3z = 6$$

$$9x - 12y + 3z = 6 \qquad \text{Add this to the second equation, unchanged}$$
$$\underline{-9x + 12y - 3z = -5}$$
$$0 = 1 \qquad \textit{A} \text{ false statement}$$
$$\text{No Solution } \varnothing \qquad \text{Our Solution}$$

Equations with three (or more) variables are not any more difficult than two variables if we are careful to keep our information organized and eliminate the same variable twice using two different pairs of equations. It is possible to solve each system several different ways. We can use different pairs of equations or eliminate variables in different orders, but as long as our information is organized and our algebra is correct, we will arrive at the same final solution.

4.4 Practice - Three Variables

Solve each of the following systems of equation.

1) $a - 2b + c = 5$
$2a + b - c = -1$
$3a + 3b - 2c = -4$

2) $2x + 3y = z - 1$
$3x = 8z - 1$
$5y + 7z = -1$

3) $3x + y - z = 11$
$x + 3y = z + 13$
$x + y - 3z = 11$

4) $x + y + z = 2$
$6x - 4y + 5z = 31$
$5x + 2y + 2z = 13$

5) $x + 6y + 3z = 4$
$2x + y + 2z = 3$
$3x - 2y + z = 0$

6) $x - y + 2z = -3$
$x + 2y + 3z = 4$
$2x + y + z = -3$

7) $x + y + z = 6$
$2x - y - z = -3$
$x - 2y + 3z = 6$

8) $x + y - z = 0$
$x + 2y - 4z = 0$
$2x + y + z = 0$

9) $x + y - z = 0$
$x - y - z = 0$
$x + y + 2z = 0$

10) $x + 2y - z = 4$
$4x - 3y + z = 8$
$5x - y = 12$

11) $-2x + y - 3z = 1$
$x - 4y + z = 6$
$4x + 16y + 4z = 24$

12) $4x + 12y + 16z = 4$
$3x + 4y + 5z = 3$
$x + 8y + 11z = 1$

13) $2x + y - 3z = 0$
$x - 4y + z = 0$
$4x + 16y + 4z = 0$

14) $4x + 12y + 16z = 0$
$3x + 4y + 5z = 0$
$x + 8y + 11z = 0$

15) $3x + 2y + 2z = 3$
$x + 2y - z = 5$
$2x - 4y + z = 0$

16) $p + q + r = 1$
$p + 2q + 3r = 4$
$4p + 5q + 6r = 7$

17) $x - 2y + 3z = 4$
$2x - y + z = -1$
$4x + y + z = 1$

18) $x + 2y - 3z = 9$
$2x - y + 2z = -8$
$3x - y - 4z = 3$

19) $x - y + 2z = 0$
$x - 2y + 3z = -1$
$2x - 2y + z = -3$

20) $4x - 7y + 3z = 1$
$3x + y - 2z = 4$
$4x - 7y + 3z = 6$

21) $4x - 3y + 2z = 40$
$5x + 9y - 7z = 47$
$9x + 8y - 3z = 97$

22) $3x + y - z = 10$
$8x - y - 6z = -3$
$5x - 2y - 5z = 1$

23) $3x + 3y - 2z = 13$
$6x + 2y - 5z = 13$
$5x - 2y - 5z = -1$

24) $2x - 3y + 5z = 1$
$3x + 2y - z = 4$
$4x + 7y - 7z = 7$

25) $3x - 4y + 2z = 1$
$2x + 3y - 3z = -1$
$x + 10y - 8z = 7$

26) $2x + y = z$
$4x + z = 4y$
$y = x + 1$

27) $m + 6n + 3p = 8$
$3m + 4n = -3$
$5m + 7n = 1$

28) $3x + 2y = z + 2$
$y = 1 - 2x$
$3z = -2y$

29) $-2w + 2x + 2y - 2z = -10$
$w + x + y + z = -5$
$3w + 2x + 2y + 4z = -11$
$w + 3x - 2y + 2z = -6$

30) $-w + 2x - 3y + z = -8$
$-w + x + y - z = -4$
$w + x + y + z = 22$
$-w + x - y - z = -14$

31) $w + x + y + z = 2$
$w + 2x + 2y + 4z = 1$
$-w + x - y - z = -6$
$-w + 3x + y - z = -2$

32) $w + x - y + z = 0$
$-w + 2x + 2y + z = 5$
$-w + 3x + y - z = -4$
$-2w + x + y - 3z = -7$

Systems of Equations - Value Problems

Objective: Solve value problems by setting up a system of equations.

One application of system of equations are known as value problems. Value problems are ones where each variable has a value attached to it. For example, if our variable is the number of nickles in a person's pocket, those nickles would have a value of five cents each. We will use a table to help us set up and solve value problems. The basic structure of the table is shown below.

	Number	Value	Total
Item 1			
Item 2			
Total			

The first column in the table is used for the number of things we have. Quite often, this will be our variables. The second column is used for the that value each item has. The third column is used for the total value which we calculate by multiplying the number by the value. For example, if we have 7 dimes, each with a value of 10 cents, the total value is $7 \cdot 10 = 70$ cents. The last row of the table is for totals. We only will use the third row (also marked total) for the totals that are given to use. This means sometimes this row may have some blanks in it. Once the table is filled in we can easily make equations by adding each column, setting it equal to the total at the bottom of the column. This is shown in the following example.

Example 186.

In a child's bank are 11 coins that have a value of $1.85. The coins are either quarters or dimes. How many coins each does child have?

	Number	Value	Total
Quarter	q	25	
Dime	d	10	
Total			

Using value table, use q for quarters, d for dimes
Each quarter's value is 25 cents, dime's is 10 cents

	Number	Value	Total
Quarter	q	25	$25q$
Dime	d	10	$10d$
Total			

Multiply number by value to get totals

	Number	Value	Total
Quarter	q	25	$25q$
Dime	d	10	$10d$
Total	11		185

We have 11 coins total. This is the number total.
We have 1.85 for the final total,
Write final total in cents (185)
Because 25 and 10 are cents

$$q + d = 11$$
$$25q + 10d = 185$$

First and last columns are our equations by adding
Solve by either addition or substitution.

$$-10(q + d) = (11)(-10)$$
$$-10q - 10d = -110$$

Using addition, multiply first equation by -10

$$-10q - 10d = -110$$
$$\underline{25q + 10d = 185}$$

Add together equations

$$\frac{15q}{15} = \frac{75}{15}$$

Divide both sides by 15

$$q = 5$$

We have our q, number of quarters is 5

$$(5) + d = 11$$
$$\underline{-5 \qquad -5}$$
$$d = 6$$

Plug into one of original equations
Subtract 5 from both sides
We have our d, number of dimes is 6

159

5 quarters and 6 dimes   Our Solution

World View Note: American coins are the only coins that do not state the value of the coin. On the back of the dime it says "one dime" (not 10 cents). On the back of the quarter it says "one quarter" (not 25 cents). On the penny it says "one cent" (not 1 cent). The rest of the world (Euros, Yen, Pesos, etc) all write the value as a number so people who don't speak the language can easily use the coins.

Ticket sales also have a value. Often different types of tickets sell for different prices (values). These problems can be solve in much the same way.

Example 187.

There were 41 tickets sold for an event. Tickets for children cost \$1.50 and tickets for adults cost \$2.00. Total receipts for the event were \$73.50. How many of each type of ticket were sold?

	Number	Value	Total
Child	c	1.5	
Adult	a	2	
Total			

Using our value table, c for child, a for adult
Child tickets have value 1.50, adult value is 2.00
(we can drop the zeros after the decimal point)

	Number	Value	Total
Child	c	1.5	$1.5c$
Adult	a	2	$2a$
Total			

Multiply number by value to get totals

	Number	Value	Total
Child	c	1.5	$1.5c$
Adult	a	2	$2a$
Total	41		73.5

We have 41 tickets sold. This is our number total
The final total was 73.50
Write in dollars as 1.5 and 2 are also dollars

$$c + a = 41$$
First and last columns are our equations by adding

$$1.5c + 2a = 73.5$$
We can solve by either addition or substitution

$$c + a = 41$$
We will solve by substitution.

$$\underline{-c \qquad -c}$$
Solve for a by subtracting c

$$a = 41 - c$$

$$1.5c + 2(41 - c) = 73.5$$
Substitute into untouched equation

$$1.5c + 82 - 2c = 73.5$$
Distribute

$$-0.5c + 82 = 73.5$$
Combine like terms

$$\underline{-82 \quad -82}$$
Subtract 82 from both sides

$$-0.5c = -8.5$$
Divide both sides by -0.5

$$\overline{-0.5} \quad \overline{-0.5}$$

$c = 17$	We have c, number of child tickets is 17
$a = 41 - (17)$	Plug into $a =$ equation to find a
$a = 24$	We have our a, number of adult tickets is 24
17 child tickets and 24 adult tickets	Our Solution

Some problems will not give us the total number of items we have. Instead they will give a relationship between the items. Here we will have statements such as "There are twice as many dimes as nickles". While it is clear that we need to multiply one variable by 2, it may not be clear which variable gets multiplied by 2. Generally the equations are backwards from the English sentence. If there are twice as many dimes, than we multiply the other variable (nickels) by two. So the equation would be $d = 2n$. This type of problem is in the next example.

Example 188.

A man has a collection of stamps made up of 5 cent stamps and 8 cent stamps. There are three times as many 8 cent stamps as 5 cent stamps. The total value of all the stamps is $3.48. How many of each stamp does he have?

	Number	Value	Total
Five	f	5	$5f$
Eight	$3f$	8	$24f$
Total			348

Use value table, f for five cent stamp, and e for eight
Also list value of each stamp under value column

	Number	Value	Total
Five	f	5	$5f$
Eight	e	8	$8e$
Total			

Multiply number by value to get total

	Number	Value	Total
Five	f	5	$5f$
Eight	e	8	$8e$
Total			348

The final total was 338 (written in cents)
We do not know the total number, this is left blank.

$e = 3f$	**3 times as many 8 cent stamples as 5 cent stamps**
$5f + 8e = 348$	Total column gives second equation
$5f + 8(3f) = 348$	Substitution, substitute first equation in second
$5f + 24f = 348$	Multiply first
$29f = 348$	Combine like terms
$\overline{29} \quad \overline{29}$	Divide both sides by 39
$f = 12$	We have f. There are 12 five cent stamps
$e = 3(12)$	Plug into first equation

$$e = 36 \qquad \text{We have } e, \text{There are 36 eight cent stamps}$$

12 five cent, 36 eight cent stamps   Our Solution

The same process for solving value problems can be applied to solving interest problems. Our table titles will be adjusted slightly as we do so.

	Invest	Rate	Interest
Account 1			
Account 2			
Total			

Our first column is for the amount invested in each account. The second column is the interest rate earned (written as a decimal - move decimal point twice left), and the last column is for the amount of interset earned. Just as before, we multiply the investment amount by the rate to find the final column, the interest earned. This is shown in the following example.

Example 189.

A woman invests $4000 in two accounts, one at 6% interest, the other at 9% interest for one year. At the end of the year she had earned $270 in interest. How much did she have invested in each account?

	Invest	Rate	Interest
Account 1	x	0.06	
Account 2	y	0.09	
Total			

Use our investment table, x and y for accounts
Fill in interest rates as decimals

	Invest	Rate	Interest
Account 1	x	0.06	$0.06x$
Account 2	y	0.09	$0.09y$
Total			

Multiply across to find interest earned.

	Invest	Rate	Interest
Account 1	x	0.06	$0.06x$
Account 2	y	0.09	$0.09y$
Total	4000		270

Total investment is 4000,
Total interest was 276

$$x + y = 4000 \qquad \text{First and last column give our two equations}$$
$$0.06x + 0.09y = 270 \qquad \text{Solve by either substitution or addition}$$

$$-0.06(x + y) = (4000)(-0.06) \qquad \text{Use Addition, multiply first equation by } -0.06$$
$$-0.06x - 0.06y = -240$$

$$-0.06x - 0.06y = -240 \qquad \text{Add equations together}$$
$$0.06x + 0.09y = 270$$
$$0.03y = 30 \qquad \text{Divide both sides by } 0.03$$
$$\overline{0.03 \quad 0.03}$$
$$y = 1000 \qquad \text{We have } y, \$1000 \text{ invested at } 9\%$$
$$x + 1000 = 4000 \qquad \text{Plug into original equation}$$
$$-1000 - 1000 \qquad \text{Subtract } 1000 \text{ from both sides}$$
$$x = 3000 \qquad \text{We have } x, \$3000 \text{ invested at } 6\%$$
$$\$1000 \text{ at } 9\% \text{ and } \$3000 \text{ at } 6\% \qquad \text{Our Solution}$$

The same process can be used to find an unknown interest rate.

Example 190.

John invests $5000 in one account and $8000 in an account paying 4% more in interest. He earned $1230 in interest after one year. At what rates did he invest?

	Invest	Rate	Interest
Account 1	5000	x	
Account 2	8000	$x + 0.04$	
Total			

Our investment table. Use x for first rate
The second rate is 4% higher, or $x + 0.04$
Be sure to write this rate as a decimal!

	Invest	Rate	Interest
Account 1	5000	x	$5000x$
Account 2	8000	$x + 0.04$	$8000x + 320$
Total			

Multiply to fill in interest column.
Be sure to distribute $8000(x + 0.04)$

	Invest	Rate	Interest
Account 2	5000	x	$5000x$
Account 2	8000	$x + 0.04$	$8000x + 320$
Total			1230

Total interest was 1230.

$$5000x + 8000x + 320 = 1230 \qquad \text{Last column gives our equation}$$
$$13000x + 320 = 1230 \qquad \text{Combine like terms}$$
$$-320 \quad -320 \qquad \text{Subtract } 320 \text{ from both sides}$$
$$13000x = 910 \qquad \text{Divide both sides by } 13000$$
$$\overline{13000 \quad 13000}$$
$$x = 0.07 \qquad \text{We have our } x, 7\% \text{ interest}$$
$$(0.07) + 0.04 \qquad \text{Second account is 4\% higher}$$
$$0.11 \qquad \text{The account with } \$8000 \text{ is at } 11\%$$
$$\$5000 \text{ at } 7\% \text{ and } \$8000 \text{ at } 11\% \qquad \text{Our Solution}$$

4.5 Practice - Value Problems

Solve.

1) A collection of dimes and quaters is worth $15.25. There are 103 coins in all. How many of each is there?

2) A collection of half dollars and nickels is worth $13.40. There are 34 coins in all. How many are there?

3) The attendance at a school concert was 578. Admission was $2.00 for adults and $1.50 for children. The total receipts were $985.00. How many adults and how many children attended?

4) A purse contains $3.90 made up of dimes and quarters. If there are 21 coins in all, how many dimes and how many quarters were there?

5) A boy has $2.25 in nickels and dimes. If there are twice as many dimes as nickels, how many of each kind has he?

6) $3.75 is made up of quarters and half dollars. If the number of quarters exceeds the number of half dollars by 3, how many coins of each denomination are there?

7) A collection of 27 coins consisting of nickels and dimes amounts to $2.25. How many coins of each kind are there?

8) $3.25 in dimes and nickels, were distributed among 45 boys. If each received one coin, how many received dimes and how many received nickels?

9) There were 429 people at a play. Admission was $1 each for adults and 75 cents each for children. The receipts were $372.50. How many children and how many adults attended?

10) There were 200 tickets sold for a women's basketball game. Tickets for students were 50 cents each and for adults 75 cents each. The total amount of money collected was $132.50. How many of each type of ticket was sold?

11) There were 203 tickets sold for a volleyball game. For activity-card holders, the price was $1.25 each and for noncard holders the price was $2 each. The total amount of money collected was $310. How many of each type of ticket was sold?

12) At a local ball game the hotdogs sold for $2.50 each and the hamburgers sold for $2.75 each. There were 131 total sandwiches sold for a total value of $342. How many of each sandwich was sold?

13) At a recent Vikings game $445 in admission tickets was taken in. The cost of a student ticket was $1.50 and the cost of a non-student ticket was $2.50. A total of 232 tickets were sold. How many students and how many non-students attented the game?

14) A bank contains 27 coins in dimes and quarters. The coins have a total value of $4.95. Find the number of dimes and quarters in the bank.

15) A coin purse contains 18 coins in nickels and dimes. The coins have a total value of $1.15. Find the number of nickels and dimes in the coin purse.

16) A business executive bought 40 stamps for $9.60. The purchase included 25¢ stamps and 20¢ stamps. How many of each type of stamp were bought?

17) A postal clerk sold some 15¢ stamps and some 25¢ stamps. Altogether, 15 stamps were sold for a total cost of $3.15. How many of each type of stamps were sold?

18) A drawer contains 15¢ stamps and 18¢ stamps. The number of 15¢ stamps is four less than three times the number of 18¢ stamps. The total value of all the stamps is $1.29. How many 15¢ stamps are in the drawer?

19) The total value of dimes and quarters in a bank is $6.05. There are six more quarters than dimes. Find the number of each type of coin in the bank.

20) A child's piggy bank contains 44 coins in quarters and dimes. The coins have a total value of $8.60. Find the number of quaters in the bank.

21) A coin bank contains nickels and dimes. The number of dimes is 10 less than twice the number of nickels. The total value of all the coins is $2.75. Find the number of each type of coin in the bank.

22) A total of 26 bills are in a cash box. Some of the bills are one dollar bills, and the rest are five dollar bills. The total amount of cash in the box is $50. Find the number of each type of bill in the cash box.

23) A bank teller cashed a check for $200 using twenty dollar bills and ten dollar bills. In all, twelve bills were handed to the customer. Find the number of twenty dollar bills and the number of ten dollar bills.

24) A collection of stamps consists of 22¢ stamps and 40¢ stamps. The number of 22¢ stamps is three more than four times the number of 40¢ stamps. The total value of the stamps is $8.34. Find the number of 22¢ stamps in the collection.

25) A total of $27000 is invested, part of it at 12% and the rest at 13%. The total interest after one year is $3385. How much was invested at each rate?

26) A total of $50000 is invested, part of it at 5% and the rest at 7.5%. The total interest after one year is $3250. How much was invested at each rate?

27) A total of $9000 is invested, part of it at 10% and the rest at 12%. The total interest after one year is $1030. How much was invested at each rate?

28) A total of $18000 is invested, part of it at 6% and the rest at 9%. The total interest after one year is $1248. How much was invested at each rate?

29) An inheritance of $10000 is invested in 2 ways, part at 9.5% and the remainder at 11%. The combined annual interest was $1038.50. How much was invested at each rate?

30) Kerry earned a total of $900 last year on his investments. If $7000 was invested at a certain rate of return and $9000 was invested in a fund with a rate that was 2% higher, find the two rates of interest.

31) Jason earned $256 interest last year on his investments. If $1600 was invested at a certain rate of return and $2400 was invested in a fund with a rate that was double the rate of the first fund, find the two rates of interest.

32) Millicent earned $435 last year in interest. If $3000 was invested at a certain rate of return and $4500 was invested in a fund with a rate that was 2% lower, find the two rates of interest.

33) A total of $8500 is invested, part of it at 6% and the rest at 3.5%. The total interest after one year is $385. How much was invested at each rate?

34) A total of $12000 was invested, part of it at 9% and the rest at 7.5%. The total interest after one year is $1005. How much was invested at each rate?

35) A total of $15000 is invested, part of it at 8% and the rest at 11%. The total interest after one year is $1455. How much was invested at each rate?

36) A total of $17500 is invested, part of it at 7.25% and the rest at 6.5%. The total interest after one year is $1227.50. How much was invested at each rate?

37) A total of $6000 is invested, part of it at 4.25% and the rest at 5.75%. The total interest after one year is $300. How much was invested at each rate?

38) A total of $14000 is invested, part of it at 5.5% and the rest at 9%. The total interest after one year is $910. How much was invested at each rate?

39) A total of $11000 is invested, part of it at 6.8% and the rest at 8.2%. The total interest after one year is $797. How much was invested at each rate?

40) An investment portfolio earned $2010 in interest last year. If $3000 was invested at a certain rate of return and $24000 was invested in a fund with a rate that was 4% lower, find the two rates of interest.

41) Samantha earned $1480 in interest last year on her investments. If $5000 was invested at a certain rate of return and $11000 was invested in a fund with a rate that was two-thirds the rate of the first fund, find the two rates of interest.

42) A man has $5.10 in nickels, dimes, and quarters. There are twice as many nickels as dimes and 3 more dimes than quarters. How many coins of each kind were there?

43) 30 coins having a value of $3.30 consists of nickels, dimes and quarters. If there are twice as many quarters as dimes, how many coins of each kind were there?

44) A bag contains nickels, dimes and quarters having a value of $3.75. If there are 40 coins in all and 3 times as many dimes as quarters, how many coins of each kind were there?

Systems of Equations - Mixture Problems

Objective: Solve mixture problems by setting up a system of equations.

One application of systems of equations are mixture problems. Mixture problems are ones where two different solutions are mixed together resulting in a new final solution. We will use the following table to help us solve mixture problems:

	Amount	Part	Total
Item 1			
Item 2			
Final			

The first column is for the amount of each item we have. The second column is labeled "part". If we mix percentages we will put the rate (written as a decimal) in this column. If we mix prices we will put prices in this column. Then we can multiply the amount by the part to find the total. Then we can get an equation by adding the amount and/or total columns that will help us solve the problem and answer the questions.

These problems can have either one or two variables. We will start with one variable problems.

Example 191.

A chemist has 70 mL of a 50% methane solution. How much of a 80% solution must she add so the final solution is 60% methane?

	Amount	Part	Total
Start	70	0.5	
Add	x	0.8	
Final			

Set up the mixture table. We start with 70, but don't know how much we add, that is x. The part is the percentages, 0.5 for start, 0.8 for add.

	Amount	Part	Total
Start	70	0.5	
Add	x	0.8	
Final	$70 + x$	0.6	

Add amount column to get final amount. The part for this amount is 0.6 because we want the final solution to be 60% methane.

	Amount	Part	Total
Start	70	0.5	35
Add	x	0.8	$0.8x$
Final	$70 + x$	0.6	$42 + 0.6x$

Multiply amount by part to get total. be sure to distribute on the last row: $(70 + x)0.6$

$$35 + 0.8x = 42 + 0.6x$$ The last column is our equation by adding

$$\underline{-0.6x \qquad -0.6x}$$ Move variables to one side, subtract $0.6x$

$$35 + 0.2x = 42$$ Subtract 35 from both sides

$$\underline{-35 \qquad\quad -35}$$

$$0.2x = 7$$ Divide both sides by 0.2

$$\overline{0.2} \quad \overline{0.2}$$

$$x = 35$$ We have our x!

$$35\,\text{mL must be added}$$ Our Solution

The same process can be used if the starting and final amount have a price attached to them, rather than a percentage.

Example 192.

A coffee mix is to be made that sells for \$2.50 by mixing two types of coffee. The cafe has 40 mL of coffee that costs \$3.00. How much of another coffee that costs \$1.50 should the cafe mix with the first?

	Amount	Part	Total
Start	40	3	
Add	x	1.5	
Final			

Set up mixture table. We know the starting amount and its cost, \$3. The added amount we do not know but we do know its cost is \$1.50.

	Amount	Part	Total
Start	40	3	
Add	x	1.5	
Final	$40 + x$	2.5	

Add the amounts to get the final amount. We want this final amount to sell for \$2.50.

	Amount	Part	Total
Start	40	3	120
Add	x	1.5	$1.5x$
Final	$40+x$	2.5	$100+2.5x$

Multiply amount by part to get the total.
Be sure to distribute on the last row $(40+x)2.5$

$$120 + 1.5x = 100 + 2.5x$$ Adding down the total column gives our equation

$$\underline{-1.5x \qquad\quad -1.5x}$$ Move variables to one side by subtracting $1.5x$

$$120 = 100 + x$$ Subtract 100 from both sides

$$\underline{-100 - 100}$$

$$20 = x$$ We have our x.

$$20\text{mL must be added.}$$ Our Solution

World View Note: Brazil is the world's largest coffee producer, producing 2.59 million metric tons of coffee a year! That is over three times as much coffee as second place Vietnam!

The above problems illustrate how we can put the mixture table together and get an equation to solve. However, here we are interested in systems of equations, with two unknown values. The following example is one such problem.

Example 193.

A farmer has two types of milk, one that is 24% butterfat and another which is 18% butterfat. How much of each should he use to end up with 42 gallons of 20% butterfat?

	Amount	Part	Total
Milk 1	x	0.24	
Milk 2	y	0.18	
Final	42	0.2	

We don't know either start value, but we do know final is 42. Also fill in part column with percentage of each type of milk including the final solution

	Amount	Part	Total
Milk 1	x	0.24	$0.24x$
Milk 2	y	0.18	$0.18y$
Final	42	0.2	8.4

Multiply amount by part to get totals.

$$x + y = 42$$ The amount column gives one equation

$$0.24x + 0.18y = 8.4$$ The total column gives a second equation.

169

$$-0.18(x+y) = (42)(-0.18) \qquad \text{Use addition. Multiply first equation by} -0.18$$
$$-0.18x - 0.18y = -7.56$$

$$
\begin{aligned}
-0.18x - 0.18y &= -7.56 \qquad && \text{Add the equations together}\\
\underline{0.24x + 0.18y} &= \underline{8.4}\\
0.06x &= 0.84 \qquad && \text{Divide both sides by } 0.06\\
\overline{0.06} &\;\; \overline{0.06}
\end{aligned}
$$

$$
\begin{aligned}
x &= 14 \qquad && \text{We have our } x,\ 14\,\text{gal of } 24\%\ \text{butterfat}\\
(14)+y &= 42 \qquad && \text{Plug into original equation to find } y\\
\underline{-14} \phantom{{}+y} &\;\; \underline{-14} \qquad && \text{Subtract 14 from both sides}\\
y &= 28 \qquad && \text{We have our } y,\ 28\,\text{gal of } 18\%\ \text{butterfat}
\end{aligned}
$$

14 gal of 24% and 28 gal of 18% \qquad Our Solution

The same process can be used to solve mixtures of prices with two unknowns.

Example 194.

In a candy shop, chocolate which sells for \$4 a pound is mixed with nuts which are sold for \$2.50 a pound are mixed to form a chocolate-nut candy which sells for \$3.50 a pound. How much of each are used to make 30 pounds of the mixture?

	Amount	Part	Total
Chocolate	c	4	
Nut	n	2.5	
Final	30	3.5	

Using our mixture table, use c and n for variables. We do know the final amount (30) and price, include this in the table

	Amount	Part	Total
Chocolate	c	4	$4c$
Nut	n	2.5	$2.5n$
Final	30	3.5	105

Multiply amount by part to get totals

$$
\begin{aligned}
c + n &= 30 \qquad && \text{First equation comes from the first column}\\
4c + 2.5n &= 105 \qquad && \text{Second equation comes from the total column}
\end{aligned}
$$

$$
\begin{aligned}
c + n &= 30 \qquad && \text{We will solve this problem with substitution}\\
\underline{-n} &\ \underline{-n} \qquad && \text{Solve for } c \text{ by subtracting } n \text{ from the first equation}\\
c &= 30 - n
\end{aligned}
$$

170

$$4(30 - n) + 2.5n = 105 \qquad \text{Substitute into untouched equation}$$
$$120 - 4n + 2.5n = 105 \qquad \text{Distribute}$$
$$120 - 1.5n = 105 \qquad \text{Combine like terms}$$
$$\underline{-120 \qquad\qquad -120} \qquad \text{Subtract 120 from both sides}$$
$$-1.5n = -15 \qquad \text{Divide both sides by} -1.5$$
$$\overline{-1.5 \quad -1.5}$$
$$n = 10 \qquad \text{We have our } n,\, 10 \text{ lbs of nuts}$$
$$c = 30 - (10) \qquad \text{Plug into } c = \text{equation to find } c$$
$$c = 20 \qquad \text{We have our } c,\, 20 \text{ lbs of chocolate}$$

10 lbs of nuts and 20 lbs of chocolate   Our Solution

With mixture problems we often are mixing with a pure solution or using water which contains none of our chemical we are interested in. For pure solutions, the percentage is 100% (or 1 in the table). For water, the percentage is 0%. This is shown in the following example.

Example 195.

A solution of pure antifreeze is mixed with water to make a 65% antifreeze solution. How much of each should be used to make 70 L?

	Amount	Part	Final
Antifreeze	a	1	
Water	w	0	
Final	70	0.65	

We use a and w for our variables. Antifreeze is pure, 100% or 1 in our table, written as a decimal. Water has no antifreeze, its percentage is 0. We also fill in the final percent

	Amount	Part	Final
Antifreeze	a	1	a
Water	w	0	0
Final	70	0.65	45.5

Multiply to find final amounts

$$a + w = 70 \qquad \text{First equation comes from first column}$$
$$a = 45.5 \qquad \text{Second equation comes from second column}$$
$$(45.5) + w = 70 \qquad \text{We have } a, \text{ plug into to other equation}$$
$$\underline{-45.5 \qquad -45.5} \qquad \text{Subtract 45.5 from both sides}$$
$$w = 24.5 \qquad \text{We have our } w$$

45.5 L of antifreeze and 24.5 L of water   Our Solution

4.6 Practice - Mixture Problems

Solve.

1) A tank contains 8000 liters of a solution that is 40% acid. How much water should be added to make a solution that is 30% acid?

2) How much antifreeze should be added to 5 quarts of a 30% mixture of antifreeze to make a solution that is 50% antifreeze?

3) Of 12 pounds of salt water 10% is salt; of another mixture 3% is salt. How many pounds of the second should be added to the first in order to get a mixture of 5% salt?

4) How much alcohol must be added to 24 gallons of a 14% solution of alcohol in order to produce a 20% solution?

5) How many pounds of a 4% solution of borax must be added to 24 pounds of a 12% solution of borax to obtain a 10% solution of borax?

6) How many grams of pure acid must be added to 40 grams of a 20% acid solution to make a solution which is 36% acid?

7) A 100 LB bag of animal feed is 40% oats. How many pounds of oats must be added to this feed to produce a mixture which is 50% oats?

8) A 20 oz alloy of platinum that costs $220 per ounce is mixed with an alloy that costs $400 per ounce. How many ounces of the $400 alloy should be used to make an alloy that costs $300 per ounce?

9) How many pounds of tea that cost $4.20 per pound must be mixed with 12 lb of tea that cost $2.25 per pound to make a mixture that costs $3.40 per pound?

10) How many liters of a solvent that costs $80 per liter must be mixed with 6 L of a solvent that costs $25 per liter to make a solvent that costs $36 per liter?

11) How many kilograms of hard candy that cost $7.50 per kilogram must be mixed with 24 kg of jelly beans that cost $3.25 per kilogram to make a mixture that sells for $4.50 per kilogram?

12) How many kilograms of soil supplement that costs $7.00 per kilogram must be mixed with 20 kg of aluminum nitrate that costs $3.50 per kilogram to make a fertilizer that costs $4.50 per kilogram?

13) How many pounds of lima beans that cost 90c per pound must be mixed with 16 lb of corn that cost 50c per pound to make a mixture of vegetables that costs 65c per pound?

14) How many liters of a blue dye that costs $1.60 per liter must be mixed with 18 L of anil that costs $2.50 per liter to make a mixture that costs $1.90 per liter?

15) Solution A is 50% acid and solution B is 80% acid. How much of each should be used to make 100cc. of a solution that is 68% acid?

16) A certain grade of milk contains 10% butter fat and a certain grade of cream 60% butter fat. How many quarts of each must be taken so as to obtain a mixture of 100 quarts that will be 45% butter fat?

17) A farmer has some cream which is 21% butterfat and some which is 15% butter fat. How many gallons of each must be mixed to produce 60 gallons of cream which is 19% butterfat?

18) A syrup manufacturer has some pure maple syrup and some which is 85% maple syrup. How many liters of each should be mixed to make 150L which is 96% maple syrup?

19) A chemist wants to make 50ml of a 16% acid solution by mixing a 13% acid solution and an 18% acid solution. How many milliliters of each solution should the chemist use?

20) A hair dye is made by blending 7% hydrogen peroxide solution and a 4% hydrogen peroxide solution. How many mililiters of each are used to make a 300 ml solution that is 5% hydrogen peroxide?

21) A paint that contains 21% green dye is mixed with a paint that contains 15% green dye. How many gallons of each must be used to make 60 gal of paint that is 19% green dye?

22) A candy mix sells for $2.20 per kilogram. It contains chocolates worth $1.80 per kilogram and other candy worth $3.00 per kilogram. How much of each are in 15 kilograms of the mixture?

23) To make a weed and feed mixture, the Green Thumb Garden Shop mixes fertilizer worth $4.00/lb. with a weed killer worth $8.00/lb. The mixture will cost $6.00/lb. How much of each should be used to prepare 500 lb. of the mixture?

24) A grocer is mixing 40 cent per lb. coffee with 60 cent per lb. coffee to make a mixture worth 54c per lb. How much of each kind of coffee should be used to make 70 lb. of the mixture?

25) A grocer wishes to mix sugar at 9 cents per pound with sugar at 6 cents per pound to make 60 pounds at 7 cents per pound. What quantity of each must he take?

26) A high-protein diet supplement that costs $6.75 per pound is mixed with a vitamin supplement that costs $3.25 per pound. How many pounds of each should be used to make 5 lb of a mixture that costs $4.65 per pound?

27) A goldsmith combined an alloy that costs $4.30 per ounce with an alloy that costs $1.80 per ounce. How many ounces of each were used to make a mixture of 200 oz costing $2.50 per ounce?

28) A grocery store offers a cheese and fruit sampler that combines cheddar cheese that costs $8 per kilogram with kiwis that cost $3 per kilogram. How many kilograms of each were used to make a 5 kg mixture that costs $4.50 per kilogram?

29) The manager of a garden shop mixes grass seed that is 60% rye grass with 70 lb of grass seed that is 80% rye grass to make a mixture that is 74% rye grass. How much of the 60% mixture is used?

30) How many ounces of water evaporated from 50 oz of a 12% salt solution to produce a 15% salt solution?

31) A caterer made an ice cream punch by combining fruit juice that cost $2.25 per gallon with ice cream that costs $3.25 per gallon. How many gallons of each were used to make 100 gal of punch costing $2.50 per pound?

32) A clothing manufacturer has some pure silk thread and some thread that is 85% silk. How many kilograms of each must be woven together to make 75 kg of cloth that is 96% silk?

33) A carpet manufacturer blends two fibers, one 20% wool and the second 50% wool. How many pounds of each fiber should be woven together to produce 600 lb of a fabric that is 28% wool?

34) How many pounds of coffee that is 40% java beans must be mixed with 80 lb of coffee that is 30% java beans to make a coffee blend that is 32% java beans?

35) The manager of a specialty food store combined almonds that cost $4.50 per pound with walnuts that cost $2.50 per pound. How many pounds of each were used to make a 100 lb mixture that cost $3.24 per pound?

36) A tea that is 20% jasmine is blended with a tea that is 15% jasmine. How many pounds of each tea are used to make 5 lb of tea that is 18% jasmine?

37) How many ounces of dried apricots must be added to 18 oz of a snack mix that contains 20% dried apricots to make a mixture that is 25% dried apricots?

38) How many mililiters of pure chocolate must be added to 150 ml of chocolate topping that is 50% chocolate to make a topping that is 75% chocolate?

39) How many ounces of pure bran flakes must be added to 50 oz of cereal that is 40% bran flakes to produce a mixture that is 50% bran flakes?

40) A ground meat mixture is formed by combining meat that costs $2.20 per pound with meat that costs $4.20 per pound. How many pounds of each were used to make a 50 lb mixture tha costs $3.00 per pound?

41) How many grams of pure water must be added to 50 g of pure acid to make a solution that is 40% acid?

42) A lumber company combined oak wood chips that cost $3.10 per pound with pine wood chips that cost $2.50 per pound. How many pounds of each were used to make an 80 lb mixture costing $2.65 per pound?

43) How many ounces of pure water must be added to 50 oz of a 15% saline solution to make a saline solution that is 10% salt?

Chapter 5 : Polynomials

Polynomials - Exponent Properties

Objective: Simplify expressions using the properties of exponents.

Problems with exponents can often be simplified using a few basic exponent properties. Exponents represent repeated multiplication. We will use this fact to discover the important properties.

World View Note: The word exponent comes from the Latin "expo" meaning out of and "ponere" meaning place. While there is some debate, it seems that the Babylonians living in Iraq were the first to do work with exponents (dating back to the 23rd century BC or earlier!)

Example 196.

$$a^3a^2 \quad \text{Expand exponents to multiplication problem}$$
$$(aaa)(aa) \quad \text{Now we have 5 } a's \text{ being multiplied together}$$
$$a^5 \quad \text{Our Solution}$$

A quicker method to arrive at our answer would have been to just add the exponents: $a^3a^2 = a^{3+2} = a^5$ This is known as the **product rule of exponents**

Product Rule of Exponents: $a^m a^n = a^{m+n}$

The product rule of exponents can be used to simplify many problems. We will add the exponent on like variables. This is shown in the following examples

Example 197.

$$3^2 \cdot 3^6 \cdot 3 \quad \text{Same base, add the exponents } 2 + 6 + 1$$
$$3^9 \quad \text{Our Solution}$$

Example 198.

$$2x^3y^5z \cdot 5xy^2z^3 \quad \text{Multiply } 2 \cdot 5, \text{ add exponents on } x, y \text{ and } z$$
$$10x^4y^7z^4 \quad \text{Our Solution}$$

Rather than multiplying, we will now try to divide with exponents

Example 199.

$$\frac{a^5}{a^2} \quad \text{Expand exponents}$$
$$\frac{aaaaa}{aa} \quad \text{Divide out two of the } a's$$
$$aaa \quad \text{Convert to exponents}$$
$$a^3 \quad \text{Our Solution}$$

A quicker method to arrive at the solution would have been to just subtract the exponents, $\frac{a^5}{a^2} = a^{5-2} = a^3$. This is known as the quotient rule of exponents.

$$\textbf{Quotient Rule of Exponents: } \frac{a^m}{a^n} = a^{m-n}$$

The quotient rule of exponents can similarly be used to simplify exponent problems by subtracting exponents on like variables. This is shown in the following examples.

Example 200.

$\dfrac{7^{13}}{7^5}$ Same base, subtract the exponents

7^8 Our Solution

Example 201.

$\dfrac{5a^3b^5c^2}{2ab^3c}$ Subtract exponents on a, b and c

$\dfrac{5}{2}a^2b^2c$ Our Solution

A third property we will look at will have an exponent problem raised to a second exponent. This is investigated in the following example.

Example 202.

$\left(a^2\right)^3$ This means we have a^2 three times

$a^2 \cdot a^2 \cdot a^2$ Add exponents

a^6 Our solution

A quicker method to arrive at the solution would have been to just multiply the exponents, $(a^2)^3 = a^{2 \cdot 3} = a^6$. This is known as the power of a power rule of exponents.

$$\textbf{Power of } a \textbf{ Power Rule of Exponents: } (a^m)^n = a^{mn}$$

This property is often combined with two other properties which we will investigate now.

Example 203.

$(ab)^3$ This means we have (ab) three times

$(ab)(ab)(ab)$ Three $a's$ and three $b's$ can be written with exponents

a^3b^3 Our Solution

A quicker method to arrive at the solution would have been to take the exponent of three and put it on each factor in parenthesis, $(ab)^3 = a^3b^3$. This is known as the power of a product rule or exponents.

$$\text{Power of } a \text{ Product Rule of Exponents: } (ab)^m = a^m b^m$$

It is important to be careful to only use the power of a product rule with multiplication inside parenthesis. This property does NOT work if there is addition or subtraction.

Warning 204.

$$(a+b)^m \neq a^m + b^m \quad \text{These are NOT equal, beware of this error!}$$

Another property that is very similar to the power of a product rule is considered next.

Example 205.

$$\left(\frac{a}{b}\right)^3 \quad \text{This means we have the fraction three timse}$$

$$\left(\frac{a}{b}\right)\left(\frac{a}{b}\right)\left(\frac{a}{b}\right) \quad \text{Multiply fractions across the top and bottom, using exponents}$$

$$\frac{a^3}{b^3} \quad \text{Our Solution}$$

A quicker method to arrive at the solution would have been to put the exponent on every factor in both the numerator and denominator, $\left(\frac{a}{b}\right)^3 = \frac{a^3}{b^3}$. This is known as the power of a quotient rule of exponents.

$$\text{Power of } a \text{ Quotient Rule of Exponents: } \left(\frac{a}{b}\right)^m = \frac{a^m}{b^m}$$

The power of a power, product and quotient rules are often used together to simplify expressions. This is shown in the following examples.

Example 206.

$$(x^3yz^2)^4 \quad \text{Put the exponent of 4 on each factor, multiplying powers}$$
$$x^{12}y^4z^8 \quad \text{Our solution}$$

Example 207.

$$\left(\frac{a^3b}{c^8d^5}\right)^2 \qquad \text{Put the exponent of 2 on each factor, multiplying powers}$$

$$\frac{a^6b^2}{c^8d^{10}} \qquad \text{Our Solution}$$

As we multiply exponents its important to remember these properties apply to exponents, not bases. An expressions such as 5^3 does not mean we multipy 5 by 3, rather we multiply 5 three times, $5 \times 5 \times 5 = 125$. This is shown in the next example.

Example 208.

$$(4x^2y^5)^3 \qquad \text{Put the exponent of 3 on each factor, multiplying powers}$$
$$4^3x^6y^{15} \qquad \text{Evaluate } 4^3$$
$$64x^6y^{15} \qquad \text{Our Solution}$$

In the previous example we did not put the 3 on the 4 and multipy to get 12, this would have been incorrect. Never multipy a base by the exponent. These properties pertain to exponents only, not bases.

In this lesson we have discussed 5 different exponent properties. These rules are summarized in the following table.

Rules of Exponents

Product Rule of Exponents	$a^m a^n = a^{m+n}$
Quotient Rule of Exponents	$\dfrac{a^m}{a^n} = a^{m-n}$
Power of a Power Rule of Exponents	$(a^m)^n = a^{mn}$
Power of a Product Rule of Exponents	$(ab)^m = a^m b^m$
Power of a Quotient Rule of Exponents	$\left(\dfrac{a}{b}\right)^m = \dfrac{a^m}{b^m}$

These five properties are often mixed up in the same problem. Often there is a bit of flexibility as to which property is used first. However, order of operations still applies to a problem. For this reason it is the suggestion of the auther to simplify inside any parenthesis first, then simplify any exponents (using power rules), and finally simplify any multiplication or division (using product and quotient rules). This is illustrated in the next few examples.

Example 209.

$$(4x^3y \cdot 5x^4y^2)^3 \qquad \text{In parenthesis simplify using product rule, adding exponents}$$
$$(20x^7y^3)^3 \qquad \text{With power rules, put three on each factor, multiplying exponents}$$
$$20^3x^{21}y^9 \qquad \text{Evaluate } 20^3$$
$$8000x^{21}y^9 \qquad \text{Our Solution}$$

Example 210.

$7a^3(2a^4)^3$	Parenthesis are already simplified, next use power rules
$7a^3(8a^{12})$	Using product rule, add exponents and multiply numbers
$56a^{15}$	Our Solution

Example 211.

$$\frac{3a^3b \cdot 10a^4b^3}{2a^4b^2}$$ Simplify numerator with product rule, adding exponents

$$\frac{30a^7b^4}{2a^4b^2}$$ Now use the quotient rule to subtract exponents

$$15a^3b^2$$ Our Solution

Example 212.

$$\frac{3m^8n^{12}}{(m^2n^3)^3}$$ Use power rule in denominator

$$\frac{3m^8n^{12}}{m^6n^9}$$ Use quotient rule

$$3m^2n^3$$ Our solution

Example 213.

$$\left(\frac{3ab^2(2a^4b^2)^3}{6a^5b^7}\right)^2$$ Simplify inside parenthesis first, using power rule in numerator

$$\left(\frac{3ab^2(8a^{12}b^6)}{6a^5b^7}\right)^2$$ Simplify numerator using product rule

$$\left(\frac{24a^{13}b^8}{6a^5b^7}\right)^2$$ Simplify using the quotient rule

$(4a^8b)^2$	Now that the parenthesis are simplified, use the power rules
$16a^{16}b^2$	Our Solution

Clearly these problems can quickly become quite involved. Remember to follow order of operations as a guide, simplify inside parenthesis first, then power rules, then product and quotient rules.

5.1 Practice - Exponent Properties

Simplify.

1) $4 \cdot 4^4 \cdot 4^4$

2) $4 \cdot 4^4 \cdot 4^2$

3) $4 \cdot 2^2$

4) $3 \cdot 3^3 \cdot 3^2$

5) $3m \cdot 4mn$

6) $3x \cdot 4x^2$

7) $2m^4n^2 \cdot 4nm^2$

8) $x^2y^4 \cdot xy^2$

9) $(3^3)^4$

10) $(4^3)^4$

11) $(4^4)^2$

12) $(3^2)^3$

13) $(2u^3v^2)^2$

14) $(xy)^3$

15) $(2a^4)^4$

16) $(2xy)^4$

17) $\frac{4^5}{4^3}$

18) $\frac{3^7}{3^3}$

19) $\frac{3^2}{3}$

20) $\frac{3^4}{3}$

21) $\frac{3nm^2}{3n}$

22) $\frac{x^2y^4}{4xy}$

23) $\frac{4x^3y^4}{3xy^3}$

24) $\frac{xy^3}{4xy}$

25) $(x^3y^4 \cdot 2x^2y^3)^2$

26) $(u^2v^2 \cdot 2u^4)^3$

27) $2x(x^4y^4)^4$

28) $\frac{3vu^5 \cdot 2v^3}{uv^2 \cdot 2u^3v}$

29) $\frac{2x^7y^5}{3x^3y \cdot 4x^2y^3}$

30) $\frac{2ba^7 \cdot 2b^4}{ba^2 \cdot 3a^3b^4}$

31) $\left(\frac{(2x)^3}{x^3}\right)^2$

32) $\frac{2a^2b^2a^7}{(ba^4)^2}$

33) $\left(\frac{2y^{17}}{(2x^2y^4)^4}\right)^3$

34) $\frac{yx^2 \cdot (y^4)^2}{2y^4}$

35) $\left(\frac{2mn^4 \cdot 2m^4n^4}{mn^4}\right)^3$

36) $\frac{n^3(n^4)^2}{2mn}$

37) $\frac{2xy^5 \cdot 2x^2y^3}{2xy^4 \cdot y^3}$

38) $\frac{(2y^3x^2)^2}{2x^2y^4 \cdot x^2}$

39) $\frac{q^3r^2 \cdot (2p^2q^2r^3)^2}{2p^3}$

40) $\frac{2x^4y^5 \cdot 2z^{10}x^2y^7}{(xy^2z^2)^4}$

41) $\left(\frac{zy^3 \cdot z^3x^4y^4}{x^3y^3z^3}\right)^4$

42) $\left(\frac{2q^3p^3r^4 \cdot 2p^3}{(qrp^3)^2}\right)^4$

43) $\frac{2x^2y^2z^6 \cdot 2zx^2y^2}{(x^2z^3)^2}$

182

Polynomials - Negative Exponents

Objective: Simplify expressions with negative exponents using the properties of exponents.

There are a few special exponent properties that deal with exponents that are not positive. The first is considered in the following example, which is worded out 2 different ways:

Example 214.

$$\frac{a^3}{a^3} \qquad \text{Use the quotient rule to subtract exponents}$$

$$a^0 \qquad \text{Our Solution, but now we consider the problem } a \text{ the second way:}$$

$$\frac{a^3}{a^3} \qquad \text{Rewrite exponents as repeated multiplication}$$

$$\frac{a\,a\,a}{a\,a\,a} \qquad \text{Reduce out all the } a's$$

$$\frac{1}{1} = 1 \qquad \text{Our Solution, when we combine the two solutions we get:}$$

$$a^0 = 1 \qquad \text{Our final result.}$$

This final result is an important property known as the zero power rule of exponents

Zero Power Rule of Exponents: $a^0 = 1$

Any number or expression raised to the zero power will always be 1. This is illustrated in the following example.

Example 215.

$$(3x^2)^0 \qquad \text{Zero power rule}$$
$$1 \qquad \text{Our Solution}$$

Another property we will consider here deals with negative exponents. Again we will solve the following example two ways.

Example 216.

$$\frac{a^3}{a^5} \qquad \text{Using the quotient rule, subtract exponents}$$

$$a^{-2} \qquad \text{Our Solution, but we will also solve this problem another way.}$$

$$\frac{a^3}{a^5} \qquad \text{Rewrite exponents as repeated multiplication}$$

$$\frac{aaa}{aaaaa} \qquad \text{Reduce three } a's \text{ out of top and bottom}$$

$$\frac{1}{aa} \qquad \text{Simplify to exponents}$$

$$\frac{1}{a^2} \qquad \text{Our Solution, putting these solutions together gives:}$$

$$a^{-2} = \frac{1}{a^2} \qquad \text{Our Final Solution}$$

This example illustrates an important property of exponents. Negative exponents yield the reciprocal of the base. Once we take the reciprical the exponent is now positive. Also, it is important to note a negative exponent does not mean the expression is negative, only that we need the reciprocal of the base. Following are the rules of negative exponents

$$a^{-m} = \frac{1}{m}$$

Rules of Negative Exponets: $\dfrac{1}{a^{-m}} = a^m$

$$\left(\frac{a}{b}\right)^{-m} = \frac{b^m}{a^m}$$

Negative exponents can be combined in several different ways. As a general rule if we think of our expression as a fraction, negative exponents in the numerator must be moved to the denominator, likewise, negative exponents in the denominator need to be moved to the numerator. When the base with exponent moves, the exponent is now positive. This is illustrated in the following example.

Example 217.

$$\frac{a^3 b^{-2} c}{2d^{-1} e^{-4} f^2} \qquad \text{Negative exponents on } b, d, \text{ and } e \text{ need to flip}$$

$$\frac{a^3 c d e^4}{2b^2 f^2} \qquad \text{Our Solution}$$

As we simplified our fraction we took special care to move the bases that had a negative exponent, but the expression itself did not become negative because of those exponents. Also, it is important to remember that exponents only effect what they are attached to. The 2 in the denominator of the above example does not have an exponent on it, so it does not move with the d.

We now have the following nine properties of exponents. It is important that we are very familiar with all of them.

Properties of Exponents

$$a^m a^n = a^{m+n} \qquad (ab)^m = a^m b^m \qquad a^{-m} = \frac{1}{a^m}$$

$$\frac{a^m}{a^n} = a^{m-n} \qquad \left(\frac{a}{b}\right)^m = \frac{a^m}{b^m} \qquad \frac{1}{a^{-m}} = a^m$$

$$(a^m)^n = a^{mn} \qquad a^0 = 1 \qquad \left(\frac{a}{b}\right)^{-m} = \frac{b^m}{a^m}$$

World View Note: Nicolas Chuquet, the French mathematician of the 15th century wrote 12^{1m} to indicate $12x^{-1}$. This was the first known use of the negative exponent.

Simplifying with negative exponents is much the same as simplifying with positive exponents. It is the advice of the author to keep the negative exponents until the end of the problem and then move them around to their correct location (numerator or denominator). As we do this it is important to be very careful of rules for adding, subtracting, and multiplying with negatives. This is illustrated in the following examples

Example 218.

$$\frac{4x^{-5}y^{-3} \cdot 3x^3y^{-2}}{6x^{-5}y^3} \qquad \text{Simplify numerator with product rule, adding exponents}$$

$$\frac{12x^{-2}y^{-5}}{6x^{-5}y^3} \qquad \text{Quotient rule to subtract exponets, be careful with negatives!}$$

$$(-2) - (-5) = (-2) + 5 = 3$$
$$(-5) - 3 = (-5) + (-3) = -8$$

$$2x^3 y^{-8} \qquad \text{Negative exponent needs to move down to denominator}$$

$$\frac{2x^3}{y^8} \qquad \text{Our Solution}$$

185

Example 219.

$$\frac{(3ab^3)^{-2}ab^3}{2a^{-4}b^0}$$ In numerator, use power rule with -2, multiplying exponents
In denominator, $b^0 = 1$

$$\frac{3^{-2}a^{-2}b^{-6}ab^{-3}}{2a^{-4}}$$ In numerator, use product rule to add exponents

$$\frac{3^{-2}a^{-1}b^{-9}}{2a^{-4}}$$ Use quotient rule to subtract exponents, be careful with negatives
$$(-1) - (-4) = (-1) + 4 = 3$$

$$\frac{3^{-2}a^3b^{-9}}{2}$$ Move 3 and b to denominator because of negative exponents

$$\frac{a^3}{3^2 2b^9}$$ Evaluate $3^2 2$

$$\frac{a^3}{18b^9}$$ Our Solution

In the previous example it is important to point out that when we simplified 3^{-2} we moved the three to the denominator and the exponent became positive. We did not make the number negative! Negative exponents never make the bases negative, they simply mean we have to take the reciprocal of the base. One final example with negative exponents is given here.

Example 220.

$$\left(\frac{3x^{-2}y^5z^3 \cdot 6x^{-6}y^{-2}z^{-3}}{9(x^2y^{-2})^{-3}}\right)^3$$ In numerator, use product rule, adding exponents
In denominator, use power rule, multiplying exponets

$$\left(\frac{18x^{-8}y^3z^0}{9x^{-6}y^6}\right)^3$$ Use quotient rule to subtract exponents,

be careful with negatives:
$$(-8) - (-6) = (-8) + 6 = -2$$
$$3 - 6 = 3 + (-6) = -3$$

$$(2x^{-2}y^{-3}z^0)^3$$ Parenthesis are done, use power rule with -3

$$2^3x^6y^9z^0$$ Move 2 with negative exponent down and $z^0 = 1$

$$\frac{x^6y^9}{2^3}$$ Evaluate 2^3

$$\frac{x^6y^9}{8}$$ Our Solution

5.2 Practice - Negative Exponents

Simplify. Your answer should contain only positive expontents.

1) $2x^4y^{-2} \cdot (2xy^3)^4$

2) $2a^{-2}b^{-3} \cdot (2a^0b^4)^4$

3) $(a^4b^{-3})^3 \cdot 2a^3b^{-2}$

4) $2x^3y^2 \cdot (2x^3)^0$

5) $(2x^2y^2)^4x^{-4}$

6) $(m^0n^3 \cdot 2m^{-3}n^{-3})^0$

7) $(x^3y^4)^3 \cdot x^{-4}y^4$

8) $2m^{-1}n^{-3} \cdot (2m^{-1}n^{-3})^4$

9) $\dfrac{2x^{-3}y^2}{3x^{-3}y^3 \cdot 3x^0}$

10) $\dfrac{3y^3}{3yx^3 \cdot 2x^4y^{-3}}$

11) $\dfrac{4xy^{-3} \cdot x^{-4}y^0}{4y^{-1}}$

12) $\dfrac{3x^3y^2}{4y^{-2} \cdot 3x^{-2}y^{-4}}$

13) $\dfrac{u^2v^{-1}}{2u^0v^4 \cdot 2uv}$

14) $\dfrac{2xy^2 \cdot 4x^3y^{-4}}{4x^{-4}y^{-4} \cdot 4x}$

15) $\dfrac{u^2}{4u^0v^3 \cdot 3v^2}$

16) $\dfrac{2x^{-2}y^2}{4yx^2}$

17) $\dfrac{2y}{(x^0y^2)^4}$

18) $\dfrac{(a^4)^4}{2b}$

19) $\left(\dfrac{2a^2b^3}{a^{-1}}\right)^4$

20) $\left(\dfrac{2y^{-4}}{x^2}\right)^{-2}$

21) $\dfrac{2nm^4}{(2m^2n^2)^4}$

22) $\dfrac{2y^2}{(x^4y^0)^{-4}}$

23) $\dfrac{(2mn)^4}{m^0n^{-2}}$

24) $\dfrac{2x^{-3}}{(x^4y^{-3})^{-1}}$

25) $\dfrac{y^3 \cdot x^{-3}y^2}{(x^4y^2)^3}$

26) $\dfrac{2x^{-2}y^0 \cdot 2xy^4}{(xy^0)^{-1}}$

27) $\dfrac{2u^{-2}v^3 \cdot (2uv^4)^{-1}}{2u^{-4}v^0}$

28) $\dfrac{2yx^2 \cdot x^{-2}}{(2x^0y^4)^{-1}}$

29) $\left(\dfrac{2x^0 \cdot y^4}{y^4}\right)^3$

30) $\dfrac{u^{-3}v^{-4}}{2v(2u^{-3}v^4)^0}$

31) $\dfrac{y(2x^4y^2)^2}{2x^4y^0}$

32) $\dfrac{b^{-1}}{(2a^4b^0)^0 \cdot 2a^{-3}b^2}$

33) $\dfrac{2yzx^2}{2x^4y^4z^{-2} \cdot (zy^2)^4}$

34) $\dfrac{2b^4c^{-2} \cdot (2b^3c^2)^{-4}}{a^{-2}b^4}$

35) $\dfrac{2kh^0 \cdot 2h^{-3}k^0}{(2kj^3)^2}$

36) $\left(\dfrac{(2x^{-3}y^0z^{-1})^3 \cdot x^{-3}y^2}{2x^3}\right)^{-2}$

37) $\dfrac{(cb^3)^2 \cdot 2a^{-3}b^2}{(a^3b^{-2}c^3)^3}$

38) $\dfrac{2q^4 \cdot m^2p^2q^4}{(2m^{-4}p^2)^3}$

39) $\dfrac{(yx^{-4}z^2)^{-1}}{z^{-3} \cdot x^2y^3z^{-1}}$

40) $\dfrac{2mpn^{-3}}{(m^0n^{-4}p^2)^3 \cdot 2n^2p^0}$

Polynomials - Scientific Notation

Objective: Multiply and divide expressions using scientific notation and exponent properties.

One application of exponent properties comes from scientific notation. Scientific notation is used to represent really large or really small numbers. An example of really large numbers would be the distance that light travels in a year in miles. An example of really small numbers would be the mass of a single hydrogen atom in grams. Doing basic operations such as multiplication and division with these numbers would normally be very combersome. However, our exponent properties make this process much simpler.

First we will take a look at what scientific notation is. Scientific notation has two parts, a number between one and ten (it can be equal to one, but not ten), and that number multiplied by ten to some exponent.

$$\text{Scientific Notation: } a \times 10^b \text{ where } 1 \leqslant a < 10$$

The exponent, b, is very important to how we convert between scientific notation and normal numbers, or standard notation. The exponent tells us how many times we will multiply by 10. Multiplying by 10 in affect moves the decimal point one place. So the exponent will tell us how many times the exponent moves between scientific notation and standard notation. To decide which direction to move the decimal (left or right) we simply need to remember that positive exponents mean in standard notation we have a big number (bigger than ten) and negative exponents mean in standard notation we have a small number (less than one).

Keeping this in mind, we can easily make conversions between standard notation and scientific notation.

Example 221.

Convert $14,200$ to scientific notation	Put decimal after first nonzero number
1.42	Exponent is how many times decimal moved, 4
$\times 10^4$	Positive exponent, standard notation is big
1.42×10^4	Our Solution

Example 222.

Convert 0.0042 to scientific notation	Put decimal after first nonzero number
4.2	Exponent is how many times decimal moved, 3
$\times 10^{-3}$	Negative exponent, standard notation is small
4.2×10^{-3}	Our Solution

Example 223.

Convert 3.21×10^5 to standard notation   Positive exponent means standard notation
                                                 big number. Move decimal right 5 places

                           $321,000$ Our Solution

Example 224.

Conver 7.4×10^{-3} to standard notation   Negative exponent means standard notation
                                                 is a small number. Move decimal left 3 places

                           0.0074 Our Solution

Converting between standard notation and scientific notation is important to understand how scientific notation works and what it does. Here our main interest is to be able to multiply and divide numbers in scientific notation using exponent properties. The way we do this is first do the operation with the front number (multiply or divide) then use exponent properties to simplify the 10's. Scientific notation is the only time where it will be allowed to have negative exponents in our final solution. The negative exponent simply informs us that we are dealing with small numbers. Consider the following examples.

Example 225.

$$(2.1 \times 10^{-7})(3.7 \times 10^5)$$ Deal with numbers and $10's$ separately

$$(2.1)(3.7) = 7.77$$ Multiply numbers

$$10^{-7} 10^5 = 10^{-2}$$ Use product rule on $10's$ and add exponents

$$7.77 \times 10^{-2}$$ Our Solution

Example 226.

$$\frac{4.96 \times 10^4}{3.1 \times 10^{-3}}$$ Deal with numbers and $10's$ separately

$$\frac{4.96}{3.1} = 1.6$$ Divide Numbers

$$\frac{10^4}{10^{-3}} = 10^7$$ Use quotient rule to subtract exponents, be careful with negatives!

                       Be careful with negatives, $4 - (-3) = 4 + 3 = 7$

$$1.6 \times 10^7$$ Our Solution

Example 227.

$$(1.8 \times 10^{-4})^3 \quad \text{Use power rule to deal with numbers and } 10's \text{ separately}$$
$$1.8^3 = 5.832 \quad \text{Evaluate } 1.8^3$$
$$(10^{-4})^3 = 10^{-12} \quad \text{Multiply exponents}$$
$$5.832 \times 10^{-12} \quad \text{Our Solution}$$

Often when we multiply or divide in scientific notation the end result is not in scientific notation. We will then have to convert the front number into scientific notation and then combine the 10's using the product property of exponents and adding the exponents. This is shown in the following examples.

Example 228.

$$(4.7 \times 10^{-3})(6.1 \times 10^9) \quad \text{Deal with numbers and } 10's \text{ separately}$$
$$(4.7)(6.1) = 28.67 \quad \text{Multiply numbers}$$
$$2.867 \times 10^1 \quad \text{Convert this number into scientific notation}$$
$$10^1 10^{-3} 10^9 = 10^7 \quad \text{Use product rule, add exponents, using } 10^1 \text{ from conversion}$$
$$2.867 \times 10^7 \quad \text{Our Solution}$$

World View Note: Archimedes (287 BC - 212 BC), the Greek mathematician, developed a system for representing large numbers using a system very similar to scientific notation. He used his system to calculate the number of grains of sand it would take to fill the universe. His conclusion was 10^{63} grains of sand because he figured the universe to have a diameter of 10^{14} stadia or about 2 light years.

Example 229.

$$\frac{2.014 \times 10^{-3}}{3.8 \times 10^{-7}} \quad \text{Deal with numbers and } 10's \text{ separately}$$

$$\frac{2.014}{3.8} = 0.53 \quad \text{Divide numbers}$$

$$0.53 = 5.3 \times 10^{-1} \quad \text{Change this number into scientific notation}$$
$$\frac{10^{-1} 10^{-3}}{10^{-7}} = 10^3 \quad \text{Use product and quotient rule, using } 10^{-1} \text{ from the conversion}$$
$$\text{Be careful with signs:}$$
$$(-1) + (-3) - (-7) = (-1) + (-3) + 7 = 3$$
$$5.3 \times 10^3 \quad \text{Our Solution}$$

5.3 Practice - Scientific Notation

Write each number in scientific notiation

1) 885

2) 0.000744

3) 0.081

4) 1.09

5) 0.039

6) 15000

Write each number in standard notation

7) 8.7×10^5

8) 2.56×10^2

9) 9×10^{-4}

10) 5×10^4

11) 2×10^0

12) 6×10^{-5}

Simplify. Write each answer in scientific notation.

13) $(7 \times 10^{-1})(2 \times 10^{-3})$

14) $(2 \times 10^{-6})(8.8 \times 10^{5})$

15) $(5.26 \times 10^{5})(3.16 \times 10^{2})$

16) $(5.1 \times 10^6)(9.84 \times 10^{-1})$

17) $(2.6 \times 10^{-2})(6 \times 10^{-2})$

18) $\frac{7.4 \times 10^4}{1.7 \times 10^{-4}}$

19) $\frac{4.9 \times 10^1}{2.7 \times 10^{-3}}$

20) $\frac{7.2 \times 10^{1}}{7.32 \times 10^{1}}$

21) $\frac{5.33 \times 10^{6}}{9.62 \times 10^{2}}$

22) $\frac{3.2 \times 10^{-3}}{5.02 \times 10^0}$

23) $(5.5 \times 10^{5})^2$

24) $(9.6 \times 10^3)^{4}$

25) $(7.8 \times 10^{-2})^5$

26) $(5.4 \times 10^6)^{-3}$

27) $(8.03 \times 10^4)^{4}$

28) $(6.88 \times 10^{-4})(4.23 \times 10^1)$

29) $\frac{6.1 \times 10^{-6}}{5.1 \times 10^{-4}}$

30) $\frac{8.4 \times 10^5}{7 \times 10^{-2}}$

31) $(3.6 \times 10^0)(6.1 \times 10^{-3})$

32) $(3.15 \times 10^3)(8 \times 10^{-1})$

33) $(1.8 \times 10^{5})^{3}$

34) $\frac{9.58 \times 10^{-2}}{1.14 \times 10^{3}}$

35) $\frac{9 \times 10^4}{7.83 \times 10^{2}}$

36) $(8.3 \times 10^1)^5$

37) $\frac{3.22 \times 10^{-3}}{7 \times 10^{-6}}$

38) $\frac{5 \times 10^6}{6.69 \times 10^2}$

39) $\frac{2.4 \times 10^{6}}{6.5 \times 10^0}$

40) $(9 \times 10^{-2})^{-3}$

41) $\frac{6 \times 10^3}{5.8 \times 10^{3}}$

42) $(2 \times 10^4)(6 \times 10^1)$

Polynomials - Introduction to Polynomials

Objective: Evaluate, add, and subtract polynomials.

Many applications in mathematics have to do with what are called polynomials. Polynomials are made up of terms. **Terms** are a product of numbers and/or variables. For example, $5x$, $2y^2$, -5, ab^3c, and x are all terms. Terms are connected to each other by addition or subtraction. Expressions are often named based on the number of terms in them. A **monomial** has one term, such as $3x^2$. A **binomial** has two terms, such as $a^2 - b^2$. A Trinomial has three terms, such as $ax^2 + bx + c$. The term **polynomial** means many terms. Monomials, binomials, trinomials, and expressions with more terms all fall under the umbrella of "polynomials".

If we know what the variable in a polynomial represents we can replace the variable with the number and evaluate the polynomial as shown in the following example.

Example 230.

$$2x^2 - 4x + 6 \text{ when } x = -4 \qquad \text{Replace variable } x \text{ with} -4$$
$$2(-4)^2 - 4(-4) + 6 \qquad \text{Exponents first}$$
$$2(16) - 4(-4) + 6 \qquad \text{Multiplication (we can do all terms at once)}$$
$$32 + 16 + 6 \qquad \text{Add}$$
$$54 \qquad \text{Our Solution}$$

It is important to be careful with negative variables and exponents. Remember the exponent only effects the number it is physically attached to. This means $-3^2 = -9$ because the exponent is only attached to the 3. Also, $(-3)^2 = 9$ because the exponent is attached to the parenthesis and effects everything inside. When we replace a variable with parenthesis like in the previous example, the substituted value is in parenthesis. So the $(-4)^2 = 16$ in the example. However, consider the next example.

Example 231.

$$-x^2 + 2x + 6 \text{ when } x = 3 \qquad \text{Replace variable } x \text{ with } 3$$
$$-(3)^2 + 2(3) + 6 \qquad \text{Exponent only on the 3, not negative}$$
$$-9 + 2(3) + 6 \qquad \text{Multiply}$$
$$-9 + 6 + 6 \qquad \text{Add}$$
$$3 \qquad \text{Our Solution}$$

World View Note: Ada Lovelace in 1842 described a Difference Engine that would be used to caluclate values of polynomials. Her work became the foundation for what would become the modern computer (the programming language Ada was named in her honor), more than 100 years after her death from cancer.

Generally when working with polynomials we do not know the value of the variable, so we will try and simplify instead. The simplest operation with polynomials is addition. When adding polynomials we are mearly combining like terms. Consider the following example

Example 232.

$$(4x^3 - 2x + 8) + (3x^3 - 9x^2 - 11) \quad \text{Combine like terms } 4x^3 + 3x^3 \text{ and } 8 - 11$$
$$7x^3 - 9x^2 - 2x - 3 \quad \text{Our Solution}$$

Generally final answers for polynomials are written so the exponent on the variable counts down. Example 3 demonstrates this with the exponent counting down 3, 2, 1, 0 (recall $x^0 = 1$). Subtracting polynomials is almost as fast. One extra step comes from the minus in front of the parenthesis. When we have a negative in front of parenthesis we distribute it through, changing the signs of everything inside. The same is done for the subtraction sign.

Example 233.

$$(5x^2 - 2x + 7) - (3x^2 + 6x - 4) \quad \text{Distribute negative through second part}$$
$$5x^2 - 2x + 7 - 3x^2 - 6x + 4 \quad \text{Combine like terms } 5x^2 - 3x^3, -2x - 6x, \text{ and } 7 + 4$$
$$2x^2 - 8x + 11 \quad \text{Our Solution}$$

Addition and subtraction can also be combined into the same problem as shown in this final example.

Example 234.

$$(2x^2 - 4x + 3) + (5x^2 - 6x + 1) - (x^2 - 9x + 8) \quad \text{Distribute negative through}$$
$$2x^2 - 4x + 3 + 5x^2 - 6x + 1 - x^2 + 9x - 8 \quad \text{Combine like terms}$$
$$6x^2 - x - 4 \quad \text{Our Solution}$$

5.4 Practice - Introduction to Polynomials

Simplify each expression.

1) $-a^3 - a^2 + 6a - 21$ when $a = -4$

2) $n^2 + 3n - 11$ when $n = -6$

3) $n^3 - 7n^2 + 15n - 20$ when $n = 2$

4) $n^3 - 9n^2 + 23n - 21$ when $n = 5$

5) $-5n^4 - 11n^3 - 9n^2 - n - 5$ when $n = -1$

6) $x^4 - 5x^3 - x + 13$ when $x = 5$

7) $x^2 + 9x + 23$ when $x = -3$

8) $-6x^3 + 41x^2 - 32x + 11$ when $x = 6$

9) $x^4 - 6x^3 + x^2 - 24$ when $x = 6$

10) $m^4 + 8m^3 + 14m^2 + 13m + 5$ when $m = -6$

11) $(5p - 5p^4) - (8p - 8p^4)$

12) $(7m^2 + 5m^3) - (6m^3 - 5m^2)$

13) $(3n^2 + n^3) - (2n^3 - 7n^2)$

14) $(x^2 + 5x^3) + (7x^2 + 3x^3)$

15) $(8n + n^4) - (3n - 4n^4)$

16) $(3v^4 + 1) + (5 - v^4)$

17) $(1 + 5p^3) - (1 - 8p^3)$

18) $(6x^3 + 5x) - (8x + 6x^3)$

19) $(5n^4 + 6n^3) + (8 - 3n^3 - 5n^4)$

20) $(8x^2 + 1) - (6 - x^2 - x^4)$

21) $(3 + b^4) + (7 + 2b + b^4)$

22) $(1 + 6r^2) + (6r^2 - 2 - 3r^4)$

23) $(8x^3 + 1) - (5x^4 - 6x^3 + 2)$

24) $(4n^4 + 6) - (4n - 1 - n^4)$

25) $(2a + 2a^4) - (3a^2 - 5a^4 + 4a)$

26) $(6v + 8v^3) + (3 + 4v^3 - 3v)$

27) $(4p^2 - 3 - 2p) - (3p^2 - 6p + 3)$

28) $(7 + 4m + 8m^4) - (5m^4 + 1 + 6m)$

29) $(4b^3 + 7b^2 - 3) + (8 + 5b^2 + b^3)$

30) $(7n + 1 - 8n^4) - (3n + 7n^4 + 7)$

31) $(3 + 2n^2 + 4n^4) + (n^3 - 7n^2 - 4n^4)$

32) $(7x^2 + 2x^4 + 7x^3) + (6x^3 - 8x^4 - 7x^2)$

33) $(n - 5n^4 + 7) + (n^2 - 7n^4 - n)$

34) $(8x^2 + 2x^4 + 7x^3) + (7x^4 - 7x^3 + 2x^2)$

35) $(8r^4 - 5r^3 + 5r^2) + (2r^2 + 2r^3 - 7r^4 + 1)$

36) $(4x^3 + x - 7x^2) + (x^2 - 8 + 2x + 6x^3)$

37) $(2n^2 + 7n^4 - 2) + (2 + 2n^3 + 4n^2 + 2n^4)$

38) $(7b^3 - 4b + 4b^4) - (8b^3 - 4b^2 + 2b^4 - 8b)$

39) $(8 - b + 7b^3) - (3b^4 + 7b - 8 + 7b^2) + (3 - 3b + 6b^3)$

40) $(1 - 3n^4 - 8n^3) + (7n^4 + 2 - 6n^2 + 3n^3) + (4n^3 + 8n^4 + 7)$

41) $(8x^4 + 2x^3 + 2x) + (2x + 2 - 2x^3 - x^4) - (x^3 + 5x^4 + 8x)$

42) $(6x - 5x^4 - 4x^2) - (2x - 7x^2 - 4x^4 - 8) - (8 - 6x^2 - 4x^4)$

Polynomials - Multiplying Polynomials

Objective: Multiply polynomials.

Multiplying polynomials can take several different forms based on what we are multiplying. We will first look at multiplying monomials, then monomials by polynomials and finish with polynomials by polynomials.

Multiplying monomials is done by multiplying the numbers or coefficients and then adding the exponents on like factors. This is shown in the next example.

Example 235.

$$(4x^3y^4z)(2x^2y^6z^3) \quad \text{Multiply numbers and add exponents for } x, y, \text{ and } z$$
$$8x^5y^{10}z^4 \quad \text{Our Solution}$$

In the previous example it is important to remember that the z has an exponent of 1 when no exponent is written. Thus for our answer the z has an exponent of $1 + 3 = 4$. Be very careful with exponents in polynomials. If we are adding or subtracting the exponnets will stay the same, but when we multiply (or divide) the exponents will be changing.

Next we consider multiplying a monomial by a polynomial. We have seen this operation before with distributing through parenthesis. Here we will see the exact same process.

Example 236.

$$4x^3(5x^2 - 2x + 5) \quad \text{Distribute the } 4x^3, \text{multiplying numbers, adding exponents}$$
$$20x^5 - 8x^4 + 20x^3 \quad \text{Our Solution}$$

Following is another example with more variables. When distributing the exponents on a are added and the exponents on b are added.

Example 237.

$$2a^3b(3ab^2 - 4a) \quad \text{Distribute, multiplying numbers and adding exponents}$$
$$6a^4b^3 - 8a^4b \quad \text{Our Solution}$$

There are several different methods for multiplying polynomials. All of which work, often students prefer the method they are first taught. Here three methods will be discussed. All three methods will be used to solve the same two multiplication problems.

Multiply by Distributing

Just as we distribute a monomial through parenthesis we can distribute an entire polynomial. As we do this we take each term of the second polynomial and put it in front of the first polynomial.

Example 238.

$$(4x + 7y)(3x - 2y) \quad \text{Distribute } (4x + 7y) \text{ through parenthesis}$$
$$3x(4x + 7y) - 2y(4x + 7y) \quad \text{Distribute the } 3x \text{ and} -2y$$
$$12x^2 + 21xy - 8xy - 14y^2 \quad \text{Combine like terms } 21xy - 8xy$$
$$12x^2 + 13xy - 14y^2 \quad \text{Our Solution}$$

This example illustrates an important point, the negative/subtraction sign stays with the $2y$. Which means on the second step the negative is also distributed through the last set of parenthesis.

Multiplying by distributing can easily be extended to problems with more terms. First distribute the front parenthesis onto each term, then distribute again!

Example 239.

$$(2x - 5)(4x^2 - 7x + 3) \quad \text{Distribute } (2x - 5) \text{ through parenthesis}$$
$$4x^2(2x - 5) - 7x(2x - 5) + 3(2x - 5) \quad \text{Distribute again through each parenthesis}$$
$$8x^3 - 20x^2 - 14x^2 + 35x + 6x - 15 \quad \text{Combine like terms}$$
$$8x^3 - 34x^2 + 41x - 15 \quad \text{Our Solution}$$

This process of multiplying by distributing can easily be reversed to do an important procedure known as factoring. Factoring will be addressed in a future lesson.

Multiply by FOIL

Another form of multiplying is known as FOIL. Using the FOIL method we multiply each term in the first binomial by each term in the second binomial. The letters of FOIL help us remember every combination. F stands for First, we multiply the first term of each binomial. O stand for Outside, we multiply the outside two terms. I stands for Inside, we multiply the inside two terms. L stands for Last, we multiply the last term of each binomial. This is shown in the next example:

Example 240.

$$(4x + 7y)(3x - 2y) \quad \text{Use FOIL to multiply}$$
$$(4x)(3x) = 12x^2 \quad F - \text{First terms } (4x)(3x)$$
$$(4x)(-2y) = -8xy \quad O - \text{Outside terms } (4x)(-2y)$$
$$(7y)(3x) = 21xy \quad I - \text{Inside terms } (7y)(3x)$$
$$(7y)(-2y) = -14y^2 \quad L - \text{Last terms } (7y)(-2y)$$
$$12x^2 - 8xy + 21xy - 14y^2 \quad \text{Combine like terms } -8xy + 21xy$$
$$12x^2 + 13xy - 14y^2 \quad \text{Our Solution}$$

Some students like to think of the FOIL method as distributing the first term $4x$ through the $(3x - 2y)$ and distributing the second term $7y$ through the $(3x - 2y)$. Thinking about FOIL in this way makes it possible to extend this method to problems with more terms.

Example 241.

$$
\begin{array}{rl}
(2x - 5)(4x^2 - 7x + 3) & \text{Distribute } 2x \text{ and} - 5 \\
(2x)(4x^2) + (2x)(-7x) + (2x)(3) - 5(4x^2) - 5(-7x) - 5(3) & \text{Multiply out each term} \\
8x^3 - 14x^2 + 6x - 20x^2 + 35x - 15 & \text{Combine like terms} \\
8x^3 - 34x^2 + 41x - 15 & \text{Our Solution}
\end{array}
$$

The second step of the FOIL method is often not written, for example, consider the previous example, a student will often go from the problem $(4x + 7y)(3x - 2y)$ and do the multiplication mentally to come up with $12x^2 - 8xy + 21xy - 14y^2$ and then combine like terms to come up with the final solution.

Multiplying in rows

A third method for multiplying polynomials looks very similar to multiplying numbers. Consider the problem:

$$
\begin{array}{rl}
35 & \\
\times\, 27 & \\
\hline
245 & \text{Multiply 7 by 5 then 3} \\
\underline{700} & \text{Use 0 for placeholder, multiply 2 by 5 then 3} \\
945 & \text{Add to get Our Solution}
\end{array}
$$

World View Note: The first known system that used place values comes from Chinese mathematics, dating back to 190 AD or earlier.

The same process can be done with polynomials. Multiply each term on the bottom with each term on the top.

Example 242.

$$
\begin{array}{rl}
(4x + 7y)(3x - 2y) & \text{Rewrite as vertical problem} \\
4x + 7y & \\
\times\, 3x - 2y & \\
\hline
-8xy - 14y^2 & \text{Multiply } -2y \text{ by } 7y \text{ then } 4x \\
\underline{12x^2 + 21xy} & \text{Multiply } 3x \text{ by } 7y \text{ then } 4x. \text{ Line up like terms} \\
12x^2 + 13xy - 14y^2 & \text{Add like terms to get Our Solution}
\end{array}
$$

This same process is easily expanded to a problem with more terms.

Example 243.

$$(2x - 5)(4x^2 - 7x + 3) \qquad \text{Rewrite as vertical problem}$$

$$4x^3 - 7x + 3 \qquad \text{Put polynomial with most terms on top}$$

$$\underline{\times\, 2x - 5}$$

$$-20x^2 + 35x - 15 \qquad \text{Multiply} -5 \text{ by each term}$$

$$\underline{8x^3 - 14x^2 + 6x } \qquad \text{Multiply } 2x \text{ by each term. Line up like terms}$$

$$8x^3 - 34x^2 + 41x - 15 \qquad \text{Add like terms to get our solution}$$

This method of multiplying in rows also works with multiplying a monomial by a polynomial!

Any of the three described methods work to multiply polynomials. It is suggested that you are very comfortable with at least one of these methods as you work through the practice problems. All three methods are shown side by side in the example.

Example 244.

$$(2x - y)(4x - 5y)$$

Distribute	FOIL	Rows
$4x(2x - y) - 5y(2x - y)$	$2x(4x) + 2x(-5y) - y(4x) - y(-5y)$	$2x - y$
$8x^2 - 4xy - 10xy - 5y^2$	$8x^2 - 10xy - 4xy + 5y^2$	$\underline{\times\, 4x - 5y}$
$8x^2 - 14xy - 5y^2$	$8x^2 - 14xy + 5y^2$	$-10xy + 5y^2$
		$\underline{8x^2 - 4xy }$
		$8x^2 - 14xy + 5y^2$

When we are multiplying a monomial by a polynomial by a polynomial we can solve by first multiplying the polynomials then distributing the coefficient last. This is shown in the last example.

Example 245.

$$3(2x - 4)(x + 5) \qquad \text{Multiply the binomials, we will use FOIL}$$

$$3(2x^2 + 10x - 4x - 20) \qquad \text{Combine like terms}$$

$$3(2x^2 + 6x - 20) \qquad \text{Distribute the 3}$$

$$6x^2 + 18x - 60 \qquad \text{Our Solution}$$

A common error students do is distribute the three at the start into both parenthesis. While we can distribute the 3 into the $(2x - 4)$ factor, distributing into both would be wrong. Be careful of this error. This is why it is suggested to multiply the binomials first, then distribute the coeffienct last.

5.5 Practice - Multiply Polynomials

Find each product.

1) $6(p - 7)$

2) $4k(8k + 4)$

3) $2(6x + 3)$

4) $3n^2(6n + 7)$

5) $5m^4(4m + 4)$

6) $3(4r - 7)$

7) $(4n + 6)(8n + 8)$

8) $(2x + 1)(x - 4)$

9) $(8b + 3)(7b - 5)$

10) $(r + 8)(4r + 8)$

11) $(4x + 5)(2x + 3)$

12) $(7n - 6)(n + 7)$

13) $(3v - 4)(5v - 2)$

14) $(6a + 4)(a - 8)$

15) $(6x - 7)(4x + 1)$

16) $(5x - 6)(4x - 1)$

17) $(5x + y)(6x - 4y)$

18) $(2u + 3v)(8u - 7v)$

19) $(x + 3y)(3x + 4y)$

20) $(8u + 6v)(5u - 8v)$

21) $(7x + 5y)(8x + 3y)$

22) $(5a + 8b)(a - 3b)$

23) $(r - 7)(6r^2 - r + 5)$

24) $(4x + 8)(4x^2 + 3x + 5)$

25) $(6n - 4)(2n^2 - 2n + 5)$

26) $(2b - 3)(4b^2 + 4b + 4)$

27) $(6x + 3y)(6x^2 - 7xy + 4y^2)$

28) $(3m - 2n)(7m^2 + 6mn + 4n^2)$

29) $(8n^2 + 4n + 6)(6n^2 - 5n + 6)$

30) $(2a^2 + 6a + 3)(7a^2 - 6a + 1)$

31) $(5k^2 + 3k + 3)(3k^2 + 3k + 6)$

32) $(7u^2 + 8uv - 6v^2)(6u^2 + 4uv + 3v^2)$

33) $3(3x - 4)(2x + 1)$

34) $5(x - 4)(2x - 3)$

35) $3(2x + 1)(4x - 5)$

36) $2(4x + 1)(2x - 6)$

37) $7(x - 5)(x - 2)$

38) $5(2x - 1)(4x + 1)$

39) $6(4x - 1)(4x + 1)$

40) $3(2x + 3)(6x + 9)$

Polynomials - Multiply Special Products

Objective: Recognize and use special product rules of a sum and difference and perfect squares to multiply polynomials.

There are a few shortcuts that we can take when multiplying polynomials. If we can recognize them the shortcuts can help us arrive at the solution much quicker. These shortcuts will also be useful to us as our study of algebra continues.

The first shortcut is often called a **sum and a difference**. A sum and a difference is easily recognized as the numbers and variables are exactly the same, but the sign in the middle is different (one sum, one difference). To illustrate the shortcut consider the following example, multiplied by the distributing method.

Example 246.

$$(a+b)(a-b) \quad \text{Distribute } (a+b)$$
$$a(a+b) - b(a+b) \quad \text{Distribute } a \text{ and } -b$$
$$a^2 + ab - ab - b^2 \quad \text{Combine like terms } ab - ab$$
$$a^2 - b^2 \quad \text{Our Solution}$$

The important part of this example is the middle terms subtracted to zero. Rather than going through all this work, when we have a sum and a difference we will jump right to our solution by squaring the first term and squaring the last term, putting a subtraction between them. This is illustrated in the following example

Example 247.

$$(x-5)(x+5) \quad \text{Recognize sum and difference}$$
$$x^2 - 25 \quad \text{Square both, put subtraction between. Our Solution}$$

This is much quicker than going through the work of multiplying and combining like terms. Often students ask if they can just multiply out using another method and not learn the shortcut. These shortcuts are going to be very useful when we get to factoring polynomials, or reversing the multiplication process. For this reason it is very important to be able to recognize these shortcuts. More examples are shown here.

Example 248.

$$(3x + 7)(3x - 7) \quad \text{Recognize sum and difference}$$
$$9x^2 - 49 \quad \text{Square both, put subtraction between. Our Solution}$$

Example 249.

$$(2x - 6y)(2x + 6y) \quad \text{Recognize sum and difference}$$
$$4x^2 - 36y^2 \quad \text{Square both, put subtraction between. Our Solution}$$

It is interesting to note that while we can multiply and get an answer like $a^2 - b^2$ (with subtraction), it is impossible to multiply real numbers and end up with a product such as $a^2 + b^2$ (with addition).

Another shortcut used to multiply is known as a **perfect square**. These are easy to recognize as we will have a binomial with a 2 in the exponent. The following example illustrates multiplying a perfect square

Example 250.

$$(a + b)^2 \quad \text{Squared is same as multiplying by itself}$$
$$(a + b)(a + b) \quad \text{Distribute } (a + b)$$
$$a(a + b) + b(a + b) \quad \text{Distribute again through final parenthesis}$$
$$a^2 + ab + ab + b^2 \quad \text{Combine like terms } ab + ab$$
$$a^2 + 2ab + b^2 \quad \text{Our Solution}$$

This problem also helps us find our shortcut for multiplying. The first term in the answer is the square of the first term in the problem. The middle term is 2 times the first term times the second term. The last term is the square of the last term. This can be shortened to square the first, twice the product, square the last. If we can remember this shortcut we can square any binomial. This is illustrated in the following example

Example 251.

$$(x - 5)^2 \quad \text{Recognize perfect square}$$
$$x^2 \quad \text{Square the first}$$
$$2(x)(-5) = -10x \quad \text{Twice the product}$$
$$(-5)^2 = 25 \quad \text{Square the last}$$
$$x^2 - 10x + 25 \quad \text{Our Solution}$$

Be very careful when we are squaring a binomial to **NOT** distribute the square through the parenthesis. A common error is to do the following: $(x-5)^2 = x^2 - 25$ (or $x^2 + 25$). Notice both of these are missing the middle term, $-10x$. This is why it is important to use the shortcut to help us find the correct solution. Another important observation is that the middle term in the solution always has the same sign as the middle term in the problem. This is illustrated in the next examples.

Example 252.

$$(2x+5)^2 \quad \text{Recognize perfect square}$$
$$(2x)^2 = 4x^2 \quad \text{Square the first}$$
$$2(2x)(5) = 20x \quad \text{Twice the product}$$
$$5^2 = 25 \quad \text{Square the last}$$
$$4x^2 + 20x + 25 \quad \text{Our Solution}$$

Example 253.

$$(3x-7y)^2 \quad \text{Recognize perfect square}$$
$$9x^2 - 42xy + 49y^2 \quad \text{Square the first, twice the product, square the last. Our Solution}$$

Example 254.

$$(5a+9b)^2 \quad \text{Recognize perfect square}$$
$$25a^2 + 90ab + 81b^2 \quad \text{Square the first, twice the product, square the last. Our Solution}$$

These two formulas will be important to commit to memory. The more familiar we are with them, the easier factoring, or multiplying in reverse, will be. The final example covers both types of problems (two perfect squares, one positive, one negative), be sure to notice the difference between the examples and how each formula is used

Example 255.

$$(4x-7)(4x+7) \qquad (4x+7)^2 \qquad (4x-7)^2$$
$$16x^2 - 49 \qquad 16x^2 + 56x + 49 \qquad 16x^2 - 56x + 49$$

World View Note: There are also formulas for higher powers of binomials as well, such as $(a+b)^3 = a^3 + 3a^2b + 3ab^2 + b^3$. While French mathematician Blaise Pascal often gets credit for working with these expansions of binomials in the 17th century, Chinese mathematicians had been working with them almost 400 years earlier!

5.6 Practice - Multiply Special Products

Find each product.

1) $(x+8)(x-8)$

2) $(a-4)(a+4)$

3) $(1+3p)(1-3p)$

4) $(x-3)(x+3)$

5) $(1-7n)(1+7n)$

6) $(8m+5)(8m-5)$

7) $(5n-8)(5n+8)$

8) $(2r+3)(2r-3)$

9) $(4x+8)(4x-8)$

10) $(b-7)(b+7)$

11) $(4y-x)(4y+x)$

12) $(7a+7b)(7a-7b)$

13) $(4m-8n)(4m+8n)$

14) $(3y-3x)(3y+3x)$

15) $(6x-2y)(6x+2y)$

16) $(1+5n)^2$

17) $(a+5)^2$

18) $(v+4)^2$

19) $(x-8)^2$

20) $(1-6n)^2$

21) $(p+7)^2$

22) $(7k-7)^2$

23) $(7-5n)^2$

24) $(4x-5)^2$

25) $(5m-8)^2$

26) $(3a+3b)^2$

27) $(5x+7y)^2$

28) $(4m-n)^2$

29) $(2x+2y)^2$

30) $(8x+5y)^2$

31) $(5+2r)^2$

32) $(m-7)^2$

33) $(2+5x)^2$

34) $(8n+7)(8n-7)$

35) $(4v-7)(4v+7)$

36) $(b+4)(b-4)$

37) $(n-5)(n+5)$

38) $(7x+7)^2$

39) $(4k+2)^2$

40) $(3a-8)(3a+8)$

Polynomials - Divide Polynomials

Objective: Divide polynomials using long division.

Dividing polynomials is a process very similar to long division of whole numbers. But before we look at that, we will first want to be able to master dividing a polynomial by a monomial. The way we do this is very similar to distributing, but the operation we distribute is the division, dividing each term by the monomial and reducing the resulting expression. This is shown in the following examples

Example 256.

$$\frac{9x^5 + 6x^4 - 18x^3 - 24x^2}{3x^2}$$ Divide each term in the numerator by $3x^2$

$$\frac{9x^5}{3x^2} + \frac{6x^4}{3x^2} - \frac{18x^3}{3x^2} - \frac{24x^2}{3x^2}$$ Reduce each fraction, subtracting exponents

$$3x^3 + 2x^2 - 6x - 8$$ Our Solution

Example 257.

$$\frac{8x^3 + 4x^2 - 2x + 6}{4x^2}$$ Divide each term in the numerator by $4x^2$

$$\frac{8x^3}{4x^2} + \frac{4x^2}{4x^2} - \frac{2x}{4x^2} + \frac{6}{4x^2}$$ Reduce each fraction, subtracting exponents

Remember negative exponents are moved to denominator

$$2x + 1 - \frac{1}{2x} + \frac{3}{2x^2}$$ Our Solution

The previous example illustrates that sometimes we will have fractions in our solution, as long as they are reduced this will be correct for our solution. Also interesting in this problem is the second term $\frac{4x^2}{4x^2}$ divided out completely. Remember that this means the reduced answer is 1 not 0.

Long division is required when we divide by more than just a monomial. Long division with polynomials works very similar to long division with whole numbers.

An example is given to review the (general) steps that are used with whole numbers that we will also use with polynomials

Example 258.

$4\overline{)631}$ Divide front numbers: $\dfrac{6}{4} = 1...$

    1
$4\overline{)631}$ Multiply this number by divisor: $1 \cdot 4 = 4$
$\underline{-4}$ Change the sign of this number (make it subtract) and combine
  23   Bring down next number

  15   Repeat, divide front numbers: $\dfrac{23}{4} = 5...$
$4\overline{)631}$
$\underline{-4}$
  23   Multiply this number by divisor: $5 \cdot 4 = 20$
$\underline{-20}$ Change the sign of this number (make it subtract) and combine
  31   Bring down next number

 157   Repeat, divide front numbers: $\dfrac{31}{4} = 7...$
$4\overline{)631}$
$\underline{-4}$
  23
$\underline{-20}$
  31   Multiply this number by divisor: $7 \cdot 4 = 28$
$\underline{-28}$ Change the sign of this number (make it subtract) and combine
   3   We will write our remainder as a fraction, over the divisor, added to the end

$157\dfrac{3}{4}$ Our Solution

This same process will be used to multiply polynomials. The only difference is we will replace the word "number" with the word "term"

Dividing Polynomials

1. Divide front terms

2. Multiply this term by the divisor

3. Change the sign of the terms and combine

4. Bring down the next term

5. Repeat

Step number 3 tends to be the one that students skip, not changing the signs of the terms would be equivalent to adding instead of subtracting on long division with whole numbers. Be sure not to miss this step! This process is illustrated in the following two examples.

Example 259.

$$\frac{3x^3 - 5x^2 - 32x + 7}{x - 4}$$ Rewrite problem as long division

$x - 4\overline{\smash{\big)}3x^3 - 5x^2 - 32x + 7}$ Divide front terms: $\dfrac{3x^3}{x} = 3x^2$

$$
\begin{array}{r}
\mathbf{3x^2} \\
x - 4\overline{\smash{\big)}3x^3 - 5x^2 - 32x + 7} \\
\underline{\mathbf{-3x^3 + 12x^2}} \\
\mathbf{7x^2 - 32x}
\end{array}
$$
Multiply this term by divisor: $3x^2(x - 4) = 3x^3 - 12x^2$
Change the signs and combine
Bring down the next term

$$
\begin{array}{r}
\mathbf{3x^2 + 7x} \\
x - 4\overline{\smash{\big)}3x^3 - 5x^2 - 32x + 7} \\
\underline{-3x^3 + 12x^2} \\
7x^2 - 32x \\
\underline{\mathbf{-7x^2 + 28x}} \\
\mathbf{-4x + 7}
\end{array}
$$
Repeat, divide front terms: $\dfrac{7x^2}{x} = 7x$

Multiply this term by divisor: $7x(x - 4) = 7x^2 - 28x$
Change the signs and combine
Bring down the next term

$$
\begin{array}{r}
\mathbf{3x^2 + 7x - 4} \\
x - 4\overline{\smash{\big)}3x^3 - 5x^2 - 32x + 7} \\
\underline{-3x^3 + 12x^2} \\
7x^2 - 32x \\
\underline{-7x^2 + 28x} \\
-4x + 7 \\
\underline{\mathbf{+4x - 16}} \\
\mathbf{-9}
\end{array}
$$
Repeat, divide front terms: $\dfrac{-4x}{x} = -4$

Multiply this term by divisor: $-4(x - 4) = -4x + 16$
Change the signs and combine
Remainder put over divisor and subtracted (due to negative)

207

$$3x^2 + 7x - 4 - \frac{9}{x-4} \qquad \text{Our Solution}$$

Example 260.

$$\frac{6x^3 - 8x^2 + 10x + 103}{2x+4} \qquad \text{Rewrite problem as long division}$$

$$2x+4\overline{)6x^3 - 8x^2 + 10x + 103} \qquad \text{Divide front terms: } \frac{6x^3}{2x} = 3x^2$$

$$
\begin{array}{l}
\mathbf{3x^2} \\
2x+4\overline{)6x^3 - 8x^2 + 10x + 103} \\
\underline{-\,\mathbf{6x^3 - 12x^2}} \\
\mathbf{-\,20x^2 + 10x}
\end{array}
\qquad
\begin{array}{l}
\text{Multiply term by divisor: } 3x^2(2x+4) = 6x^3 + 12x^2 \\
\text{Change the signs and combine} \\
\text{Bring down the next term}
\end{array}
$$

$$
\begin{array}{l}
\mathbf{3x^2 - 10x} \\
2x+4\overline{)6x^3 - 8x^2 + 10x + 103} \\
\underline{-\,6x^3 - 12x^2} \\
-\,20x^2 + 10x \\
\underline{+\,\mathbf{20x^2 + 40x}} \\
\mathbf{50x + 103}
\end{array}
\qquad
\begin{array}{l}
\text{Repeat, divide front terms: } \frac{-20x^2}{2x} = -10x \\
\text{Multiply this term by divisor:} \\
-10x(2x+4) = -20x^2 - 40x \\
\text{Change the signs and combine} \\
\text{Bring down the next term}
\end{array}
$$

$$
\begin{array}{l}
\mathbf{3x^2 - 10x + 25} \\
2x+4\overline{)6x^3 - 8x^2 + 10x + 103} \\
\underline{-\,6x^3 - 12x^2} \\
-\,20x^2 + 10x \\
\underline{+\,20x^2 + 40x} \\
50x + 103 \\
\underline{-\,\mathbf{50x - 100}} \\
3
\end{array}
\qquad
\begin{array}{l}
\text{Repeat, divide front terms: } \frac{50x}{2x} = 25 \\[2ex]
\text{Multiply this term by divisor: } 25(2x+4) = 50x + 100 \\
\text{Change the signs and combine} \\
\text{Remainder is put over divsor and added (due to positive)}
\end{array}
$$

$$3x^2 - 10x + 25 + \frac{3}{2x+4} \qquad \text{Our Solution}$$

In both of the previous example the dividends had the exponents on our variable counting down, no exponent skipped, third power, second power, first power, zero power (remember $x^0 = 1$ so there is no variable on zero power). This is very important in long division, the variables must count down and no exponent can be skipped. If they don't count down we must put them in order. If an exponent is skipped we will have to add a term to the problem, with zero for its coefficient. This is demonstrated in the following example.

Example 261.

$$\frac{2x^3 + 42 - 4x}{x + 3}$$ Reorder dividend, need x^2 term, add $0x^2$ for this

$x + 3 \overline{)2x^3 + 0x^2 - 4x + 42}$ Divide front terms: $\dfrac{2x^3}{x} = 2x^2$

$$\begin{array}{r} 2x^2 \\ x + 3 \overline{)2x^3 + 0x^2 - 4x + 42} \\ \underline{-2x^3 - 6x^2} \\ -6x^2 - 4x \end{array}$$

Multiply this term by divisor: $2x^2(x + 3) = 2x^3 + 6x^2$
Change the signs and combine
Bring down the next term

$$\begin{array}{r} 2x^2 - 6x \\ x + 3 \overline{)2x^3 + 0x^2 - 4x + 42} \\ \underline{-2x^3 - 6x^2} \\ -6x^2 - 4x \\ \underline{+6x^2 + 18x} \\ 14x + 42 \end{array}$$

Repeat, divide front terms: $\dfrac{-6x^2}{x} = -6x$

Multiply this term by divisor: $-6x(x + 3) = -6x^2 - 18x$
Change the signs and combine
Bring down the next term

$$\begin{array}{r} 2x^2 - 6x + 14 \\ x + 3 \overline{)2x^3 + 0x^2 - 4x + 42} \\ \underline{-2x^3 - 6x^2} \\ -6x^2 - 4x \\ \underline{+6x^2 + 18x} \\ 14x + 42 \\ \underline{-14x - 42} \\ 0 \end{array}$$

Repeat, divide front terms: $\dfrac{14x}{x} = 14$

Multiply this term by divisor: $14(x + 3) = 14x + 42$
Change the signs and combine
No remainder

$2x^2 - 6x + 14$ Our Solution

It is important to take a moment to check each problem to verify that the exponents count down and no exponent is skipped. If so we will have to adjust the problem. Also, this final example illustrates, just as in regular long division, sometimes we have no remainder in a problem.

World View Note: Paolo Ruffini was an Italian Mathematician of the early 19th century. In 1809 he was the first to describe a process called synthetic division which could also be used to divide polynomials.

5.7 Practice - Divide Polynomials

Divide.

1) $\dfrac{20x^4 + x^3 + 2x^2}{4x^3}$

2) $\dfrac{5x^4 + 45x^3 + 4x^2}{9x}$

3) $\dfrac{20n^4 + n^3 + 40n^2}{10n}$

4) $\dfrac{3k^3 + 4k^2 + 2k}{8k}$

5) $\dfrac{12x^4 + 24x^3 + 3x^2}{6x}$

6) $\dfrac{5p^4 + 16p^3 + 16p^2}{4p}$

7) $\dfrac{10n^4 + 50n^3 + 2n^2}{10n^2}$

8) $\dfrac{3m^4 + 18m^3 + 27m^2}{9m^2}$

9) $\dfrac{x^2 - 2x - 71}{x + 8}$

10) $\dfrac{r^2 - 3r - 53}{r \quad 9}$

11) $\dfrac{n^2 + 13n + 32}{n + 5}$

12) $\dfrac{b^2 - 10b + 16}{b \quad 7}$

13) $\dfrac{v^2 - 2v - 89}{v \quad 10}$

14) $\dfrac{x^2 + 4x \quad 26}{x + 7}$

15) $\dfrac{a^2 - 4a - 38}{a \quad 8}$

16) $\dfrac{x^2 \quad 10x + 22}{x - 4}$

17) $\dfrac{45p^2 + 56p + 19}{9p + 4}$

18) $\dfrac{48k^2 \quad 70k + 16}{6k - 2}$

19) $\dfrac{10x^2 \quad 32x + 9}{10x - 2}$

20) $\dfrac{n^2 + 7n + 15}{n + 4}$

21) $\dfrac{4r^2 - r - 1}{4r + 3}$

22) $\dfrac{3m^2 + 9m \quad 9}{3m - 3}$

23) $\dfrac{n^2 - 4}{n - 2}$

24) $\dfrac{2x^2 - 5x - 8}{2x + 3}$

25) $\dfrac{27b^2 + 87b + 35}{3b + 8}$

26) $\dfrac{3v^2 - 32}{3v \quad 9}$

27) $\dfrac{4x^2 - 33x + 28}{4x \quad 5}$

28) $\dfrac{4n^2 - 23n - 38}{4n + 5}$

29) $\dfrac{a^3 + 15a^2 + 49a - 55}{a + 7}$

30) $\dfrac{8k^3 - 66k^2 + 12k + 37}{k \quad 8}$

31) $\dfrac{x^3 - 26x - 41}{x + 4}$

32) $\dfrac{x^3 \quad 16x^2 + 71x \quad 56}{x - 8}$

33) $\dfrac{3n^3 + 9n^2 - 64n - 68}{n + 6}$

34) $\dfrac{k^3 \quad 4k^2 \quad 6k + 4}{k - 1}$

35) $\dfrac{x^3 - 46x + 22}{x + 7}$

36) $\dfrac{2n^3 + 21n^2 + 25n}{2n + 3}$

37) $\dfrac{9p^3 + 45p^2 + 27p - 5}{9p + 9}$

38) $\dfrac{8m^3 \quad 57m^2 + 42}{8m + 7}$

39) $\dfrac{r^3 \quad r^2 \quad 16r + 8}{r \quad 4}$

40) $\dfrac{2x^3 + 12x^2 + 4x \quad 37}{2x + 6}$

41) $\dfrac{12n^3 + 12n^2 \quad 15n \quad 4}{2n + 3}$

42) $\dfrac{24b^3 - 38b^2 + 29b - 60}{4b - 7}$

43) $\dfrac{4v^3 - 21v^2 + 6v + 19}{4v + 3}$

Chapter 6 : Factoring

Factoring - Greatest Common Factor

Objective: Find the greatest common factor of a polynomial and factor it out of the expression.

The opposite of multiplying polynomials together is factoring polynomials. There are many benifits of a polynomial being factored. We use factored polynomials to help us solve equations, learn behaviors of graphs, work with fractions and more. Because so many concepts in algebra depend on us being able to factor polynomials it is very important to have very strong factoring skills.

In this lesson we will focus on factoring using the greatest common factor or GCF of a polynomial. When we multiplied polynomials, we multiplied monomials by polynomials by distributing, solving problems such as $4x^2(2x^2 - 3x + 8) = 8x^4 - 12x^3 + 32x$. In this lesson we will work the same problem backwards. We will start with $8x^2 - 12x^3 + 32x$ and try and work backwards to the $4x^2(2x - 3x + 8)$.

To do this we have to be able to first identify what is the GCF of a polynomial. We will first introduce this by looking at finding the GCF of several numbers. To find a GCF of sevearal numbers we are looking for the largest number that can be divided by each of the numbers. This can often be done with quick mental math and it is shown in the following example

Example 262.

$$\text{Find the GCF of } 15, 24, \text{and } 27$$
$$\frac{15}{3} = 5, \ \frac{24}{3} = 6, \ \frac{27}{3} = 9 \quad \text{Each of the numbers can be divided by 3}$$
$$\text{GCF} = 3 \quad \text{Our Solution}$$

When there are variables in our problem we can first find the GCF of the numbers using mental math, then we take any variables that are in common with each term, using the lowest exponent. This is shown in the next example

Example 263.

$$\text{GCF of } 24x^4y^2z, \ 18x^2y^4, \text{ and } 12x^3yz^5$$

$$\frac{24}{6} = 4, \ \frac{18}{6} = 3, \ \frac{12}{6} = 2 \qquad \text{Each number can be divided by 6}$$

$$x^2y \qquad x \text{ and } y \text{ are in all 3, using lowest exponets}$$

$$\text{GCF} = 6x^2y \qquad \text{Our Solution}$$

To factor out a GCF from a polynomial we first need to identify the GCF of all the terms, this is the part that goes in front of the parenthesis, then we divide each term by the GCF, the answer is what is left inside the parenthesis. This is shown in the following examples

Example 264.

$$4x^2 - 20x + 16 \qquad \text{GCF is 4, divide each term by 4}$$

$$\frac{4x^2}{4} = x^2, \ \frac{-20x}{4} = -5x, \ \frac{16}{4} = 4 \qquad \text{This is what is left inside the parenthesis}$$

$$4(x^2 - 5x + 4) \qquad \text{Our Solution}$$

With factoring we can always check our solutions by multiplying (distributing in this case) out the answer and the solution should be the original equation.

Example 265.

$$25x^4 - 15x^3 + 20x^2 \qquad \text{GCF is } 5x^2, \text{ divide each term by this}$$

$$\frac{25x^4}{5x^2} = 5x^2, \ \frac{-15x^3}{5x^2} = -3x, \ \frac{20x^2}{5x^2} = 4 \qquad \text{This is what is left inside the parenthesis}$$

$$5x^2(5x^2 - 3x + 4) \qquad \text{Our Solution}$$

Example 266.

$$3x^3y^2z + 5x^4y^3z^5 - 4xy^4 \qquad \text{GCF is } xy^2, \text{ divide each term by this}$$

$$\frac{3x^3y^2z}{xy^2} = 3x^2z, \frac{5x^4y^3z^5}{xy^2} = 5x^3yz^5, \frac{-4xy^4}{xy^2} = -4y^2 \qquad \text{This is what is left in parenthesis}$$
$$xy^2(3x^2z + 5x^3yz^5 - 4y^2) \qquad \text{Our Solution}$$

World View Note: The first recorded algorithm for finding the greatest common factor comes from Greek mathematician Euclid around the year 300 BC!

Example 267.

$$21x^3 + 14x^2 + 7x \qquad \text{GCF is } 7x, \text{divide each term by this}$$
$$\frac{21x^3}{7x} = 3x^2, \frac{14x^2}{7x} = 2x, \frac{7x}{7x} = 1 \qquad \text{This is what is left inside the parenthesis}$$
$$7x(3x^2 + 2x + 1) \qquad \text{Our Solution}$$

It is important to note in the previous example, that when the GCF was $7x$ and $7x$ was one of the terms, dividing gave an answer of 1. Students often try to factor out the $7x$ and get zero which is incorrect, factoring will never make terms dissapear. Anything divided by itself is 1, be sure to not forget to put the 1 into the solution.

Often the second line is not shown in the work of factoring the GCF. We can simply identify the GCF and put it in front of the parenthesis as shown in the following two examples.

Example 268.

$$12x^5y^2 - 6x^4y^4 + 8x^3y^5 \qquad \text{GCF is } 2x^3y^2, \text{put this in front of parenthesis and divide}$$
$$2x^3y^2(6x^2 - 3xy^2 + 4y^3) \qquad \text{Our Solution}$$

Example 269.

$$18a^4b^3 - 27a^3b^3 + 9a^2b^3 \qquad \text{GCF is } 9a^2b^3, \text{divide each term by this}$$
$$9a^2b^3(2a^2 - 3a + 1) \qquad \text{Our Solution}$$

Again, in the previous problem, when dividing $9a^2b^3$ by itself, the answer is 1, not zero. Be very careful that each term is accounted for in your final solution.

6.1 Practice - Greatest Common Factor

Factor the common factor out of each expression.

1) $9 + 8b^2$

2) $x - 5$

3) $45x^2 - 25$

4) $1 + 2n^2$

5) $56 - 35p$

6) $50x - 80y$

7) $7ab - 35a^2b$

8) $27x^2y^5 - 72x^3y^2$

9) $-3a^2b + 6a^3b^2$

10) $8x^3y^2 + 4x^3$

11) $-5x^2 - 5x^3 - 15x^4$

12) $-32n^9 + 32n^6 + 40n^5$

13) $20x^4 - 30x + 30$

14) $21p^6 + 30p^2 + 27$

15) $28m^4 + 40m^3 + 8$

16) $-10x^4 + 20x^2 + 12x$

17) $30b^9 + 5ab - 15a^2$

18) $27y^7 + 12y^2x + 9y^2$

19) $-48a^2b^2 - 56a^3b - 56a^5b$

20) $30m^6 + 15mn^2 - 25$

21) $20x^8y^2z^2 + 15x^5y^2z + 35x^3y^3z$

22) $3p + 12q - 15q^2r^2$

23) $50x^2y + 10y^2 + 70xz^2$

24) $30y^4z^3x^5 + 50y^4z^5 - 10y^4z^3x$

25) $30qpr - 5qp + 5q$

26) $28b + 14b^2 + 35b^3 + 7b^5$

27) $-18n^5 + 3n^3 - 21n + 3$

28) $30a^8 + 6a^5 + 27a^3 + 21a^2$

29) $-40x^{11} - 20x^{12} + 50x^{13} - 50x^{14}$

30) $-24x^6 - 4x^4 + 12x^3 + 4x^2$

31) $-32mn^8 + 4m^6n + 12mn^4 + 16mn$

32) $-10y^7 + 6y^{10} - 4y^{10}x - 8y^8x$

Factoring - Grouping

Objective: Factor polynomials with four terms using grouping.

The first thing we will always do when factoring is try to factor out a GCF. This GCF is often a monomial like in the problem $5xy + 10xz$ the GCF is the monomial $5x$, so we would have $5x(y + 2z)$. However, a GCF does not have to be a monomial, it could be a binomial. To see this, consider the following two example.

Example 270.

$$3ax - 7bx \quad \text{Both have } x \text{ in common, factor it out}$$
$$x(3a - 7b) \quad \text{Our Solution}$$

Now the same problem, but instead of x we have $(2a + 5b)$.

Example 271.

$$3a(2a + 5b) - 7b(2a + 5b) \quad \text{Both have } (2a + 5b) \text{ in common, factor it out}$$
$$(2a + 5b)(3a - 7b) \quad \text{Our Solution}$$

In the same way we factored out a GCF of x we can factor out a GCF which is a binomial, $(2a + 5b)$. This process can be extended to factor problems where there is no GCF to factor out, or after the GCF is factored out, there is more factoring that can be done. Here we will have to use another strategy to factor. We will use a process known as grouping. Grouping is how we will factor if there are four terms in the problem. Remember, factoring is like multiplying in reverse, so first we will look at a multiplication problem and then try to reverse the process.

Example 272.

$$(2a + 3)(5b + 2) \quad \text{Distribute } (2a + 3) \text{ into second parenthesis}$$
$$5b(2a + 3) + 2(2a + 3) \quad \text{Distribute each monomial}$$
$$10ab + 15b + 4a + 6 \quad \text{Our Solution}$$

The solution has four terms in it. We arrived at the solution by looking at the two parts, $5b(2a + 3)$ and $2(2a + 3)$. When we are factoring by grouping we will always divide the problem into two parts, the first two terms and the last two terms. Then we can factor the GCF out of both the left and right sides. When we do this our hope is what is left in the parenthesis will match on both the left and right. If they match we can pull this matching GCF out front, putting the rest in parenthesis and we will be factored. The next example is the same problem worked backwards, factoring instead of multiplying.

Example 273.

$$10ab + 15b + 4a + 6 \qquad \text{Split problem into two groups}$$

$$\boxed{10ab + 15b} + 4a + 6 \qquad \text{GCF on left is } 5b, \text{ on the right is } 2$$

$$\boxed{5b(2a + 3)} + 2(2a + 3) \qquad (2a + 3) \text{ is the same! Factor out this GCF}$$

$$(2a + 3)(5b + 2) \qquad \text{Our Solution}$$

The key for grouping to work is after the GCF is factored out of the left and right, the two binomials must match exactly. If there is any difference between the two we either have to do some adjusting or it can't be factored using the grouping method. Consider the following example.

Example 274.

$$6x^2 + 9xy - 14x - 21y \qquad \text{Split problem into two groups}$$

$$\boxed{6x^2 + 9xy} - 14x - 21y \qquad \text{GCF on left is } 3x, \text{ on right is } 7$$

$$\boxed{3x(2x + 3y)} + 7(-2x - 3y) \qquad \text{The signs in the parenthesis don}'t \text{ match!}$$

when the signs don't match on both terms we can easily make them match by factoring the opposite of the GCF on the right side. Instead of 7 we will use -7. This will change the signs inside the second parenthesis.

$$\boxed{3x(2x + 3y)} - 7(2x + 3y) \qquad (2x + 3y) \text{ is the same! Factor out this GCF}$$

$$(2x + 3y)(3x - 7) \qquad \text{Our Solution}$$

Often we can recognize early that we need to use the opposite of the GCF when factoring. If the first term of the first binomial is positive in the problem, we will also want the first term of the second binomial to be positive. If it is negative then we will use the opposite of the GCF to be sure they match.

Example 275.

$$5xy - 8x - 10y + 16 \qquad \text{Split the problem into two groups}$$

$$\boxed{5xy - 8x} - 10y + 16 \qquad \text{GCF on left is } x, \text{ on right we need } a \text{ negative,}$$
$$\qquad\qquad\qquad\qquad\qquad\qquad \text{so we use} - 2$$

$$\boxed{x(5y - 8)} - 2(5y - 8) \qquad (5y - 8) \text{ is the same! Factor out this GCF}$$

$$(5y - 8)(x - 2) \qquad \text{Our Solution}$$

Sometimes when factoring the GCF out of the left or right side there is no GCF to factor out. In this case we will use either the GCF of 1 or -1. Often this is all we need to be sure the two binomials match.

Example 276.

$$12ab - 14a - 6b + 7 \quad \text{Split the problem into two groups}$$

$$\boxed{12ab - 14a}\boxed{-6b + 7} \quad \text{GCF on left is } 2a, \text{ on right, no GCF, use } -1$$

$$\boxed{2a(6b - 7)}\boxed{-1(6b - 7)} \quad (6b - 7) \text{ is the same! Factor out this GCF}$$

$$(6b - 7)(2a - 1) \quad \text{Our Solution}$$

Example 277.

$$6x^3 - 15x^2 + 2x - 5 \quad \text{Split problem into two groups}$$

$$\boxed{6x^3 - 15x^2}\boxed{+2x - 5} \quad \text{GCF on left is } 3x^2, \text{ on right, no GCF, use } 1$$

$$\boxed{3x^2(2x - 5)}\boxed{+1(2x - 5)} \quad (2x - 5) \text{ is the same! Factor out this GCF}$$

$$(2x - 5)(3x^2 + 1) \quad \text{Our Solution}$$

Another problem that may come up with grouping is after factoring out the GCF on the left and right, the binomials don't match, more than just the signs are different. In this case we may have to adjust the problem slightly. One way to do this is to change the order of the terms and try again. To do this we will move the second term to the end of the problem and see if that helps us use grouping.

Example 278.

$$4a^2 - 21b^3 + 6ab - 14ab^2 \quad \text{Split the problem into two groups}$$

$$\boxed{4a^2 - 21b^3}\boxed{+6ab - 14ab^2} \quad \text{GCF on left is } 1, \text{ on right is } 2ab$$

$$\boxed{1(4a^2 - 21b^3)}\boxed{+2ab(3 - 7b)} \quad \text{Binomials don't match! Move second term to end}$$

$$4a^2 + 6ab - 14ab^2 - 21b^3 \quad \text{Start over, split the problem into two groups}$$

$$\boxed{4a^2 + 6ab}\boxed{-14ab^2 - 21b^3} \quad \text{GCF on left is } 2a, \text{ on right is } -7b^2$$

$$\boxed{2a(2a + 3b)}\boxed{-7b^2(2a + 3b)} \quad (2a + 3b) \text{ is the same! Factor out this GCF}$$

$$(2a + 3b)(2a - 7b^2) \quad \text{Our Solution}$$

When rearranging terms the problem can still be out of order. Sometimes after factoring out the GCF the terms are backwards. There are two ways that this can happen, one with addition, one with subtraction. If it happens with addition, for example the binomials are $(a + b)$ and $(b + a)$, we don't have to do any extra work. This is because addition is the same in either order $(5 + 3 = 3 + 5 = 8)$.

Example 279.

$$7 + y - 3xy - 21x \qquad \text{Split the problem into two groups}$$

$$\boxed{7 + y} \; \boxed{-3xy - 21x} \qquad \text{GCF on left is 1, on the right is } -3x$$

$$\boxed{1(7 + y)} \; \boxed{-3x(y + 7)} \qquad y + 7 \text{ and } 7 + y \text{ are the same, use either one}$$

$$(y + 7)(1 - 3x) \qquad \text{Our Solution}$$

However, if the binomial has subtraction, then we need to be a bit more careful. For example, if the binomials are $(a - b)$ and $(b - a)$, we will factor out the opposite of the GCF on one part, usually the second. Notice what happens when we factor out -1.

Example 280.

$$(b - a) \qquad \text{Factor out } -1$$

$$-1(-b + a) \qquad \text{Addition can be in either order, switch order}$$

$$-1(a - b) \qquad \text{The order of the subtraction has been switched!}$$

Generally we won't show all the above steps, we will simply factor out the opposite of the GCF and switch the order of the subtraction to make it match the other binomial.

Example 281.

$$8xy - 12y + 15 - 10x \qquad \text{Split the problem into two groups}$$

$$\boxed{8xy - 12y} \; \boxed{15 - 10x} \qquad \text{GCF on left is } 4y, \text{ on right, } 5$$

$$\boxed{4y(2x - 3)} \; \boxed{+5(3 - 2x)} \qquad \text{Need to switch subtraction order, use } -5 \text{ in middle}$$

$$\boxed{4y(2y - 3)} \; \boxed{-5(2x - 3)} \qquad \text{Now } 2x - 3 \text{ match on both! Factor out this GCF}$$

$$(2x - 3)(4y - 5) \qquad \text{Our Solution}$$

World View Note: Sofia Kovalevskaya of Russia was the first woman on the editorial staff of a mathematical journal in the late 19th century. She also did research on how the rings of Saturn rotated.

6.2 Practice - Grouping

Factor each completely.

1) $40r^3 - 8r^2 - 25r + 5$

2) $35x^3 - 10x^2 - 56x + 16$

3) $3n^3 - 2n^2 - 9n + 6$

4) $14v^3 + 10v^2 - 7v - 5$

5) $15b^3 + 21b^2 - 35b - 49$

6) $6x^3 - 48x^2 + 5x - 40$

7) $3x^3 + 15x^2 + 2x + 10$

8) $28p^3 + 21p^2 + 20p + 15$

9) $35x^3 - 28x^2 - 20x + 16$

10) $7n^3 + 21n^2 - 5n - 15$

11) $7xy - 49x + 5y - 35$

12) $42r^3 - 49r^2 + 18r - 21$

13) $32xy + 40x^2 + 12y + 15x$

14) $15ab - 6a + 5b^3 - 2b^2$

15) $16xy - 56x + 2y - 7$

16) $3mn - 8m + 15n - 40$

17) $2xy - 8x^2 + 7y^3 - 28y^2x$

18) $5mn + 2m - 25n - 10$

19) $40xy + 35x - 8y^2 - 7y$

20) $8xy + 56x - y - 7$

21) $32uv - 20u + 24v - 15$

22) $4uv + 14u^2 + 12v + 42u$

23) $10xy + 30 + 25x + 12y$

24) $24xy + 25y^2 - 20x - 30y^3$

25) $3uv + 14u - 6u^2 - 7v$

26) $56ab + 14 - 49a - 16b$

27) $16xy - 3x - 6x^2 + 8y$

Factoring - Trinomials where a = 1

Objective: Factor trinomials where the coefficient of x^2 is one.

Factoring with three terms, or trinomials, is the most important type of factoring to be able to master. As factoring is multiplication backwards we will start with a multipication problem and look at how we can reverse the process.

Example 282.

$$
\begin{array}{ll}
(x+6)(x-4) & \text{Distribute } (x+6) \text{ through second parenthesis} \\
x(x+6)-4(x+6) & \text{Distribute each monomial through parenthesis} \\
x^2+6x-4x-24 & \text{Combine like terms} \\
x^2+2x-24 & \text{Our Solution}
\end{array}
$$

You may notice that if you reverse the last three steps the process looks like grouping. This is because it is grouping! The GCF of the left two terms is x and the GCF of the second two terms is -4. The way we will factor trinomials is to make them into a polynomial with four terms and then factor by grouping. This is shown in the following example, the same problem worked backwards

Example 283.

$$
\begin{array}{ll}
x^2+2x-24 & \text{Split middle term into} +6x-4x \\
x^2+6x-4x-24 & \text{Grouping: GCF on left is } x, \text{on right is} -4 \\
x(x+6)-4(x+6) & (x+6) \text{ is the same, factor out this GCF} \\
(x+6)(x-4) & \text{Our Solution}
\end{array}
$$

The trick to make these problems work is how we split the middle term. Why did we pick $+6x-4x$ and not $+5x-3x$? The reason is because $6x-4x$ is the only combination that works! So how do we know what is the one combination that works? To find the correct way to split the middle term we will use what is called the ac method. In the next lesson we will discuss why it is called the ac method. The way the ac method works is we find a pair of numers that multiply to a certain number and add to another number. Here we will try to multiply to get the last term and add to get the coefficient of the middle term. In the previous example that would mean we wanted to multiply to -24 and add to $+2$. The only numbers that can do this are 6 and -4 $(6 \cdot -4 = -24$ and $6 + (-4) = 2)$. This process is shown in the next few examples

Example 284.

$$
\begin{array}{ll}
x^2 + 9x + 18 & \text{Want to multiply to 18, add to 9} \\
x^2 + 6x + 3x + 18 & \text{6 and 3, split the middle term} \\
x(x+6) + 3(x+6) & \text{Factor by grouping} \\
(x+6)(x+3) & \text{Our Solution}
\end{array}
$$

Example 285.

$$
\begin{array}{ll}
x^2 - 4x + 3 & \text{Want to multiply to 3, add to } -4 \\
x^2 - 3x - x + 3 & -3 \text{ and } -1 \text{, split the middle term} \\
x(x-3) - 1(x-3) & \text{Factor by grouping} \\
(x-3)(x-1) & \text{Our Solution}
\end{array}
$$

Example 286.

$$
\begin{array}{ll}
x^2 - 8x - 20 & \text{Want to multiply to } -20 \text{, add to } -8 \\
x^2 - 10x + 2x - 20 & -10 \text{ and 2, split the middle term} \\
x(x-10) + 2(x-10) & \text{Factor by grouping} \\
(x-10)(x+2) & \text{Our Solution}
\end{array}
$$

Often when factoring we have two variables. These problems solve just like problems with one variable, using the coefficients to decide how to split the middle term

Example 287.

$$
\begin{array}{ll}
a^2 - 9ab + 14b^2 & \text{Want to multiply to 14, add to } -9 \\
a^2 - 7ab - 2ab + 14b^2 & -7 \text{ and } -2 \text{, split the middle term} \\
a(a-7b) - 2b(a-7b) & \text{Factor by grouping} \\
(a-7b)(a-2b) & \text{Our Solution}
\end{array}
$$

As the past few examples illustrate, it is very important to be aware of negatives as we find the pair of numbers we will use to split the middle term. Consier the following example, done incorrectly, ignoring negative signs

Warning 288.

$$x^2 + 5x - 6 \qquad \text{Want to multiply to 6, add 5}$$
$$x^2 + 2x + 3x - 6 \qquad \text{2 and 3, split the middle term}$$
$$x(x+2) + 3(x-2) \qquad \text{Factor by grouping}$$
$$??? \qquad \text{Binomials do not match!}$$

Because we did not use the negative sign with the six to find our pair of numbers, the binomials did not match and grouping was not able to work at the end. Now the problem will be done correctly.

Example 289.

$$x^2 + 5x - 6 \qquad \text{Want to multiply to} -6, \text{add to 5}$$
$$x^2 + 6x - x - 6 \qquad \text{6 and} -1, \text{split the middle term}$$
$$x(x+6) - 1(x+6) \qquad \text{Factor by grouping}$$
$$(x+6)(x-1) \qquad \text{Our Solution}$$

You may have noticed a shortcut for factoring these problems. Once we identify the two numbers that are used to split the middle term, these are the two numbers in our factors! In the previous example, the numbers used to split the middle term were 6 and -1, our factors turned out to be $(x+6)(x-1)$. This pattern does not always work, so be careful getting in the habit of using it. We can use it however, when we have no number (technically we have a 1) in front of x^2. In all the problems we have factored in this lesson there is no number in front of x^2. If this is the case then we can use this shortcut. This is shown in the next few examples.

Example 290.

$$x^2 - 7x - 18 \qquad \text{Want to multiply to} -18, \text{add to} -7$$
$$-9 \text{ and } 2, \text{write the factors}$$
$$(x-9)(x+2) \qquad \text{Our Solution}$$

Example 291.

$$m^2 - mn - 30n^2 \quad \text{Want to multiply to} - 30, \text{add to} - 1$$
$$\quad\quad\quad\quad\quad\quad 5 \text{ and} - 6, \text{write the factors, don't forget second variable}$$
$$(m + 5n)(m - 6n) \quad \text{Our Solution}$$

It is possible to have a problem that does not factor. If there is no combination of numbers that multiplies and adds to the correct numbers, then we say we cannot factor the polynomial, or we say the polynomial is prime. This is shown in the following example.

Example 292.

$$x^2 + 2x + 6 \quad \text{Want to multiply to 6, add to 2}$$
$$1 \cdot 6 \text{ and } 2 \cdot 3 \quad \text{Only possibilities to multiply to six, none add to 2}$$
$$\text{Prime, can't factor} \quad \text{Our Solution}$$

When factoring it is important not to forget about the GCF. If all the terms in a problem have a common factor we will want to first factor out the GCF before we factor using any other method.

Example 293.

$$3x^2 - 24x + 45 \quad \text{GCF of all terms is 3, factor this out}$$
$$3(x^2 - 8x + 15) \quad \text{Want to multiply to 15, add to} - 8$$
$$\quad\quad\quad\quad\quad\quad - 5 \text{ and} - 3, \text{write the factors}$$
$$3(x - 5)(x - 3) \quad \text{Our Solution}$$

Again it is important to comment on the shortcut of jumping right to the factors, this only works if there is no coefficient on x^2. In the next lesson we will look at how this process changes slightly when we have a number in front of x^2. Be careful not to use this shortcut on all factoring problems!

World View Note: The first person to use letters for unknown values was Francois Vieta in 1591 in France. He used vowels to represent variables we are solving for, just as codes used letters to represent an unknown message.

6.3 Practice - Trinomials where a = 1

Factor each completely.

1) $p^2 + 17p + 72$

2) $x^2 + x - 72$

3) $n^2 - 9n + 8$

4) $x^2 + x - 30$

5) $x^2 - 9x - 10$

6) $x^2 + 13x + 40$

7) $b^2 + 12b + 32$

8) $b^2 - 17b + 70$

9) $x^2 + 3x - 70$

10) $x^2 + 3x - 18$

11) $n^2 - 8n + 15$

12) $a^2 - 6a - 27$

13) $p^2 + 15p + 54$

14) $p^2 + 7p - 30$

15) $n^2 - 15n + 56$

16) $m^2 - 15mn + 50n^2$

17) $u^2 - 8uv + 15v^2$

18) $m^2 - 3mn - 40n^2$

19) $m^2 + 2mn - 8n^2$

20) $x^2 + 10xy + 16y^2$

21) $x^2 - 11xy + 18y^2$

22) $u^2 - 9uv + 14v^2$

23) $x^2 + xy - 12y^2$

24) $x^2 + 14xy + 45y^2$

25) $x^2 + 4xy - 12y^2$

26) $4x^2 + 52x + 168$

27) $5a^2 + 60a + 100$

28) $5n^2 - 45n + 40$

29) $6a^2 + 24a - 192$

30) $5v^2 + 20v - 25$

31) $6x^2 + 18xy + 12y^2$

32) $5m^2 + 30mn - 90n^2$

33) $6x^2 + 96xy + 378y^2$

34) $6m^2 - 36mn - 162n^2$

Factoring - Trinomials where a $\neq$ 1

Objective: Factor trinomials using the ac method when the coefficient of x^2 is not one.

When factoring trinomials we used the ac method to split the middle term and then factor by grouping. The ac method gets it's name from the general trinomial equation, $ax^2 + bx + c$, where a, b, and c are the numbers in front of x^2, x and the constant at the end respectively.

World View Note: It was French philosopher Rene Descartes who first used letters from the beginning of the alphabet to represent values we know (a, b, c) and letters from the end to represent letters we don't know and are solving for (x, y, z).

The ac method is named ac because we multiply $a \cdot c$ to find out what we want to multiply to. In the previous lesson we always multiplied to just c because there was no number in front of x^2. This meant the number was 1 and we were multiplying to $1c$ or just c. Now we will have a number in front of x^2 so we will be looking for numbers that multiply to ac and add to b. Other than this, the process will be the same.

Example 294.

$$3x^2 + 11x + 6 \qquad \text{Multiply to } ac \text{ or } (3)(6) = 18, \text{ add to } 11$$
$$3x^2 + 9x + 2x + 6 \qquad \text{The numbers are 9 and 2, split the middle term}$$
$$3x(x+3) + 2(x+3) \qquad \text{Factor by grouping}$$
$$(x+3)(3x+2) \qquad \text{Our Solution}$$

When $a = 1$, or no coefficient in front of x^2, we were able to use a shortcut, using the numbers that split the middle term in the factors. The previous example illustrates an important point, the shortcut does not work when $a \neq 1$. We must go through all the steps of grouping in order to factor the problem.

Example 295.

$$8x^2 - 2x - 15 \qquad \text{Multiply to } ac \text{ or } (8)(-15) = -120, \text{ add to } -2$$
$$8x^2 - 12x + 10x - 15 \qquad \text{The numbers are } -12 \text{ and } 10, \text{ split the middle term}$$
$$4x(2x-3) + 5(2x-3) \qquad \text{Factor by grouping}$$
$$(2x-3)(4x+5) \qquad \text{Our Solution}$$

Example 296.

$$10x^2 - 27x + 5 \quad \text{Multiply to } ac \text{ or } (10)(5) = 50, \text{ add to } -27$$
$$10x^2 - 25x - 2x + 5 \quad \text{The numbers are } -25 \text{ and } -2, \text{ split the middle term}$$
$$5x(2x - 5) - 1(2x - 5) \quad \text{Factor by grouping}$$
$$(2x - 5)(5x - 1) \quad \text{Our Solution}$$

The same process works with two variables in the problem

Example 297.

$$4x^2 - xy - 5y^2 \quad \text{Multiply to } ac \text{ or } (4)(-5) = -20, \text{ add to } -1$$
$$4x^2 + 4xy - 5xy - 5y^2 \quad \text{The numbers are } 4 \text{ and } -5, \text{ split the middle term}$$
$$4x(x + y) - 5y(x + y) \quad \text{Factor by grouping}$$
$$(x + y)(4x - 5y) \quad \text{Our Solution}$$

As always, when factoring we will first look for a GCF before using any other method, including the ac method. Factoring out the GCF first also has the added bonus of making the numbers smaller so the ac method becomes easier.

Example 298.

$$18x^3 + 33x^2 - 30x \quad \text{GCF} = 3x, \text{ factor this out first}$$
$$3x[6x^2 + 11x - 10] \quad \text{Multiply to } ac \text{ or } (6)(-10) = -60, \text{ add to } 11$$
$$3x[6x^2 + 15x - 4x - 10] \quad \text{The numbers are } 15 \text{ and } -4, \text{ split the middle term}$$
$$3x[3x(2x + 5) - 2(2x + 5)] \quad \text{Factor by grouping}$$
$$3x(2x + 5)(3x - 2) \quad \text{Our Solution}$$

As was the case with trinomials when $a = 1$, not all trinomials can be factored. If there is no combinations that multiply and add correctly then we can say the trinomial is prime and cannot be factored.

Example 299.

$$3x^2 + 2x - 7 \quad \text{Multiply to } ac \text{ or } (3)(-7) = -21, \text{ add to } 2$$
$$-3(7) \text{ and } -7(3) \quad \text{Only two ways to multiply to } -21, \text{ it doesn}'t \text{ add to } 2$$
$$\text{Prime, cannot be factored} \quad \text{Our Solution}$$

6.4 Practice - Trinomials where $a \neq 1$

Factor each completely.

1) $7x^2 - 48x + 36$

2) $7n^2 - 44n + 12$

3) $7b^2 + 15b + 2$

4) $7v^2 - 24v - 16$

5) $5a^2 - 13a - 28$

6) $5n^2 - 4n - 20$

7) $2x^2 - 5x + 2$

8) $3r^2 - 4r - 4$

9) $2x^2 + 19x + 35$

10) $7x^2 + 29x - 30$

11) $2b^2 - b - 3$

12) $5k^2 - 26k + 24$

13) $5k^2 + 13k + 6$

14) $3r^2 + 16r + 21$

15) $3x^2 - 17x + 20$

16) $3u^2 + 13uv - 10v^2$

17) $3x^2 + 17xy + 10y^2$

18) $7x^2 - 2xy - 5y^2$

19) $5x^2 + 28xy - 49y^2$

20) $5u^2 + 31uv - 28v^2$

21) $6x^2 - 39x - 21$

22) $10a^2 - 54a - 36$

23) $21k^2 - 87k - 90$

24) $21n^2 + 45n - 54$

25) $14x^2 - 60x + 16$

26) $4r^2 + r - 3$

27) $6x^2 + 29x + 20$

28) $6p^2 + 11p - 7$

29) $4k^2 - 17k + 4$

30) $4r^2 + 3r - 7$

31) $4x^2 + 9xy + 2y^2$

32) $4m^2 + 6mn + 6n^2$

33) $4m^2 - 9mn - 9n^2$

34) $4x^2 - 6xy + 30y^2$

35) $4x^2 + 13xy + 3y^2$

36) $18u^2 - 3uv - 36v^2$

37) $12x^2 + 62xy + 70y^2$

38) $16x^2 + 60xy + 36y^2$

39) $24x^2 - 52xy + 8y^2$

40) $12x^2 + 50xy + 28y^2$

Factoring - Factoring Special Products

Objective: Identify and factor special products including a difference of squares, perfect squares, and sum and difference of cubes.

When factoring there are a few special products that, if we can recognize them, can help us factor polynomials. The first is one we have seen before. When multiplying special products we found that a sum and a difference could multiply to a difference of squares. Here we will use this special product to help us factor

$$\textbf{Difference of Squares: } a^2 - b^2 = (a + b)(a - b)$$

If we are subtracting two perfect squares then it will always factor to the sum and difference of the square roots.

Example 300.

$$x^2 - 16 \qquad \text{Subtracting two perfect squares, the square roots are } x \text{ and } 4$$
$$(x + 4)(x - 4) \qquad \text{Our Solution}$$

Example 301.

$$9a^2 - 25b^2 \qquad \text{Subtracting two perfect squares, the square roots are } 3a \text{ and } 5b$$
$$(3a + 5b)(3a - 5b) \qquad \text{Our Solution}$$

It is important to note, that a sum of squares will never factor. It is always prime. This can be seen if we try to use the ac method to factor $x^2 + 36$.

Example 302.

$$x^2 + 36 \qquad \text{No } bx \text{ term, we use } 0x.$$
$$x^2 + 0x + 36 \qquad \text{Multiply to } 36, \text{ add to } 0$$
$$1 \cdot 36, 2 \cdot 18, 3 \cdot 12, 4 \cdot 9, 6 \cdot 6 \qquad \text{No combinations that multiply to } 36 \text{ add to } 0$$
$$\text{Prime, cannot factor} \qquad \text{Our Solution}$$

It turns out that a sum of squares is always prime.

$$\text{Sum of Squares: } a^2 + b^2 = \text{Prime}$$

A great example where we see a sum of squares comes from factoring a difference of 4th powers. Because the square root of a fourth power is a square ($\sqrt{a^4} = a^2$), we can factor a difference of fourth powers just like we factor a difference of squares, to a sum and difference of the square roots. This will give us two factors, one which will be a prime sum of squares, and a second which will be a difference of squares which we can factor again. This is shown in the following examples.

Example 303.

$$
\begin{array}{ll}
a^4 - b^4 & \text{Difference of squares with roots } a^2 \text{ and } b^2 \\
(a^2 + b^2)(a^2 - b^2) & \text{The first factor is prime, the second is a difference of squares!} \\
(a^2 + b^2)(a + b)(a - b) & \text{Our Solution}
\end{array}
$$

Example 304.

$$
\begin{array}{ll}
x^4 - 16 & \text{Difference of squares with roots } x^2 \text{ and } 4 \\
(x^2 + 4)(x^2 - 4) & \text{The first factor is prime, the second is a difference of squares!} \\
(x^2 + 4)(x + 2)(x - 2) & \text{Our Solution}
\end{array}
$$

Another factoring shortcut is the perfect square. We had a shortcut for multiplying a perfect square which can be reversed to help us factor a perfect square

$$\text{Perfect Square: } a^2 + 2ab + b^2 = (a + b)^2$$

A perfect square can be difficult to recognize at first glance, but if we use the ac method and get two of the same numbers we know we have a perfect square. Then we can just factor using the square roots of the first and last terms and the sign from the middle. This is shown in the following examples.

Example 305.

$$
\begin{array}{ll}
x^2 - 6x + 9 & \text{Multiply to } 9, \text{ add to } -6 \\
& \text{The numbers are } -3 \text{ and } -3, \text{ the same! Perfect square} \\
(x - 3)^2 & \text{Use square roots from first and last terms and sign from the middle}
\end{array}
$$

Example 306.

$$4x^2 + 20xy + 25y^2 \quad \text{Multiply to 100, add to 20}$$

Multiply to 100, add to 20

The numbers are 10 and 10, the same! Perfect square

$$(2x + 5y)^2 \quad \text{Use square roots from first and last terms and sign from the middle}$$

World View Note: The first known record of work with polynomials comes from the Chinese around 200 BC. Problems would be written as "three sheafs of a good crop, two sheafs of a mediocre crop, and one sheaf of a bad crop sold for 29 dou. This would be the polynomial (trinomial) $3x + 2y + z = 29$.

Another factoring shortcut has cubes. With cubes we can either do a sum or a difference of cubes. Both sum and difference of cubes have very similar factoring formulas

$$\textbf{Sum of Cubes:} \, a^3 + b^3 = (a + b)(a^2 - ab + b^2)$$

$$\textbf{Difference of Cubes:} \, a^3 - b^3 = (a - b)(a^2 + ab + b^2)$$

Comparing the formulas you may notice that the only difference is the signs in between the terms. One way to keep these two formulas straight is to think of SOAP. S stands for Same sign as the problem. If we have a sum of cubes, we add first, a difference of cubes we subtract first. O stands for Opposite sign. If we have a sum, then subtraction is the second sign, a difference would have addition for the second sign. Finally, AP stands for Always Positive. Both formulas end with addition. The following examples show factoring with cubes.

Example 307.

$$m^3 - 27 \quad \text{We have cube roots } m \text{ and 3}$$

$$(m \quad 3)(m^2 \quad 3m \quad 9) \quad \text{Use formula, use SOAP to fill in signs}$$

$$(m - 3)(m^2 + 3m + 9) \quad \text{Our Solution}$$

Example 308.

$$125p^3 + 8r^3 \quad \text{We have cube roots } 5p \text{ and } 2r$$

$$(5p \quad 2r)(25p^2 \quad 10r \quad 4r^2) \quad \text{Use formula, use SOAP to fill in signs}$$

$$(5p + 2r)(25p^2 - 10r + 4r^2) \quad \text{Our Solution}$$

The previous example illustrates an important point. When we fill in the trinomial's first and last terms we square the cube roots $5p$ and $2r$. Often students forget to square the number in addition to the variable. Notice that when done correctly, both get cubed.

Often after factoring a sum or difference of cubes, students want to factor the second factor, the trinomial further. As a general rule, this factor will always be prime (unless there is a GCF which should have been factored out before using cubes rule).

The following table sumarizes all of the shortcuts that we can use to factor special products

Factoring Special Products

Difference of Squares	$a^2 - b^2 = (a+b)(a-b)$
Sum of Squares	$a^2 + b^2 = \text{Prime}$
Perfect Square	$a^2 + 2ab + b^2 = (a+b)^2$
Sum of Cubes	$a^3 + b^3 = (a+b)(a^2 - ab + b^2)$
Difference of Cubes	$a^3 - b^3 = (a-b)(a^2 + ab + b^2)$

As always, when factoring special products it is important to check for a GCF first. Only after checking for a GCF should we be using the special products. This is shown in the following examples

Example 309.

$$72x^2 - 2 \qquad \text{GCF is 2}$$
$$2(36x^2 - 1) \qquad \text{Difference of Squares, square roots are } 6x \text{ and } 1$$
$$2(6x + 1)(6x - 1) \qquad \text{Our Solution}$$

Example 310.

$$48x^2y - 24xy + 3y \qquad \text{GCF is } 3y$$
$$3y(16x^2 - 8x + 1) \qquad \text{Multiply to 16 add to 8}$$
$$\qquad \text{The numbers are 4 and 4, the same! Perfect Square}$$
$$3y(4x - 1)^2 \qquad \text{Our Solution}$$

Example 311.

$$128a^4b^2 + 54ab^5 \qquad \text{GCF is } 2ab^2$$
$$2ab^2(64a^3 + 27b^3) \qquad \text{Sum of cubes! Cube roots are } 4a \text{ and } 3b$$
$$2ab^2(4a + 3b)(16a^2 - 12ab + 9b^2) \qquad \text{Our Solution}$$

6.5 Practice - Factoring Special Products

Factor each completely.

1) $r^2 - 16$

2) $x^2 - 9$

3) $v^2 - 25$

4) $x^2 - 1$

5) $p^2 - 4$

6) $4v^2 - 1$

7) $9k^2 - 4$

8) $9a^2 - 1$

9) $3x^2 - 27$

10) $5n^2 - 20$

11) $16x^2 - 36$

12) $125x^2 + 45y^2$

13) $18a^2 - 50b^2$

14) $4m^2 + 64n^2$

15) $a^2 - 2a + 1$

16) $k^2 + 4k + 4$

17) $x^2 + 6x + 9$

18) $n^2 - 8n + 16$

19) $x^2 - 6x + 9$

20) $k^2 - 4k + 4$

21) $25p^2 - 10p + 1$

22) $x^2 + 2x + 1$

23) $25a^2 + 30ab + 9b^2$

24) $x^2 + 8xy + 16y^2$

25) $4a^2 - 20ab + 25b^2$

26) $18m^2 - 24mn + 8n^2$

27) $8x^2 - 24xy + 18y^2$

28) $20x^2 + 20xy + 5y^2$

29) $8 - m^3$

30) $x^3 + 64$

31) $x^3 - 64$

32) $x^3 + 8$

33) $216 - u^3$

34) $125x^3 - 216$

35) $125a^3 - 64$

36) $64x^3 - 27$

37) $64x^3 + 27y^3$

38) $32m^3 - 108n^3$

39) $54x^3 + 250y^3$

40) $375m^3 + 648n^3$

41) $a^4 - 81$

42) $x^4 - 256$

43) $16 - z^4$

44) $n^4 - 1$

45) $x^4 - y^4$

46) $16a^4 - b^4$

47) $m^4 - 81b^4$

48) $81c^4 - 16d^4$

Factoring - Factoring Strategy

Objective: Idenfity and use the correct method to factor various polynomials.

With so many different tools used to factor, it is easy to get lost as to which tool to use when. Here we will attempt to organize all the different factoring types we have seen. A large part of deciding how to solve a problem is based on how many terms are in the problem. For all problem types we will always try to factor out the GCF first.

Factoring Strategy (GCF First!!!!!)

- **2 terms:** sum or difference of squares or cubes:

  $a^2 - b^2 = (a+b)(a-b)$

  $a^2 + b^2 = \text{Prime}$

  $a^3 + b^3 = (a+b)(a^2 - ab + b^2)$

  $a^3 - b^3 = (a-b)(a^2 + ab + b^2)$

- **3 terms:** ac method, watch for perfect square!

  $a^2 + 2ab + b^2 = (a+b)^2$

  Multiply to ac and add to b

- **4 terms:** grouping

We will use the above strategy to factor each of the following examples. Here the emphasis will be on which strategy to use rather than the steps used in that method.

Example 312.

$$4x^2 + 56xy + 196y^2 \quad \text{GCF first, 4}$$
$$4(x^2 + 14xy + 49y^2) \quad \text{Three terms, try ac method, multiply to 49, add to 14}$$
$$\qquad\qquad\qquad\qquad\qquad 7 \text{ and } 7, \text{ perfect square!}$$

$$4(x+7y)^2 \quad \text{Our Solution}$$

Example 313.

$$\begin{aligned}
5x^2y + 15xy - 35x^2 - 105x \quad &\text{GCF first, } 5x \\
5x(xy + 3y - 7x - 21) \quad &\text{Four terms, try grouping} \\
5x[y(x+3) - 7(x+3)] \quad &(x+3) \text{ match!} \\
5x(x+3)(y-7) \quad &\text{Our Solution}
\end{aligned}$$

Example 314.

$$\begin{aligned}
100x^2 - 400 \quad &\text{GCF first, } 100 \\
100(x^2 - 4) \quad &\text{Two terms, difference of squares} \\
100(x+4)(x-4) \quad &\text{Our Solution}
\end{aligned}$$

Example 315.

$$\begin{aligned}
108x^3y^2 - 39x^2y^2 + 3xy^2 \quad &\text{GCF first, } 3xy^2 \\
3xy^2(36x^2 - 13x + 1) \quad &\text{Thee terms, ac method, multiply to 36, add to } -13 \\
3xy^2(36x^2 - 9x - 4x + 1) \quad &-9 \text{ and } -4, \text{ split middle term} \\
3xy^2[9x(4x-1) - 1(4x-1)] \quad &\text{Factor by grouping} \\
3xy^2(4x-1)(9x-1) \quad &\text{Our Solution}
\end{aligned}$$

World View Note: Variables originated in ancient Greece where Aristotle would use a single capital letter to represent a number.

Example 316.

$$\begin{aligned}
5 + 625y^3 \quad &\text{GCF first, } 5 \\
5(1 + 125y^3) \quad &\text{Two terms, sum of cubes} \\
5(1 + 5y)(1 - 5y + 25y^2) \quad &\text{Our Solution}
\end{aligned}$$

It is important to be comfortable and confident not just with using all the factoring methods, but decided on which method to use. This is why practice is very important!

6.6 Practice - Factoring Strategy

Factor each completely.

1) $24az - 18ah + 60yz - 45yh$

2) $2x^2 - 11x + 15$

3) $5u^2 - 9uv + 4v^2$

4) $16x^2 + 48xy + 36y^2$

5) $-2x^3 + 128y^3$

6) $20uv - 60u^3 - 5xv + 15xu^2$

7) $5n^3 + 7n^2 - 6n$

8) $2x^3 + 5x^2y + 3y^2x$

9) $54u^3 - 16$

10) $54 - 128x^3$

11) $n^2 - n$

12) $5x^2 - 22x - 15$

13) $x^2 - 4xy + 3y^2$

14) $45u^2 - 150uv + 125v^2$

15) $9x^2 - 25y^2$

16) $x^3 - 27y^3$

17) $m^2 - 4n^2$

18) $12ab - 18a + 6nb - 9n$

19) $36b^2c - 16xd - 24b^2d + 24xc$

20) $3m^3 - 6m^2n - 24n^2m$

21) $128 + 54x^3$

22) $64m^3 + 27n^3$

23) $2x^3 + 6x^2y - 20y^2x$

24) $3ac + 15ad^2 + x^2c + 5x^2d^2$

25) $n^3 + 7n^2 + 10n$

26) $64m^3 - n^3$

27) $27x^3 - 64$

28) $16a^2 - 9b^2$

29) $5x^2 + 2x$

30) $2x^2 - 10x + 12$

31) $3k^3 - 27k^2 + 60k$

32) $32x^2 - 18y^2$

33) $mn - 12x + 3m - 4xn$

34) $2k^2 + k - 10$

35) $16x^2 - 8xy + y^2$

36) $v^2 + v$

37) $27m^2 - 48n^2$

38) $x^3 + 4x^2$

39) $9x^3 + 21x^2y - 60y^2x$

40) $9n^3 - 3n^2$

41) $2m^2 + 6mn - 20n^2$

42) $2u^2v^2 - 11uv^3 + 15v^4$

6.7

Factoring - Solve by Factoring

Objective: Solve quadratic equation by factoring and using the zero product rule.

When solving linear equations such as $2x - 5 = 21$ we can solve for the variable directly by adding 5 and dividing by 2 to get 13. However, when we have x^2 (or a higher power of x) we cannot just isolate the variable as we did with the linear equations. One method that we can use to solve for the varaible is known as the zero product rule

Zero Product Rule: If $ab = 0$ then either $a = 0$ or $b = 0$

The zero product rule tells us that if two factors are multiplied together and the answer is zero, then one of the factors must be zero. We can use this to help us solve factored polynomials as in the following example.

Example 317.

$$(2x - 3)(5x + 1) = 0 \quad \text{One factor must be zero}$$
$$2x - 3 = 0 \text{ or } 5x + 1 = 0 \quad \text{Set each factor equal to zero}$$
$$\underline{+3 + 3} \qquad \underline{-1 - 1} \quad \text{Solve each equation}$$
$$\frac{2x = 3}{2 \quad 2} \text{ or } \frac{5x = -1}{5 \quad 5}$$
$$x = \frac{3}{2} \text{ or } \frac{-1}{5} \quad \text{Our Solution}$$

For the zero product rule to work we must have factors to set equal to zero. This means if the problem is not already factored we will factor it first.

Example 318.

$$4x^2 + x - 3 = 0 \quad \text{Factor using the ac method, multiply to} -12, \text{add to } 1$$
$$4x^2 - 3x + 4x - 3 = 0 \quad \text{The numbers are} -3 \text{ and } 4, \text{split the middle term}$$
$$x(4x - 3) + 1(4x - 3) = 0 \quad \text{Factor by grouping}$$
$$(4x - 3)(x + 1) = 0 \quad \text{One factor must be zero}$$
$$4x - 3 = 0 \text{ or } x + 1 = 0 \quad \text{Set each factor equal to zero}$$

$$\underline{+3+3} \qquad \underline{-1-1} \qquad \text{Solve each equation}$$
$$4x = 3 \ \text{or} \ x = -1$$
$$\overline{4} \ \overline{4}$$
$$x = \frac{3}{4} \ \text{or} \ -1 \qquad \text{Our Solution}$$

Another important part of the zero product rule is that before we factor, the equation must equal zero. If it does not, we must move terms around so it does equal zero. Generally we like the x^2 term to be positive.

Example 319.

$$x^2 = 8x - 15 \qquad \text{Set equal to zero by moving terms to the left}$$
$$\underline{-8x + 15} \quad \underline{-8x + 15}$$
$$x^2 - 8x + 15 = 0 \qquad \text{Factor using the ac method, multiply to 15, add to} -8$$
$$(x - 5)(x - 3) = 0 \qquad \text{The numbers are} -5 \text{ and} -3$$
$$x - 5 = 0 \ \text{or} \ x - 3 = 0 \qquad \text{Set each factor equal to zero}$$
$$\underline{+5 + 5} \qquad \underline{+3 + 3} \qquad \text{Solve each equation}$$
$$x = 5 \ \text{or} \ x = 3 \qquad \text{Our Solution}$$

Example 320.

$$(x - 7)(x + 3) = -9 \qquad \text{Not equal to zero, multiply first, use FOIL}$$
$$x^2 - 7x + 3x - 21 = -9 \qquad \text{Combine like terms}$$
$$x^2 - 4x - 21 = -9 \qquad \text{Move} -9 \text{ to other side so equation equals zero}$$
$$\underline{+9 \quad +9}$$
$$x^2 - 4x - 12 = 0 \qquad \text{Factor using the ac method, mutiply to} -12, \text{add to} -4$$
$$(x - 6)(x + 2) = 0 \qquad \text{The numbers are 6 and} -2$$
$$x - 6 = 0 \ \text{or} \ x + 2 = 0 \qquad \text{Set each factor equal to zero}$$
$$\underline{+6 + 6} \qquad \underline{-2 - 2} \qquad \text{Solve each equation}$$
$$x = 6 \ \text{or} \ -2 \qquad \text{Our Solution}$$

Example 321.

$$3x^2 + 4x - 5 = 7x^2 + 4x - 14 \qquad \text{Set equal to zero by moving terms to the right}$$
$$\underline{-3x^2 - 4x + 5} \quad \underline{-3x^2 - 4x + 5}$$
$$0 = 4x^2 - 9 \qquad \text{Factor using difference of squares}$$

$$0 = (2x+3)(2x-3) \quad \text{One factor must be zero}$$
$$2x+3=0 \ \text{ or } \ 2x-3=0 \quad \text{Set each factor equal to zero}$$
$$\underline{-3-3} \qquad \underline{+3+3} \quad \text{Solve each equation}$$
$$\frac{2x=-3}{2 \quad 2} \ \text{ or } \ \frac{2x=3}{2 \quad 2}$$
$$x = \frac{-3}{2} \ \text{ or } \ \frac{3}{2} \quad \text{Our Solution}$$

Most problems with x^2 will have two unique solutions. However, it is possible to have only one solution as the next example illustrates.

Example 322.

$$4x^2 = 12x - 9 \qquad \text{Set equal to zero by moving terms to left}$$
$$\underline{-12x+9 \quad -12x+9}$$
$$4x^2 - 12x + 9 = 0 \qquad \text{Factor using the ac method, multiply to 36, add to} -12$$
$$(2x-3)^2 = 0 \qquad -6 \text{ and } -6, a \text{ perfect square!}$$
$$2x - 3 = 0 \qquad \text{Set this factor equal to zero}$$
$$\underline{+3+3} \qquad \text{Solve the equation}$$
$$\frac{2x=3}{2 \quad 2}$$
$$x = \frac{3}{2} \qquad \text{Our Solution}$$

As always it will be important to factor out the GCF first if we have one. This GCF is also a factor and must also be set equal to zero using the zero product rule. This may give us more than just two solution. The next few examples illustrate this.

Example 323.

$$4x^2 = 8x \qquad \text{Set equal to zero by moving the terms to left}$$
$$\underline{-8x -8x} \qquad \text{Be careful, on the right side, they are not like terms!}$$
$$4x^2 - 8x = 0 \qquad \text{Factor out the GCF of } 4x$$
$$4x(x-2) = 0 \qquad \text{One factor must be zero}$$
$$4x = 0 \ \text{ or } \ x - 2 = 0 \qquad \text{Set each factor equal to zero}$$
$$\overline{4 \quad 4} \qquad \underline{+2+2} \qquad \text{Solve each equation}$$
$$x = 0 \ \text{ or } \ 2 \qquad \text{Our Solution}$$

Example 324.

$$2x^3 - 14x^2 + 24x = 0 \qquad \text{Factor out the GCF of } 2x$$
$$2x(x^2 - 7x + 12) = 0 \qquad \text{Factor with ac method, multiply to 12, add to } -7$$
$$2x(x - 3)(x - 4) = 0 \qquad \text{The numbers are } -3 \text{ and } -4$$

$$2x = 0 \text{ or } x - 3 = 0 \text{ or } x - 4 = 0 \qquad \text{Set each factor equal to zero}$$
$$\overline{2} \quad \overline{2} \qquad +3+3 \qquad +4+4 \qquad \text{Solve each equation}$$
$$x = 0 \text{ or } 3 \text{ or } 4 \qquad \text{Our Solutions}$$

Example 325.

$$6x^2 + 21x - 27 = 0 \qquad \text{Factor out the GCF of 3}$$
$$3(2x^2 + 7x - 9) = 0 \qquad \text{Factor with ac method, multiply to } -18, \text{ add to 7}$$
$$3(2x^2 + 9x - 2x - 9) = 0 \qquad \text{The numbers are 9 and } -2$$
$$3[x(2x + 9) - 1(2x + 9)] = 0 \qquad \text{Factor by grouping}$$
$$3(2x + 9)(x - 1) = 0 \qquad \text{One factor must be zero}$$

$$3 = 0 \text{ or } 2x + 9 = 0 \text{ or } x - 1 = 0 \qquad \text{Set each factor equal to zero}$$
$$3 \neq 0 \qquad -9-9 \qquad +1+1 \qquad \text{Solve each equation}$$
$$2x = -9 \text{ or } x = 1$$
$$\overline{2} \quad \overline{2}$$
$$x = -\frac{9}{2} \text{ or } 1 \qquad \text{Our Solution}$$

In the previous example, the GCF did not have a variable in it. When we set this factor equal to zero we got a false statement. No solutions come from this factor. Often a student will skip setting the GCF factor equal to zero if there is no variables in the GCF.

Just as not all polynomials cannot factor, all equations cannot be solved by factoring. If an equation does not factor we will have to solve it using another method. These other methods are saved for another section.

World View Note: While factoring works great to solve problems with x^2, Tartaglia, in 16th century Italy, developed a method to solve problems with x^3. He kept his method a secret until another mathematician, Cardan, talked him out of his secret and published the results. To this day the formula is known as Cardan's Formula.

A question often asked is if it is possible to get rid of the square on the variable by taking the square root of both sides. While it is possible, there are a few properties of square roots that we have not covered yet and thus it is common to break a rule of roots that we are not aware of at this point. The short reason we want to avoid this for now is because taking a square root will only give us one of the two answers. When we talk about roots we will come back to problems like these and see how we can solve using square roots in a method called completing the square. For now, **never** take the square root of both sides!

6.7 Practice - Solve by Factoring

Solve each equation by factoring.

1) $(k-7)(k+2)=0$

2) $(a+4)(a-3)=0$

3) $(x-1)(x+4)=0$

4) $(2x+5)(x-7)=0$

5) $6x^2-150=0$

6) $p^2+4p-32=0$

7) $2n^2+10n-28=0$

8) $m^2-m-30=0$

9) $7x^2+26x+15=0$

10) $40r^2-285r-280=0$

11) $5n^2-9n-2=0$

12) $2b^2-3b-2=0$

13) $x^2-4x-8=-8$

14) $v^2-8v-3=-3$

15) $x^2-5x-1=-5$

16) $a^2-6a+6=-2$

17) $49p^2+371p-163=5$

18) $7k^2+57k+13=5$

19) $7x^2+17x-20=-8$

20) $4n^2-13n+8=5$

21) $7r^2+84=-49r$

22) $7m^2-224=28m$

23) $x^2-6x=16$

24) $7n^2-28n=0$

25) $3v^2+7v=40$

26) $6b^2=5+7b$

27) $35x^2+120x=-45$

28) $9n^2+39n=-36$

29) $4k^2+18k-23=6k-7$

30) $a^2+7a-9=-3+6a$

31) $9x^2-46+7x=7x+8x^2+3$

32) $x^2+10x+30=6$

33) $2m^2+19m+40=-2m$

34) $5n^2+41n+40=-2$

35) $40p^2+183p-168=p+5p^2$

36) $24x^2+11x-80=3x$

Chapter 7 : Rational Expressions

Rational Expressions - Reduce Rational Expressions

Objective: Reduce rational expressions by dividing out common factors.

Rational expressions are expressions written as a quotient of polynomials. Examples of rational expressions include:

$$\frac{x^2 - x - 12}{x^2 - 9x + 20} \quad \text{and} \quad \frac{3}{x - 2} \quad \text{and} \quad \frac{a - b}{b - a} \quad \text{and} \quad \frac{3}{2}$$

As rational expressions are a special type of fraction, it is important to remember with fractions we cannot have zero in the denominator of a fraction. For this reason, rational expressions may have one more excluded values, or values that the variable cannot be or the expression would be undefined.

Example 326.

$$\text{State the excluded value(s): } \frac{x^2 - 1}{3x^2 + 5x} \qquad \text{Denominator cannot be zero}$$

$$3x^2 + 5x \neq 0 \qquad \text{Factor}$$

$$x(3x + 5) \neq 0 \qquad \text{Set each factor not equal to zero}$$

$$x \neq 0 \text{ or } 3x + 5 \neq 0 \qquad \text{Subtract 5 from second equation}$$

$$\underline{-5 \quad -5}$$

$$\frac{3x \neq -5}{3 \quad 3} \qquad \text{Divide by 3}$$

$$x \neq \frac{-5}{3} \qquad \text{Second equation is solved}$$

$$x \neq 0 \text{ or } \frac{-5}{3} \qquad \text{Our Solution}$$

This means we can use any value for x in the equation except for 0 and $-\frac{5}{3}$. We can however, evaluate any other value in the expression.

World View Note: The number zero was not widely accepted in mathematical thought around the world for many years. It was the Mayans of Central America who first used zero to aid in the use of their base 20 system as a place holder!

Rational expressions are easily evaluated by simply substituting the value for the variable and using order of operations.

Example 327.

$$\frac{x^2-4}{x^2+6x+8} \quad \text{when} \quad x=-6 \qquad \text{Substitute} - 5 \text{ in for each variable}$$

$$\frac{(-6)^2-4}{(-6)^2+6(-6)+8} \qquad \text{Exponents first}$$

$$\frac{36-4}{36+6(-6)+8} \qquad \text{Multiply}$$

$$\frac{36-4}{36-36+8} \qquad \text{Add and subtract}$$

$$\frac{32}{8} \qquad \text{Reduce}$$

$$4 \qquad \text{Our Solution}$$

Just as we reduced the previous example, often a rational expression can be reduced, even without knowing the value of the variable. When we reduce we divide out common factors. We have already seen this with monomials when we discussed properties of exponents. If the problem only has monomials we can reduce the coefficients, and subtract exponents on the variables.

Example 328.

$$\frac{15x^4y^2}{25x^2y^6} \qquad \text{Reduce , subtract exponents. Negative exponents move to denominator}$$

$$\frac{3x^2}{5y^4} \qquad \text{Our Solution}$$

However, if there is more than just one term in either the numerator or denominator, we can't divide out common factors unless we first factor the numerator and denominator.

Example 329.

$$\frac{28}{8x^2 - 16} \qquad \text{Denominator has } a \text{ common factor of } 8$$

$$\frac{28}{8(x^2 - 2)} \qquad \text{Reduce by dividing } 24 \text{ and } 8 \text{ by } 4$$

$$\frac{7}{2(x^2 - 2)} \qquad \text{Our Solution}$$

Example 330.

$$\frac{9x - 3}{18x - 6} \qquad \text{Numerator has } a \text{ common factor of } 3, \text{ denominator of } 6$$

$$\frac{3(3x - 1)}{6(3x - 1)} \qquad \text{Divide out common factor } (3x - 1) \text{ and divide } 3 \text{ and } 6 \text{ by } 3$$

$$\frac{1}{2} \qquad \text{Our Solution}$$

Example 331.

$$\frac{x^2 - 25}{x^2 + 8x + 15} \qquad \text{Numerator is difference of squares, denominator is factored using ac}$$

$$\frac{(x + 5)(x - 5)}{(x + 3)(x + 5)} \qquad \text{Divide out common factor } (x + 5)$$

$$\frac{x - 5}{x + 3} \qquad \text{Our Solution}$$

It is important to remember we cannot reduce terms, only factors. This means if there are any + or − between the parts we want to reduce we cannot. In the previous example we had the solution $\frac{x \quad 5}{x+3}$, we cannot divide out the x's because they are terms (separated by + or −) not factors (separated by multiplication).

7.1 Practice - Reduce Rational Expressions

Evaluate

1) $\frac{4v+2}{6}$ when $v=4$

2) $\frac{b \quad 3}{3b-9}$ when $b=-2$

3) $\frac{x-3}{x^2-4x+3}$ when $x=-4$

4) $\frac{a+2}{a^2+3a+2}$ when $a=-1$

5) $\frac{b+2}{b^2+4b+4}$ when $b=0$

6) $\frac{n^2 \quad n \quad 6}{n-3}$ when $n=4$

State the excluded values for each.

7) $\frac{3k^2+30k}{k+10}$

8) $\frac{27p}{18p^2-36p}$

9) $\frac{15n^2}{10n+25}$

10) $\frac{x+10}{8x^2+80x}$

11) $\frac{10m^2+8m}{10m}$

12) $\frac{10x+16}{6x+20}$

13) $\frac{r^2+3r+2}{5r+10}$

14) $\frac{6n^2 \quad 21n}{6n^2+3n}$

15) $\frac{b^2+12b+32}{b^2+4b-32}$

16) $\frac{10v^2+30v}{35v^2-5v}$

Simplify each expression.

17) $\frac{21x^2}{18x}$

18) $\frac{12n}{4n^2}$

19) $\frac{24a}{40a^2}$

20) $\frac{21k}{24k^2}$

21) $\frac{32x^3}{8x^4}$

22) $\frac{90x^2}{20x}$

23) $\frac{18m \quad 24}{60}$

24) $\frac{10}{81n^3+36n^2}$

25) $\frac{20}{4p+2}$

26) $\frac{n-9}{9n \quad 81}$

27) $\frac{x+1}{x^2+8x+7}$

28) $\frac{28m+12}{36}$

29) $\frac{32x^2}{28x^2+28x}$

30) $\frac{49r+56}{56r}$

31) $\frac{n^2+4n-12}{n^2-7n+10}$

32) $\frac{b^2+14b+48}{b^2+15b+56}$

33) $\frac{9v+54}{v^2 \quad 4v \quad 60}$

34) $\frac{30x \quad 90}{50x+40}$

35) $\frac{12x^2-42x}{30x^2-42x}$

36) $\frac{k^2-12k+32}{k^2-64}$

37) $\frac{6a-10}{10a+4}$

38) $\frac{9p+18}{p^2+4p+4}$

39) $\frac{2n^2+19n \quad 10}{9n+90}$

40) $\frac{3x^2 \quad 29x+40}{5x^2 \quad 30x \quad 80}$

41) $\dfrac{8m+16}{20m\quad 12}$

42) $\dfrac{56x-48}{24x^2+56x+32}$

43) $\dfrac{2x^2\quad 10x+8}{3x^2-7x+4}$

44) $\dfrac{50b\quad 80}{50b+20}$

45) $\dfrac{7n^2-32n+16}{4n-16}$

46) $\dfrac{35v+35}{21v+7}$

47) $\dfrac{n^2-2n+1}{6n+6}$

48) $\dfrac{56x\quad 48}{24x^2+56x+32}$

49) $\dfrac{7a^2-26a-45}{6a^2-34a+20}$

50) $\dfrac{4k^3-2k^2-2k}{9k^3-18k^2+9k}$

Rational Expressions - Multiply & Divide

Objective: Multiply and divide rational expressions.

Multiplying and dividing rational expressions is very similar to the process we use to multiply and divide fractions.

Example 332.

$$\frac{15}{49} \cdot \frac{14}{45} \qquad \text{First reduce common factors from numerator and denominator (15 and 7)}$$

$$\frac{1}{7} \cdot \frac{2}{3} \qquad \text{Multiply numerators across and denominators across}$$

$$\frac{2}{21} \qquad \text{Our Solution}$$

The process is identical for division with the extra first step of multiplying by the reciprocal. When multiplying with rational expressions we follow the same process, first divide out common factors, then multiply straight across.

Example 333.

$$\frac{25x^2}{9y^8} \cdot \frac{24y^4}{55x^7} \qquad \text{Reduce coefficients by dividing out common factors (3 and 5)}$$
$$\text{Reduce, subtracting exponents, negative exponents in denominator}$$

$$\frac{5}{3y^4} \cdot \frac{8}{11x^5} \qquad \text{Multiply across}$$

$$\frac{40}{33x^5y^4} \qquad \text{Our Solution}$$

Division is identical in process with the extra first step of multiplying by the reciprocal.

Example 334.

$$\frac{a^4b^2}{a} \div \frac{b^4}{4} \qquad \text{Multiply by the reciprocal}$$

$$\frac{a^4b^2}{a} \cdot \frac{4}{b^4} \qquad \text{Subtract exponents on variables, negative exponents in denominator}$$

$$\frac{a^3}{1} \cdot \frac{4}{b^2} \qquad \text{Multiply across}$$

$$\frac{4a^3}{b^2} \qquad \text{Our Solution}$$

Just as with reducing rational expressions, before we reduce a multiplication problem, it must be factored first.

Example 335.

$$\frac{x^2-9}{x^2+x-20} \cdot \frac{x^2-8x+16}{3x+9} \qquad \text{Factor each numerator and denominator}$$

$$\frac{(x+3)(x-3)}{(x-4)(x+5)} \cdot \frac{(x-4)(x-4)}{3(x+3)} \qquad \text{Divide out common factors } (x+3) \text{ and } (x-4)$$

$$\frac{x-3}{x+5} \cdot \frac{x-4}{3} \qquad \text{Multiply across}$$

$$\frac{(x-3)(x-4)}{3(x+5)} \quad \text{Our Solution}$$

Again we follow the same pattern with division with the extra first step of multiplying by the reciprocal.

Example 336.

$$\frac{x^2-x-12}{x^2-2x-8} \div \frac{5x^2+15x}{x^2+x-2} \quad \text{Multiply by the reciprocal}$$

$$\frac{x^2-x-12}{x^2-2x-8} \cdot \frac{x^2+x-2}{5x^2+15x} \quad \text{Factor each numerator and denominator}$$

$$\frac{(x-4)(x+3)}{(x+2)(x-4)} \cdot \frac{(x+2)(x-1)}{5x(x+3)} \quad \text{Divide out common factors:}$$

$$(x-4) \text{ and } (x+3) \text{ and } (x+2)$$

$$\frac{1}{1} \cdot \frac{x-1}{5x} \quad \text{Multiply across}$$

$$\frac{x-1}{5x} \quad \text{Our Solution}$$

We can combine multiplying and dividing of fractions into one problem as shown below. To solve we still need to factor, and we use the reciprocal of the divided fraction.

Example 337.

$$\frac{a^2+7a+10}{a^2+6a+5} \cdot \frac{a+1}{a^2+4a+4} \div \frac{a-1}{a+2} \quad \text{Factor each expression}$$

$$\frac{(a+5)(a+2)}{(a+5)(a+1)} \cdot \frac{(a+1)}{(a+2)(a+2)} \div \frac{(a-1)}{(a+2)} \quad \text{Reciprocal of last fraction}$$

$$\frac{(a+5)(a+2)}{(a+5)(a+1)} \cdot \frac{(a+1)}{(a+2)(a+2)} \cdot \frac{(a+2)}{(a-1)} \quad \text{Divide out common factors}$$

$$(a+2), (a+2), (a+1), (a+5)$$

$$\frac{1}{a-1} \quad \text{Our Solution}$$

World View Note: Indian mathematician Aryabhata, in the 6th century, published a work which included the rational expression $\frac{n(n+1)(n+2)}{6}$ for the sum of the first n squares $(1^1 + 2^2 + 3^2 + \ldots + n^2)$

7.2 Practice - Multiply and Divide

Simplify each expression.

1) $\dfrac{8x^2}{9} \cdot \dfrac{9}{2}$

2) $\dfrac{8x}{3x} \div \dfrac{4}{7}$

3) $\dfrac{9n}{2n} \cdot \dfrac{7}{5n}$

4) $\dfrac{9m}{5m^2} \cdot \dfrac{7}{2}$

5) $\dfrac{5x^2}{4} \cdot \dfrac{6}{5}$

6) $\dfrac{10p}{5} \div \dfrac{8}{10}$

7) $\dfrac{7(m-6)}{m-6} \cdot \dfrac{5m(7m-5)}{7(7m-5)}$

8) $\dfrac{7}{10(n+3)} \div \dfrac{n-2}{(n+3)(n-2)}$

9) $\dfrac{7r}{7r(r+10)} \div \dfrac{r-6}{(r-6)^2}$

10) $\dfrac{6x(x+4)}{x-3} \cdot \dfrac{(x-3)(x-6)}{6x(x-6)}$

11) $\dfrac{25n+25}{5} \cdot \dfrac{4}{30n+30}$

12) $\dfrac{9}{b^2-b-12} \div \dfrac{b-5}{b^2-b-12}$

13) $\dfrac{x-10}{35x+21} \div \dfrac{7}{35x+21}$

14) $\dfrac{v-1}{4} \cdot \dfrac{4}{v^2-11v+10}$

15) $\dfrac{x^2-6x-7}{x+5} \cdot \dfrac{x+5}{x-7}$

16) $\dfrac{1}{a-6} \cdot \dfrac{8a+80}{8}$

17) $\dfrac{8k}{24k^2-40k} \div \dfrac{1}{15k-25}$

18) $\dfrac{p-8}{p^2-12p+32} \div \dfrac{1}{p-10}$

19) $(n-8) \cdot \dfrac{6}{10n-80}$

20) $\dfrac{x^2-7x+10}{x-2} \cdot \dfrac{x+10}{x^2-x-20}$

21) $\dfrac{4m+36}{m+9} \cdot \dfrac{m-5}{5m^2}$

22) $\dfrac{2r}{r+6} \div \dfrac{2r}{7r+42}$

23) $\dfrac{3x-6}{12x-24}(x+3)$

24) $\dfrac{2n^2-12n-54}{n+7} \div (2n+6)$

25) $\dfrac{b+2}{40b^2-24b}(5b-3)$

26) $\dfrac{21v^2+16v-16}{3v+4} \div \dfrac{35v-20}{v-9}$

27) $\dfrac{n-7}{6n-12} \cdot \dfrac{12-6n}{n^2-13n+42}$

28) $\dfrac{x^2+11x+24}{6x^3+18x^2} \cdot \dfrac{6x^3+6x^2}{x^2+5x-24}$

29) $\dfrac{27a+36}{9a+63} \div \dfrac{6a+8}{2}$

30) $\dfrac{k-7}{k^2-k-12} \cdot \dfrac{7k^2-28k}{8k^2-56k}$

31) $\dfrac{x^2-12x+32}{x^2-6x-16} \cdot \dfrac{7x^2+14x}{7x^2+21x}$

32) $\dfrac{9x^3+54x^2}{x^2+5x-14} \cdot \dfrac{x^2+5x-14}{10x^2}$

33) $(10m^2+100m) \cdot \dfrac{18m^3-36m^2}{20m^2-40m}$

34) $\dfrac{n-7}{n^2-2n-35} \div \dfrac{9n+54}{10n+50}$

35) $\dfrac{7p^2+25p+12}{6p+48} \cdot \dfrac{3p-8}{21p^2-44p-32}$

36) $\dfrac{7x^2-66x+80}{49x^2+7x-72} \div \dfrac{7x^2+39x-70}{49x^2+7x-72}$

37) $\dfrac{10b^2}{30b+20} \cdot \dfrac{30b+20}{2b^2+10b}$

38) $\dfrac{35n^2-12n-32}{49n^2-91n+40} \cdot \dfrac{7n^2+16n-15}{5n+4}$

39) $\dfrac{7r^2-53r-24}{7r+2} \div \dfrac{49r+21}{49r+14}$

40) $\dfrac{12x+24}{10x^2+34x+28} \cdot \dfrac{15x+21}{5}$

41) $\dfrac{x^2-1}{2x-4} \cdot \dfrac{x^2-4}{x^2-x-2} \div \dfrac{x^2+x-2}{3x-6}$

42) $\dfrac{a^3+b^3}{a^2+3ab+2b^2} \cdot \dfrac{3a-6b}{3a^2-3ab+3b^2} \div \dfrac{a^2-4b^2}{a+2b}$

43) $\dfrac{x^2+3x+9}{x^2+x-12} \cdot \dfrac{x^2+2x-8}{x^3-27} \div \dfrac{x^2-4}{x^2-6x+9}$

44) $\dfrac{x^2+3x-10}{x^2+6x+5} \cdot \dfrac{2x^2-x-3}{2x^2+x-6} \div \dfrac{8x+20}{6x+15}$

Rational Expressions - Least Common Denominators

Objective: Idenfity the least common denominator and build up denominators to match this common denominator.

As with fractions, the least common denominator or LCD is very important to working with rational expressions. The process we use to find the LCD is based on the process used to find the LCD of intergers.

Example 338.

Find the LCD of 8 and 6	Consider multiples of the larger number
$8, 16, 24....$	24 is the first multiple of 8 that is also divisible by 6
24	Our Solution

When finding the LCD of several monomials we first find the LCD of the coefficients, then use all variables and attach the highest exponent on each variable.

Example 339.

Find the LCD of $4x^2y^5$ and $6x^4y^3z^6$

	First find the LCD of coefficients 4 and 6
12	12 is the LCD of 4 and 6
$x^4y^5z^6$	Use all variables with highest exponents on each variable
$12x^4y^5z^6$	Our Solution

The same pattern can be used on polynomials that have more than one term. However, we must first factor each polynomial so we can identify all the factors to be used (attaching highest exponent if necessary).

Example 340.

Find the LCD of $x^2 + 2x - 3$ and $x^2 - x - 12$	Factor each polynomial
$(x-1)(x+3)$ and $(x-4)(x+3)$	LCD uses all unique factors
$(x-1)(x+3)(x-4)$	Our Solution

Notice we only used $(x+3)$ once in our LCD. This is because it only appears as a factor once in either polynomial. The only time we need to repeat a factor or use an exponent on a factor is if there are exponents when one of the polynomials is factored

Example 341.

Find the LCD of $x^2 - 10x + 25$ and $x^2 - 14x + 45$

$$(x-5)^2 \text{ and } (x-5)(x-9) \qquad \text{Factor each polynomial}$$
$$\text{LCD uses all unique factors with highest exponent}$$
$$(x-5)^2(x-9) \qquad \text{Our Solution}$$

The previous example could have also been done with factoring the first polynomial to $(x - 5)(x - 5)$. Then we would have used $(x - 5)$ twice in the LCD because it showed up twice in one of the polynomials. However, it is the author's suggestion to use the exponents in factored form so as to use the same pattern (highest exponent) as used with monomials.

Once we know the LCD, our goal will be to build up fractions so they have matching denominators. In this lesson we will not be adding and subtracting fractions, just building them up to a common denominator. We can build up a fraction's denominator by multipliplying the numerator and denoinator by any factors that are not already in the denominator.

Example 342.

$$\frac{5a}{3a^2b} = \frac{?}{6a^5b^3} \qquad \text{Idenfity what factors we need to match denominators}$$

$$2a^3b^2 \qquad 3 \cdot 2 = 6 \text{ and we need three more } a's \text{ and two more } b's$$

$$\frac{5a}{3a^2b}\left(\frac{2a^3b^2}{2a^3b^2}\right) \qquad \text{Multiply numerator and denominator by this}$$

$$\frac{10a^4b^2}{6a^5b^3} \qquad \text{Our Solution}$$

Example 343.

$$\frac{x-2}{x+4} = \frac{?}{x^2+7x+12} \qquad \text{Factor to idenfity factors we need to match denominators}$$
$$(x+4)(x+3)$$

$$(x+3) \qquad \text{The missing factor}$$

$$\frac{x-2}{x+4}\left(\frac{x+3}{x+3}\right) \qquad \text{Multiply numerator and denominator by missing factor,}$$
$$\text{FOIL numerator}$$

$$\frac{x^2+x-6}{(x+4)(x+3)} \qquad \text{Our Solution}$$

As the above example illustrates, we will multiply out our numerators, but keep our denominators factored. The reason for this is to add and subtract fractions we will want to be able to combine like terms in the numerator, then when we reduce at the end we will want our denominators factored.

Once we know how to find the LCD and how to build up fractions to a desired denominator we can combine them together by finding a common denominator and building up those fractions.

Example 344.

Build up each fraction so they have a common denominator

$$\frac{5a}{4b^3c} \quad \text{and} \quad \frac{3c}{6a^2b} \qquad \text{First identify LCD}$$

$$12a^2b^3c \qquad \text{Determine what factors each fraction is missing}$$

$$\text{First: } 3a^2 \quad \text{Second: } 2b^2c \qquad \text{Multiply each fraction by missing factors}$$

$$\frac{5a}{4b^3c}\left(\frac{3a^2}{3a^2}\right) \quad \text{and} \quad \frac{3c}{6a^2b}\left(\frac{2b^2c}{2b^2c}\right)$$

$$\frac{15a^3}{12a^2b^3c} \quad \text{and} \quad \frac{6b^2c^2}{12a^2b^3c} \qquad \text{Our Solution}$$

Example 345.

Build up each fraction so they have a common denominator

$$\frac{5x}{x^2-5x-6} \quad \text{and} \quad \frac{x-2}{x^2+4x+3} \qquad \text{Factor to find LCD}$$

$$(x-6)(x+1) \qquad (x+1)(x+3) \qquad \text{Use factors to find LCD}$$

$$\text{LCD: } (x-6)(x+1)(x+3) \qquad \text{Identify which factors are missing}$$

$$\text{First: } (x+3) \quad \text{Second: } (x-6) \qquad \text{Multiply fractions by missing factors}$$

$$\frac{5x}{(x-6)(x+1)}\left(\frac{x+3}{x+3}\right) \quad \text{and} \quad \frac{x-2}{(x+1)(x+3)}\left(\frac{x-6}{x-6}\right) \qquad \text{Multiply numerators}$$

$$\frac{5x^2+15x}{(x-6)(x+1)(x+3)} \quad \text{and} \quad \frac{x^2-8x+12}{(x-6)(x+1)(x+3)} \qquad \text{Our Solution}$$

World View Note: When the Egyptians began working with fractions, they expressed all fractions as a sum of unit fraction. Rather than $\frac{4}{5}$, they would write the fraction as the sum, $\frac{1}{2}+\frac{1}{4}+\frac{1}{20}$. An interesting problem with this system is this is not a unique solution, $\frac{4}{5}$ is also equal to the sum $\frac{1}{3}+\frac{1}{5}+\frac{1}{6}+\frac{1}{10}$.

7.3 Practice - Least Common Denominator

Build up denominators.

1) $\frac{3}{8} = \frac{?}{48}$

2) $\frac{a}{5} = \frac{?}{5a}$

3) $\frac{a}{x} = \frac{?}{xy}$

4) $\frac{5}{2x^2} = \frac{?}{8x^3y}$

5) $\frac{2}{3a^3b^2c} = \frac{?}{9a^5b^2c^4}$

6) $\frac{4}{3a^5b^2c^4} = \frac{?}{9a^5b^2c^4}$

7) $\frac{2}{x+4} = \frac{?}{x^2\ 16}$

8) $\frac{x+1}{x\ 3} = \frac{?}{x^2\ 6x+9}$

9) $\frac{x-4}{x+2} = \frac{?}{x^2+5x+6}$

10) $\frac{x-6}{x+3} = \frac{?}{x^2-2x-15}$

Find Least Common Denominators

11) $2a^3, 6a^4b^2, 4a^3b^5$

12) $5x^2y, 25x^3y^5z$

13) $x^2-3x, x-3, x$

14) $4x-8, x-2, 4$

15) $x+2, x-4$

16) $x, x-7, x+1$

17) $x^2-25, x+5$

18) x^2-9, x^2-6x+9

19) x^2+3x+2, x^2+5x+6

20) $x^2-7x+10, x^2-2x-15, x^2+x-6$

Find LCD and build up each fraction

21) $\frac{3a}{5b^2}, \frac{2}{10a^3b}$

22) $\frac{3x}{x\ 4}, \frac{2}{x+2}$

23) $\frac{x+2}{x-3}, \frac{x\ 3}{x+2}$

24) $\frac{5}{x^2\ 6x}, \frac{2}{x}, \frac{3}{x-6}$

25) $\frac{x}{x^2\ 16}, \frac{3x}{x^2\ 8x+16}$

26) $\frac{5x+1}{x^2-3x-10}, \frac{4}{x-5}$

27) $\frac{x+1}{x^2-36}, \frac{2x+3}{x^2+12x+36}$

28) $\frac{3x+1}{x^2\ x\ 12}, \frac{2x}{x^2+4x+3}$

29) $\frac{4x}{x^2\ x\ 6}, \frac{x+2}{x-3}$

30) $\frac{3x}{x^2\ 6x+8}, \frac{x\ 2}{x^2+x\ 20}, \frac{5}{x^2+3x\ 10}$

Rational Expressions - Add & Subtract

Objective: Add and subtract rational expressions with and without common denominators.

Adding and subtracting rational expressions is identical to adding and subtracting with integers. Recall that when adding with a common denominator we add the numerators and keep the denominator. This is the same process used with rational expressions. Remember to reduce, if possible, your final answer.

Example 346.

$$\frac{x-4}{x^2-2x-8} + \frac{x+8}{x^2-2x-8}$$ Same denominator, add numerators, combine like terms

$$\frac{2x+4}{x^2-2x-8}$$ Factor numerator and denominator

$$\frac{2(x+2)}{(x+2)(x-4)}$$ Divide out $(x+2)$

$$\frac{2}{x-4}$$ Our Solution

Subtraction with common denominator follows the same pattern, though the subtraction can cause problems if we are not careful with it. To avoid sign errors we will first distribute the subtraction through the numerator. Then we can treat it like an addition problem. This process is the same as "add the opposite" we saw when subtracting with negatives.

Example 347.

$$\frac{6x-12}{3x-6} - \frac{15x-6}{3x-6}$$ Add the opposite of the second fraction (distribute negative)

$$\frac{6x-12}{3x-6} + \frac{-15x+6}{3x-6}$$ Add numerators, combine like terms

$$\frac{-9x-6}{3x-6}$$ Factor numerator and denominator

$$\frac{-3(3x+2)}{3(x-2)}$$ Divide out common factor of 3

$$\frac{-(3x+2)}{x-2}$$ Our Solution

World View Note: The Rhind papyrus of Egypt from 1650 BC gives some of the earliest known symbols for addition and subtraction, a pair of legs walking in the direction one reads for addition, and a pair of legs walking in the opposite direction for subtraction..

When we don't have a common denominator we will have to find the least common denominator (LCD) and build up each fraction so the denominators match. The following example shows this process with integers.

Example 348.

$$\frac{5}{6} + \frac{1}{4}$$ The LCD is 12. Build up, multiply 6 by 2 and 4 by 3

$$\left(\frac{2}{2}\right)\frac{5}{6} + \frac{1}{4}\left(\frac{3}{3}\right)$$ Multiply

$$\frac{10}{12} + \frac{3}{12}$$ Add numerators

$$\frac{13}{12} \quad \text{Our Solution}$$

The same process is used with variables.

Example 349.

$$\frac{7a}{3a^2b} + \frac{4b}{6ab^4} \qquad \text{The LCD is } 6a^2b^4. \text{ We will then build up each fraction}$$

$$\left(\frac{2b^3}{2b^3}\right)\frac{7a}{3a^2b} + \frac{4b}{6ab^4}\left(\frac{a}{a}\right) \qquad \text{Multiply first fraction by } 2b^3 \text{ and second by } a$$

$$\frac{14ab^3}{6a^2b^4} + \frac{4ab}{6a^2b^4} \qquad \text{Add numerators, no like terms to combine}$$

$$\frac{14ab^3 + 4ab}{6a^2b^4} \qquad \text{Factor numerator}$$

$$\frac{2ab(7b^3 + 2)}{6a^2b^4} \qquad \text{Reduce, dividing out factors } 2, a, \text{ and } b$$

$$\frac{7b^3 + 2}{3ab^3} \qquad \text{Our Solution}$$

The same process can be used for subtraction, we will simply add the first step of adding the opposite.

Example 350.

$$\frac{4}{5a} - \frac{7b}{4a^2} \qquad \text{Add the opposite}$$

$$\frac{4}{5a} + \frac{-7b}{4a^2} \qquad \text{LCD is } 20a^2. \text{ Build up denominators}$$

$$\left(\frac{4a}{4a}\right)\frac{4}{5a} + \frac{-7b}{4a^2}\left(\frac{5}{5}\right) \qquad \text{Multiply first fraction by } 4a, \text{ second by } 5$$

$$\frac{16a - 35b}{20a^2} \qquad \text{Our Solution}$$

If our denominators have more than one term in them we will need to factor first to find the LCD. Then we build up each denominator using the factors that are missing on each fraction.

Example 351.

$$\frac{6}{8a+4} + \frac{3a}{8} \qquad \text{Factor denominators to find LCD}$$
$$4(2a+1) \quad 8 \qquad \text{LCD is } 8(2a+1), \text{build up each fraction}$$

$$\left(\frac{2}{2}\right)\frac{6}{4(2a+1)} + \frac{3a}{8}\left(\frac{2a+1}{2a+1}\right) \qquad \text{Multiply first fraction by 2, second by } 2a+1$$

$$\frac{12}{8(2a+1)} + \frac{6a^2+3a}{8(2a+1)} \qquad \text{Add numerators}$$

$$\frac{6a^2+3a+12}{8(2a+1)} \qquad \text{Our Solution}$$

With subtraction remember to add the opposite.

Example 352.

$$\frac{x+1}{x-4} - \frac{x+1}{x^2-7x+12} \qquad \text{Add the opposite (distribute negative)}$$

$$\frac{x+1}{x-4} + \frac{-x-1}{x^2-7x+12} \qquad \text{Factor denominators to find LCD}$$
$$x-4 \quad (x-4)(x-3) \qquad \text{LCD is } (x-4)(x-3), \text{build up each fraction}$$

$$\left(\frac{x-3}{x-3}\right)\frac{x+1}{x-4} + \frac{-x-1}{x^2-7x+12} \qquad \text{Only first fraction needs to be multiplied by } x-3$$

$$\frac{x^2-2x-3}{(x-3)(x-4)} + \frac{-x-1}{(x-3)(x-4)} \qquad \text{Add numerators, combine like terms}$$

$$\frac{x^2-3x-4}{(x-3)(x-4)} \qquad \text{Factor numerator}$$

$$\frac{(x-4)(x+1)}{(x-3)(x-4)} \qquad \text{Divide out } x-4 \text{ factor}$$

$$\frac{x+1}{x-3} \qquad \text{Our Solution}$$

7.4 Practice - Add and Subtract

Add or subtract the rational expressions. Simplify your answers whenever possible.

1) $\dfrac{2}{a+3} + \dfrac{4}{a+3}$

2) $\dfrac{x^2}{x-2} - \dfrac{6x\ 8}{x-2}$

3) $\dfrac{t^2+4t}{t\ 1} + \dfrac{2t-7}{t\ 1}$

4) $\dfrac{a^2+3a}{a^2+5a-6} - \dfrac{4}{a^2+5a-6}$

5) $\dfrac{2x^2+3}{x^2-6x+5} - \dfrac{x^2-5x+9}{x^2-6x+5}$

6) $\dfrac{3}{x} + \dfrac{4}{x^2}$

7) $\dfrac{5}{6r} - \dfrac{5}{8r}$

8) $\dfrac{7}{xy^2} + \dfrac{3}{x^2y}$

9) $\dfrac{8}{9t^3} + \dfrac{5}{6t^2}$

10) $\dfrac{x+5}{8} + \dfrac{x-3}{12}$

11) $\dfrac{a+2}{2} - \dfrac{a\ 4}{4}$

12) $\dfrac{2a-1}{3a^2} + \dfrac{5a+1}{9a}$

13) $\dfrac{x\ 1}{4x} - \dfrac{2x+3}{x}$

14) $\dfrac{2c-d}{c^2d} - \dfrac{c+d}{cd^2}$

15) $\dfrac{5x+3y}{2x^2y} - \dfrac{3x+4y}{xy^2}$

16) $\dfrac{2}{x-1} + \dfrac{2}{x+1}$

17) $\dfrac{2z}{z\ 1} - \dfrac{3z}{z+1}$

18) $\dfrac{2}{x\ 5} + \dfrac{3}{4x}$

19) $\dfrac{8}{x^2\ 4} - \dfrac{3}{x+2}$

20) $\dfrac{4x}{x^2\ 25} + \dfrac{x}{x+5}$

21) $\dfrac{t}{t-3} - \dfrac{5}{4t-12}$

22) $\dfrac{2}{x+3} + \dfrac{4}{(x+3)^2}$

23) $\dfrac{2}{5x^2+5x} - \dfrac{4}{3x+3}$

24) $\dfrac{3a}{4a\ 20} + \dfrac{9a}{6a\ 30}$

25) $\dfrac{t}{y\ t} - \dfrac{y}{y+t}$

26) $\dfrac{x}{x-5} + \dfrac{x\ 5}{x}$

27) $\dfrac{x}{x^2+5x+6} - \dfrac{2}{x^2+3x+2}$

28) $\dfrac{2x}{x^2-1} - \dfrac{3}{x^2+5x+4}$

29) $\dfrac{x}{x^2+15x+56} - \dfrac{7}{x^2+13x+42}$

30) $\dfrac{2x}{x^2-9} + \dfrac{5}{x^2+x-6}$

31) $\dfrac{5x}{x^2-x-6} - \dfrac{18}{x^2-9}$

32) $\dfrac{4x}{x^2-2x-3} - \dfrac{3}{x^2-5x+6}$

33) $\dfrac{2x}{x^2-1} - \dfrac{4}{x^2+2x-3}$

34) $\dfrac{x-1}{x^2+3x+2} + \dfrac{x+5}{x^2+4x+3}$

35) $\dfrac{x+1}{x^2-2x-35} + \dfrac{x+6}{x^2+7x+10}$

36) $\dfrac{3x+2}{3x+6} + \dfrac{x}{4-x^2}$

37) $\dfrac{4\ a^2}{a^2-9} - \dfrac{a\ 2}{3\ a}$

38) $\dfrac{4y}{y^2\ 1} - \dfrac{2}{y} - \dfrac{2}{y+1}$

39) $\dfrac{2z}{1\ 2z} + \dfrac{3z}{2z+1} - \dfrac{3}{4z^2-1}$

40) $\dfrac{2r}{r^2-s^2} + \dfrac{1}{r+s} - \dfrac{1}{r-s}$

41) $\dfrac{2x-3}{x^2+3x+2} + \dfrac{3x-1}{x^2+5x+6}$

42) $\dfrac{x+2}{x^2-4x+3} + \dfrac{4x+5}{x^2+4x-5}$

43) $\dfrac{2x+7}{x^2-2x-3} - \dfrac{3x-2}{x^2+6x+5}$

44) $\dfrac{3x-8}{x^2+6x+8} + \dfrac{2x-3}{x^2+3x+2}$

Rational Expressions - Complex Fractions

Objective: Simplify complex fractions by multiplying each term by the least common denominator.

Complex fractions have fractions in either the numerator, or denominator, or usually both. These fractions can be simplified in one of two ways. This will be illustrated first with integers, then we will consider how the process can be expanded to include expressions with variables.

The first method uses order of operations to simplify the numerator and denominator first, then divide the two resulting fractions by multiplying by the reciprocal.

Example 353.

$$\frac{\frac{2}{3} - \frac{1}{4}}{\frac{5}{6} + \frac{1}{2}} \qquad \text{Get common denominator in top and bottom fractions}$$

$$\frac{\frac{8}{12} - \frac{3}{12}}{\frac{5}{6} + \frac{3}{6}} \qquad \text{Add and subtract fractions, reducing solutions}$$

$$\frac{\frac{5}{12}}{\frac{4}{3}} \qquad \text{To divide fractions we multiply by the reciprocal}$$

$$\left(\frac{5}{12}\right)\left(\frac{3}{4}\right) \qquad \text{Reduce}$$

$$\left(\frac{5}{4}\right)\left(\frac{1}{4}\right) \qquad \text{Multiply}$$

$$\frac{5}{16} \qquad \text{Our Solution}$$

The process above works just fine to simplify, but between getting common denominators, taking reciprocals, and reducing, it can be a very involved process. Generally we prefer a different method, to multiply the numerator and denominator of the large fraction (in effect each term in the complex fraction) by the least common denominator (LCD). This will allow us to reduce and clear the small fractions. We will simplify the same problem using this second method.

Example 354.

$$\frac{\frac{2}{3} - \frac{1}{4}}{\frac{5}{6} + \frac{1}{2}} \qquad \text{LCD is 12, multiply each term}$$

$$\frac{\frac{2(12)}{3} - \frac{1(12)}{4}}{\frac{5(12)}{6} + \frac{1(12)}{2}} \qquad \text{Reduce each fraction}$$

$$\frac{2(4) - 1(3)}{5(2) + 1(6)} \qquad \text{Multiply}$$

$$\frac{8 - 3}{10 + 6} \qquad \text{Add and subtract}$$

$$\frac{5}{16} \qquad \text{Our Solution}$$

Clearly the second method is a much cleaner and faster method to arrive at our solution. It is the method we will use when simplifying with variables as well. We will first find the LCD of the small fractions, and multiply each term by this LCD so we can clear the small fractions and simplify.

Example 355.

$$\frac{1 - \frac{1}{x^2}}{1 - \frac{1}{x}} \qquad \text{Identify LCD (use highest exponent)}$$

$$\text{LCD} = x^2 \qquad \text{Multiply each term by LCD}$$

$$\frac{1(x^2) - \frac{1(x^2)}{x^2}}{1(x^2) - \frac{1(x^2)}{x}} \qquad \text{Reduce fractions (subtract exponents)}$$

$$\frac{1(x^2) - 1}{1(x^2) - x} \qquad \text{Multiply}$$

$$\frac{x^2 - 1}{x^2 - x} \qquad \text{Factor}$$

$$\frac{(x + 1)(x - 1)}{x(x - 1)} \qquad \text{Divide out } (x - 1) \text{ factor}$$

$$\frac{x + 1}{x} \qquad \text{Our Solution}$$

The process is the same if the LCD is a binomial, we will need to distribute

$$\frac{\frac{3}{x+4} - 2}{5 + \frac{2}{x+4}} \qquad \text{Multiply each term by LCD, } (x + 4)$$

$$\frac{\frac{3(x+4)}{x+4} - 2(x+4)}{5(x+4) + \frac{2(x+4)}{x+4}} \qquad \text{Reduce fractions}$$

$$\frac{3 - 2(x+4)}{5(x+4) + 2} \qquad \text{Distribute}$$

$$\frac{3 - 2x - 8}{5x + 20 + 2} \qquad \text{Combine like terms}$$

$$\frac{-2x - 5}{5x + 22} \qquad \text{Our Solution}$$

The more fractions we have in our problem, the more we repeat the same process.

Example 356.

$$\frac{\frac{2}{ab^2} - \frac{3}{ab^3} + \frac{1}{ab}}{\frac{4}{a^2b} + ab - \frac{1}{ab}} \qquad \text{Idenfity LCD (highest exponents)}$$

$$LCD = a^2b^3 \qquad \text{Multiply each term by LCD}$$

$$\frac{\frac{2(a^2b^3)}{ab^2} - \frac{3(a^2b^3)}{ab^3} + \frac{1(a^2b^3)}{ab}}{\frac{4(a^2b^3)}{a^2b} + ab(a^2b^3) - \frac{1(a^2b^3)}{ab}} \qquad \text{Reduce each fraction (subtract exponents)}$$

$$\frac{2ab - 3a + ab^2}{4b^2 + a^3b^4 - ab^2} \qquad \text{Our Solution}$$

World View Note: Sophie Germain is one of the most famous women in mathematics, many primes, which are important to finding an LCD, carry her name. Germain primes are prime numbers where one more than double the prime number is also prime, for example 3 is prime and so is $2 \cdot 3 + 1 = 7$ prime. The largest known Germain prime (at the time of printing) is $183027 \cdot 2^{265440} - 1$ which has 79911 digits!

Some problems may require us to FOIL as we simplify. To avoid sign errors, if there is a binomial in the numerator, we will first distribute the negative through the numerator.

Example 357.

$$\frac{\frac{x}{x+3} - \frac{3}{x} - \frac{x+3}{3}}{\frac{x-3}{x+3} + \frac{x+3}{x} \cdot 3} \qquad \text{Distribute the subtraction to numerator}$$

$$\frac{\frac{x-3}{x+3} + \frac{-x-3}{x} \cdot 3}{\frac{x}{x+3} \cdot 3 + \frac{x+3}{x-3}} \qquad \text{Identify LCD}$$

264

$$\text{LCD} = (x+3)(x-3) \qquad \text{Multiply each term by LCD}$$

$$\cfrac{\dfrac{(x-3)(x+3)(x-3)}{x+3} + \dfrac{(-x-3)(x+3)(x-3)}{x \quad 3}}{\dfrac{(x \quad 3)(x+3)(x \quad 3)}{x+3} + \dfrac{(x+3)(x+3)(x \quad 3)}{x-3}} \qquad \text{Reduce fractions}$$

$$\frac{(x-3)(x-3) + (-x-3)(x+3)}{(x-3)(x-3) + (x+3)(x+3)} \qquad \text{FOIL}$$

$$\frac{x^2 - 6x + 9 - x^2 - 6x - 9}{x^2 - 6x + 9 + x^2 + 6x - 9} \qquad \text{Combine like terms}$$

$$\frac{-12x}{2x^2 + 18} \qquad \text{Factor out 2 in denominator}$$

$$\frac{-12x}{2(x^2 + 9)} \qquad \text{Divide out common factor 2}$$

$$\frac{-6x}{x^2 - 9} \qquad \text{Our Solution}$$

If there are negative exponents in an expression we will have to first convert these negative exponents into fractions. Remember, the exponent is only on the factor it is attached to, not the whole term.

Example 358.

$$\frac{m^{-2} + 2m^{-1}}{m + 4m^{-2}} \qquad \text{Make each negative exponent into } a \text{ fraction}$$

$$\cfrac{\dfrac{1}{m^2} + \dfrac{2}{m}}{m + \dfrac{4}{m^2}} \qquad \text{Multiply each term by LCD, } m^2$$

$$\cfrac{\dfrac{1(m^2)}{m^2} + \dfrac{2(m^2)}{m}}{m(m^2) + \dfrac{4(m^2)}{m^2}} \qquad \text{Reduce the fractions}$$

$$\frac{1 + 2m}{m^3 + 4} \qquad \text{Our Solution}$$

Once we convert each negative exponent into a fraction, the problem solves exactly like the other complex fraction problems.

265

7.5 Practice - Complex Fractions

Solve.

1) $\dfrac{1+\frac{1}{x}}{1-\frac{1}{x^2}}$

2) $\dfrac{\frac{1}{y^2}-1}{1+\frac{1}{y}}$

3) $\dfrac{a-2}{\frac{4}{a}-a}$

4) $\dfrac{\frac{25}{a}-a}{5+a}$

5) $\dfrac{\frac{1}{a^2}-\frac{1}{a}}{\frac{1}{a^2}+\frac{1}{a}}$

6) $\dfrac{\frac{1}{b}+\frac{1}{2}}{\frac{4}{b^2}\ \ 1}$

7) $\dfrac{2-\frac{4}{x+2}}{5-\frac{10}{x+2}}$

8) $\dfrac{4+\frac{12}{2x\ \ 3}}{5+\frac{15}{2x-3}}$

9) $\dfrac{\frac{3}{2a-3}+2}{\frac{6}{2a\ \ 3}-4}$

10) $\dfrac{\frac{5}{b-5}-3}{\frac{10}{b\ \ 5}+6}$

11) $\dfrac{\frac{x}{x+1}-\frac{1}{x}}{\frac{x}{x+1}+\frac{1}{x}}$

12) $\dfrac{\frac{2a}{a\ \ 1}-\frac{3}{a}}{\frac{6}{a-1}-4}$

13) $\dfrac{\frac{3}{x}}{\frac{9}{x^2}}$

14) $\dfrac{\frac{x}{3x\ \ 2}}{\frac{x}{9x^2\ \ 4}}$

15) $\dfrac{\frac{a^2\ \ b^2}{4a^2b}}{\frac{a+b}{16ab^2}}$

16) $\dfrac{1-\frac{1}{x}-\frac{6}{x^2}}{1-\frac{4}{x}+\frac{3}{x^2}}$

17) $\dfrac{1-\frac{3}{x}-\frac{10}{x^2}}{1+\frac{11}{x}+\frac{18}{x^2}}$

18) $\dfrac{\frac{15}{x^2}-\frac{2}{x}-1}{\frac{4}{x^2}-\frac{5}{x}+4}$

19) $\dfrac{1-\frac{2x}{3x-4}}{x-\frac{32}{3x-4}}$

20) $\dfrac{1-\frac{12}{3x+10}}{x-\frac{8}{3x+10}}$

21) $\dfrac{x-1+\frac{2}{x\ \ 4}}{x+3+\frac{6}{x-4}}$

22) $\dfrac{x-5-\frac{18}{x+2}}{x+7+\frac{6}{x+2}}$

23) $\dfrac{x-4+\dfrac{9}{2x+3}}{x+3-\dfrac{5}{2x+3}}$

24) $\dfrac{\dfrac{1}{a}-\dfrac{3}{a-2}}{\dfrac{2}{a}+\dfrac{5}{a-2}}$

25) $\dfrac{\dfrac{2}{b}-\dfrac{5}{b+3}}{\dfrac{3}{b}+\dfrac{3}{b+3}}$

26) $\dfrac{\dfrac{1}{y^2}-\dfrac{1}{xy}-\dfrac{2}{x^2}}{\dfrac{1}{y^2}-\dfrac{3}{xy}+\dfrac{2}{x^2}}$

27) $\dfrac{\dfrac{2}{b^2}-\dfrac{5}{ab}-\dfrac{3}{a^2}}{\dfrac{2}{b^2}+\dfrac{7}{ab}+\dfrac{3}{a^2}}$

28) $\dfrac{\dfrac{x-1}{x+1}-\dfrac{x+1}{x-1}}{\dfrac{x-1}{x+1}+\dfrac{x+1}{x-1}}$

29) $\dfrac{\dfrac{y}{y+2}-\dfrac{y}{y-2}}{\dfrac{y}{y+2}+\dfrac{y}{y-2}}$

30) $\dfrac{\dfrac{x+1}{x-1}-\dfrac{1-x}{1+x}}{\dfrac{1}{(x+1)^2}+\dfrac{1}{(x-1)^2}}$

Simplify each of the following fractional expressions.

31) $\dfrac{x^{-2}-y^{-2}}{x^{-1}+y^{-1}}$

32) $\dfrac{x^{-2}y+xy^{-2}}{x^{-2}-y^{-2}}$

33) $\dfrac{x^{-3}y-xy^{-3}}{x^{-2}-y^{-2}}$

34) $\dfrac{4-4x^{-1}+x^{-2}}{4-x^{-2}}$

35) $\dfrac{x^{-2}-6x^{-1}+9}{x^2-9}$

36) $\dfrac{x^{-3}+y^{-3}}{x^{-2}-x^{-1}y^{-1}+y^{-2}}$

Rational Expressions - Proportions

Objective: Solve proportions using the cross product and use proportions to solve application problems

When two fractions are equal, they are called a proportion. This definition can be generalized to two equal rational expressions. Proportions have an important property called the cross-product.

$$\text{Cross Product: If } \frac{a}{b} = \frac{c}{d} \text{ then ad} = \text{bc}$$

The cross product tells us we can multiply diagonally to get an equation with no fractions that we can solve.

Example 359.

$$\frac{20}{6} = \frac{x}{9} \qquad \text{Calculate cross product}$$

$$(20)(9) = 6x \quad \text{Multiply}$$
$$\frac{180 = 6x}{6 \quad \ 6} \quad \text{Divide both sides by 6}$$
$$30 = x \quad \text{Our Solution}$$

World View Note: The first clear definition of a proportion and the notation for a proportion came from the German Leibniz who wrote, "I write $dy: x = \mathrm{dt}: a$; for dy is to x as dt is to a, is indeed the same as, dy divided by x is equal to dt divided by a. From this equation follow then all the rules of proportion."

If the proportion has more than one term in either numerator or denominator, we will have to distribute while calculating the cross product.

Example 360.

$$\frac{x+3}{4} = \frac{2}{5} \quad \text{Calculate cross product}$$
$$5(x+3) = (4)(2) \quad \text{Multiply and distribute}$$
$$5x + 15 = 8 \quad \text{Solve}$$
$$\frac{-15 \ -15}{} \quad \text{Subtract 15 from both sides}$$
$$\frac{5x = -7}{5 \quad \ 5} \quad \text{Divide both sides by 5}$$
$$x = -\frac{7}{5} \quad \text{Our Solution}$$

This same idea can be seen when the variable appears in several parts of the proportion.

Example 361.

$$\frac{4}{x} = \frac{6}{3x+2} \quad \text{Calculate cross product}$$
$$4(3x+2) = 6x \quad \text{Distribute}$$
$$12x + 8 = 6x \quad \text{Move variables to one side}$$
$$\frac{-12x \quad -12x}{} \quad \text{Subtract } 12x \text{ from both sides}$$
$$\frac{8 = -6x}{-6 \quad -6} \quad \text{Divide both sides by } -6$$
$$-\frac{4}{3} = x \quad \text{Our Solution}$$

Example 362.

$$\frac{2x-3}{7x+4} = \frac{2}{5} \qquad \text{Calculate cross product}$$

$$5(2x-3) = 2(7x+4) \qquad \text{Distribute}$$

$$10x - 15 = 14x + 8 \qquad \text{Move variables to one side}$$

$$\underline{-10x \qquad\quad -10x} \qquad \text{Subtract } 10x \text{ from both sides}$$

$$-15 = 4x + 8 \qquad \text{Subtract } 8 \text{ from both sides}$$

$$\underline{-8 \qquad\quad -8}$$

$$\frac{-23 = 4x}{4 \quad\ 4} \qquad \text{Divide both sides by } 4$$

$$-\frac{23}{4} = x \qquad \text{Our Solution}$$

As we solve proportions we may end up with a quadratic that we will have to solve. We can solve this quadratic in the same way we solved quadratics in the past, either factoring, completing the square or the quadratic formula. As with solving quadratics before, we will generally end up with two solutions.

Example 363.

$$\frac{k+3}{3} = \frac{8}{k-2} \qquad \text{Calculate cross product}$$

$$(k+3)(k-2) = (8)(3) \qquad \text{FOIL and multiply}$$

$$k^2 + k - 6 = 24 \qquad \text{Make equation equal zero}$$

$$\underline{\quad -24 - 24} \qquad \text{Subtract } 24 \text{ from both sides}$$

$$k^2 + k - 30 = 0 \qquad \text{Factor}$$

$$(k+6)(k-5) = 0 \qquad \text{Set each factor equal to zero}$$

$$k+6 = 0 \ \text{ or } \ k - 5 = 0 \qquad \text{Solve each equation}$$

$$\underline{-6-6 \qquad\quad +5 = 5} \qquad \text{Add or subtract}$$

$$k = -6 \ \text{ or } \ k = 5 \qquad \text{Our Solutions}$$

Proportions are very useful in how they can be used in many different types of applications. We can use them to compare different quantities and make conclusions about how quantities are related. As we set up these problems it is important to remember to stay organized, if we are comparing dogs and cats, and the number of dogs is in the numerator of the first fraction, then the numerator of the second fraction should also refer to the dogs. This consistency of the numerator and denominator is essential in setting up our proportions.

Example 364.

A six foot tall man casts a shadow that is 3.5 feet long. If the shadow of a flag pole is 8 feet long, how tall is the flag pole?

$$\frac{\text{shadow}}{\text{height}} \qquad \text{We will put shadows in numerator, heights in denomintor}$$

$$\frac{3.5}{6}$$ The man has *a* shadow of 3.5 feet and *a* height of 6 feet

$$\frac{8}{x}$$ The flagpole has *a* shadow of 8 feet, but we don't know the height

$$\frac{3.5}{6} = \frac{8}{x}$$ This gives us our proportion, calculate cross product

$3.5x = (8)(6)$ Multiply

$3.5x = 48$ Divide both sides by 3.5

$$\overline{3.5} \quad \overline{3.5}$$

$x = 13.7\,\text{ft}$ Our Solution

Example 365.

In a basketball game, the home team was down by 9 points at the end of the game. They only scored 6 points for every 7 points the visiting team scored. What was the final score of the game?

$$\frac{\text{home}}{\text{visiter}}$$ We will put home in numerator, visitor in denominator

$$\frac{x-9}{x}$$ Don't know visitor score, but home is 9 points less

$$\frac{6}{7}$$ Home team scored 6 for every 7 the visitor scored

$$\frac{x-9}{x} = \frac{6}{7}$$ This gives our proportion, calculate the cross product

$7(x-9) = 6x$ Distribute

$7x - 63 = 6x$ Move variables to one side

$-7x \qquad -7x$ Subtract $7x$ from both sides

$-63 = -x$ Divide both sides by -1

$$\overline{-1} \quad \overline{-1}$$

$63 = x$ We used x for the visitor score.

$63 - 9 = 54$ Subtract 9 to get the home score

$63\,\text{to}\,54$ Our Solution

7.6 Practice - Proportions

Solve each proportion.

1) $\frac{10}{a} = \frac{6}{8}$

2) $\frac{7}{9} = \frac{n}{6}$

3) $\frac{7}{6} = \frac{2}{k}$

4) $\frac{8}{x} = \frac{4}{8}$

5) $\frac{6}{x} = \frac{8}{2}$

6) $\frac{n-10}{8} = \frac{9}{3}$

7) $\frac{m\ 1}{5} = \frac{8}{2}$

8) $\frac{8}{5} = \frac{3}{x-8}$

9) $\frac{2}{9} = \frac{10}{p-4}$

10) $\frac{9}{n+2} = \frac{3}{9}$

11) $\frac{b\ 10}{7} = \frac{b}{4}$

12) $\frac{9}{4} = \frac{r}{r-4}$

13) $\frac{x}{5} = \frac{x+2}{9}$

14) $\frac{n}{8} = \frac{n\ 4}{3}$

15) $\frac{3}{10} = \frac{a}{a+2}$

16) $\frac{x+1}{9} = \frac{x+2}{2}$

17) $\frac{v\ 5}{v+6} = \frac{4}{9}$

18) $\frac{n+8}{10} = \frac{n-9}{4}$

19) $\frac{7}{x-1} = \frac{4}{x-6}$

20) $\frac{k+5}{k\ 6} = \frac{8}{5}$

21) $\frac{x+5}{5} = \frac{6}{x\ 2}$

22) $\frac{4}{x-3} = \frac{x+5}{5}$

23) $\frac{m+3}{4} = \frac{11}{m-4}$

24) $\frac{x-5}{8} = \frac{4}{x-1}$

25) $\frac{2}{p+4} = \frac{p+5}{3}$

26) $\frac{5}{n+1} = \frac{n-4}{10}$

27) $\frac{n+4}{3} = \frac{-3}{n\ 2}$

28) $\frac{1}{n+3} = \frac{n+2}{2}$

29) $\frac{3}{x+4} = \frac{x+2}{5}$

30) $\frac{x\ 5}{4} = \frac{3}{x+3}$

Answer each question. Round your answer to the nearest tenth. Round dollar amounts to the nearest cent.

31) The currency in Western Samoa is the Tala. The exchange rate is approximately $0.70 to 1 Tala. At this rate, how many dollars would you get if you exchanged 13.3 Tala?

32) If you can buy one plantain for $0.49 then how many can you buy with $7.84?

33) Kali reduced the size of a painting to a height of 1.3 in. What is the new width if it was originally 5.2 in. tall and 10 in. wide?

34) A model train has a scale of 1.2 in : 2.9 ft. If the model train is 5 in tall then how tall is the real train?

35) A bird bath that is 5.3 ft tall casts a shadow that is 25.4 ft long. Find the length of the shadow that a 8.2 ft adult elephant casts.

36) Victoria and Georgetown are 36.2 mi from each other. How far apart would the cities be on a map that has a scale of 0.9 in : 10.5 mi?

37) The Vikings led the Timberwolves by 19 points at the half. If the Vikings scored 3 points for every 2 points the Timberwolves scored, what was the half time score?

38) Sarah worked 10 more hours than Josh. If Sarah worked 7 hr for every 2 hr Josh worked, how long did they each work?

39) Computer Services Inc. charges $8 more for a repair than Low Cost Computer Repair. If the ratio of the costs is 3 : 6, what will it cost for the repair at Low Cost Computer Repair?

40) Kelsey's commute is 15 minutes longer than Christina's. If Christina drives 12 minutes for every 17 minutes Kelsey drives, how long is each commute?

Rational Expressions - Solving Rational Equations

Objective: Solve rational equations by identifying and multiplying by the least common denominator.

When solving equations that are made up of rational expressions we will solve them using the same strategy we used to solve linear equations with fractions. When we solved problems like the next example, we cleared the fraction by multiplying by the least common denominator (LCD)

Example 366.

$$\frac{2}{3}x - \frac{5}{6} = \frac{3}{4} \qquad \text{Multiply each term by LCD, 12}$$

$$\frac{2(12)}{3}x - \frac{5(12)}{6} = \frac{3(12)}{4} \qquad \text{Reduce fractions}$$

$$2(4)x - 5(2) = 3(3) \qquad \text{Multiply}$$
$$8x - 10 = 9 \qquad \text{Solve}$$
$$\underline{+10 + 10} \qquad \text{Add 10 to both sides}$$
$$\frac{8x = 19}{8 \quad 8} \qquad \text{Divide both sides by 8}$$
$$x = \frac{19}{8} \qquad \text{Our Solution}$$

We will use the same process to solve rational equations, the only difference is our

LCD will be more involved. We will also have to be aware of domain issues. If our LCD equals zero, the solution is undefined. We will always check our solutions in the LCD as we may have to remove a solution from our solution set.

Example 367.

$$\frac{5x+5}{x+2} + 3x = \frac{x^2}{x+2} \qquad \text{Multiply each term by LCD, } (x+2)$$

$$\frac{(5x+5)(x+2)}{x+2} + 3x(x+2) = \frac{x^2(x+2)}{x+2} \qquad \text{Reduce fractions}$$

$$5x+5+3x(x+2) = x^2 \qquad \text{Distribute}$$
$$5x+5+3x^2+6x = x^2 \qquad \text{Combine like terms}$$
$$3x^2+11x+5 = x^2 \qquad \text{Make equation equal zero}$$
$$\underline{-x^2 \qquad\qquad -x^2} \qquad \text{Subtract } x^2 \text{ from both sides}$$
$$2x^2+11x+5 = 0 \qquad \text{Factor}$$
$$(2x+1)(x+5) = 0 \qquad \text{Set each factor equal to zero}$$
$$2x+1 = 0 \quad \text{or} \quad x+5 = 0 \qquad \text{Solve each equation}$$
$$\underline{-1-1 \qquad\qquad -5-5}$$
$$\frac{2x = -1}{2} \quad \text{or} \quad x = -5$$
$$x = -\frac{1}{2} \quad \text{or} \quad -5 \qquad \text{Check solutions, LCD can't be zero}$$
$$-\frac{1}{2} + 2 = \frac{3}{2} \qquad -5+2 = -3 \qquad \text{Neither make LCD zero, both are solutions}$$
$$x = -\frac{1}{2} \quad \text{or} \quad -5 \qquad \text{Our Solution}$$

The LCD can be several factors in these problems. As the LCD gets more complex, it is important to remember the process we are using to solve is still the same.

Example 368.

$$\frac{x}{x+2} + \frac{1}{x+1} = \frac{5}{(x+1)(x+2)} \qquad \text{Multiply terms by LCD, } (x+1)(x+2)$$

$$\frac{x(x+1)(x+2)}{x+2} + \frac{1(x+1)(x+2)}{x+1} = \frac{5(x+1)(x+2)}{(x+1)(x+2)} \qquad \text{Reduce fractions}$$

$$x(x+1)+1(x+2)=5 \qquad \text{Distribute}$$
$$x^2+x+x+2=5 \qquad \text{Combine like terms}$$
$$x^2+2x+2=5 \qquad \text{Make equatino equal zero}$$
$$\underline{-5-5} \qquad \text{Subtract 6 from both sides}$$
$$x^2+2x-3=0 \qquad \text{Factor}$$
$$(x+3)(x-1)=0 \qquad \text{Set each factor equal to zero}$$
$$x+3=0 \ \text{ or } \ x-1=0 \qquad \text{Solve each equation}$$
$$\underline{-3-3} \qquad \underline{+1+1}$$
$$x=-3 \ \text{ or } \ x=1 \qquad \text{Check solutions, LCD can't be zero}$$
$$(-3+1)(-3+2)=(-2)(-1)=2 \qquad \text{Check}-3 \text{ in } (x+1)(x+2), \text{ it works}$$
$$(1+1)(1+2)=(2)(3)=6 \qquad \text{Check 1 in } (x+1)(x+2), \text{ it works}$$
$$x=-3 \ \text{ or } \ 1 \qquad \text{Our Solution}$$

In the previous example the denominators were factored for us. More often we will need to factor before finding the LCD

Example 369.

$$\frac{x}{x-1}-\frac{1}{x-2}=\frac{11}{x^2-3x+2} \qquad \text{Factor denominator}$$
$$(x-1)(x-2)$$
$$\text{LCD}=(x-1)(x-2) \qquad \text{Identify LCD}$$

$$\frac{x(x-1)(x-2)}{x-1}-\frac{1(x-1)(x-2)}{x-2}=\frac{11(x-1)(x-2)}{(x-1)(x-2)} \qquad \text{Multiply each term by LCD, reduce}$$

$$x(x-2)-1(x-1)=11 \qquad \text{Distribute}$$
$$x^2-2x-x+1=11 \qquad \text{Combine like terms}$$
$$x^2-3x+1=11 \qquad \text{Make equation equal zero}$$
$$\underline{-11-11} \qquad \text{Subtract 11 from both sides}$$
$$x^2-3x-10=0 \qquad \text{Factor}$$
$$(x-5)(x+2)=0 \qquad \text{Set each factor equal to zero}$$
$$x-5=0 \ \text{ or } \ x+2=0 \qquad \text{Solve each equation}$$
$$\underline{+5+5} \qquad \underline{-2-2}$$
$$x=5 \ \text{ or } \ x=-2 \qquad \text{Check answers, LCD can't be 0}$$
$$(5-1)(5-2)=(4)(3)=12 \qquad \text{Check 5 in } (x-1)(x-2), \text{ it works}$$
$$(-2-1)(-2-2)=(-3)(-4)=12 \qquad \text{Check}-2 \text{ in } (x-1)(x-2), \text{ it works}$$

$$x = 5 \text{ or } -2 \quad \text{Our Solution}$$

World View Note: Maria Agnesi was the first women to publish a math textbook in 1748, it took her over 10 years to write! This textbook covered everything from arithmetic thorugh differential equations and was over 1,000 pages!

If we are subtracting a fraction in the problem, it may be easier to avoid a future sign error by first distributing the negative through the numerator.

Example 370.

$$\frac{x-2}{x-3} - \frac{x+2}{x+2} = \frac{5}{8} \quad \text{Distribute negative through numerator}$$

$$\frac{x-2}{x-3} + \frac{-x-2}{x+2} = \frac{5}{8} \quad \text{Identify LCD}, 8(x-3)(x+2), \text{multiply each term}$$

$$\frac{(x-2)8(x-3)(x+2)}{x-3} + \frac{(-x-2)8(x-3)(x+2)}{x+2} = \frac{5\cdot 8(x-3)(x+2)}{8} \quad \text{Reduce}$$

$$8(x-2)(x+2) + 8(-x-2)(x-3) = 5(x-3)(x+2) \quad \text{FOIL}$$

$$8(x^2-4) + 8(-x^2+x+6) = 5(x^2-x-6) \quad \text{Distribute}$$

$$8x^2 - 32 - 8x^2 + 8x + 48 = 5x^2 - 5x - 30 \quad \text{Combine like terms}$$

$$8x + 16 = 5x^2 - 5x - 30 \quad \text{Make equation equal zero}$$

$$\underline{-8x - 16 \qquad\qquad -8x - 16} \quad \text{Subtract } 8x \text{ and } 16$$

$$0 = 5x^2 - 13x - 46 \quad \text{Factor}$$

$$0 = (5x - 23)(x + 2) \quad \text{Set each factor equal to zero}$$

$$5x - 23 = 0 \text{ or } x + 2 = 0 \quad \text{Solve each equation}$$

$$\underline{+23 + 23 \qquad\qquad -2 - 2}$$

$$\frac{5x = 23}{5 \quad 5} \text{ or } x = -2$$

$$x = \frac{23}{5} \text{ or } -2 \quad \text{Check solutions, LCD can't be 0}$$

$$8\left(\frac{23}{5} - 3\right)\left(\frac{23}{5} + 2\right) = 8\left(\frac{8}{5}\right)\left(\frac{33}{5}\right) = \frac{2112}{25} \quad \text{Check } \frac{23}{5} \text{ in } 8(x-3)(x+2), \text{it works}$$

$$8(-2-3)(-2+2) = 8(-5)(0) = 0 \quad \text{Check} -2 \text{ in } 8(x-3)(x+2), \text{can't be 0!}$$

$$x = \frac{23}{5} \quad \text{Our Solution}$$

In the previous example, one of the solutions we found made the LCD zero. When this happens we ignore this result and only use the results that make the rational expressions defined.

7.7 Practice - Solving Rational Equations

Solve the following equations for the given variable:

1) $3x - \frac{1}{2} - \frac{1}{x} = 0$

2) $x + 1 = \frac{4}{x+1}$

3) $x + \frac{20}{x\ 4} = \frac{5x}{x\ 4} - 2$

4) $\frac{x^2+6}{x-1} + \frac{x\ 2}{x-1} = 2x$

5) $x + \frac{6}{x\ 3} = \frac{2x}{x\ 3}$

6) $\frac{x\ 4}{x\ 1} = \frac{12}{3\ x} + 1$

7) $\frac{2x}{3x\ 4} = \frac{4x+5}{6x\ 1} - \frac{3}{3x\ 4}$

8) $\frac{6x+5}{2x^2\ 2x} - \frac{2}{1\ x^2} = \frac{3x}{x^2\ 1}$

9) $\frac{3m}{2m-5} - \frac{7}{3m+1} = \frac{3}{2}$

10) $\frac{4x}{2x\ 6} - \frac{4}{5x\ 15} = \frac{1}{2}$

11) $\frac{4\ x}{1-x} = \frac{12}{3-x}$

12) $\frac{7}{3\ x} + \frac{1}{2} = \frac{3}{4\ x}$

13) $\frac{7}{y-3} - \frac{1}{2} = \frac{y\ 2}{y-4}$

14) $\frac{2}{3-x} - \frac{6}{8-x} = 1$

15) $\frac{1}{x+2} - \frac{1}{2\ x} = \frac{3x+8}{x^2-4}$

16) $\frac{x+2}{3x-1} - \frac{1}{x} = \frac{3x\ 3}{3x^2-x}$

17) $\frac{x+1}{x-1} - \frac{x\ 1}{x+1} = \frac{5}{6}$

18) $\frac{x\ 1}{x-3} + \frac{x+2}{x+3} = \frac{3}{4}$

19) $\frac{3}{2x+1} + \frac{2x+1}{1-2x} = 1 - \frac{8x^2}{4x^2\ 1}$

20) $\frac{3x\ 5}{5x\ 5} + \frac{5x\ 1}{7x\ 7} - \frac{x\ 4}{1\ x} = 2$

21) $\frac{x\ 2}{x+3} - \frac{1}{x\ 2} = \frac{1}{x^2+x\ 6}$

22) $\frac{x\ 1}{x\ 2} + \frac{x+4}{2x+1} = \frac{1}{2x^2\ 3x\ 2}$

23) $\frac{3}{x+2} + \frac{x\ 1}{x+5} = \frac{5x+20}{6x+24}$

24) $\frac{x}{x+3} - \frac{4}{x\ 2} = \frac{-5x^2}{x^2+x-6}$

25) $\frac{x}{x-1} - \frac{2}{x+1} = \frac{4x^2}{x^2\ 1}$

26) $\frac{2x}{x+2} + \frac{2}{x-4} = \frac{3x}{x^2\ 2x\ 8}$

27) $\frac{2x}{x+1} - \frac{3}{x+5} = \frac{-8x^2}{x^2+6x+5}$

28) $\frac{x}{x+1} - \frac{3}{x+3} = \frac{-2x^2}{x^2+4x+3}$

29) $\frac{x\ 5}{x-9} + \frac{x+3}{x-3} = \frac{4x^2}{x^2\ 12x+27}$

30) $\frac{x\ 3}{x+6} + \frac{x\ 2}{x-3} = \frac{x^2}{x^2+3x\ 18}$

31) $\frac{x-3}{x\ 6} + \frac{x+5}{x+3} = \frac{-2x^2}{x^2-3x-18}$

32) $\frac{x+3}{x\ 2} + \frac{x-2}{x+1} = \frac{9x^2}{x^2-x-2}$

33) $\frac{4x+1}{x+3} + \frac{5x\ 3}{x-1} = \frac{8x^2}{x^2+2x\ 3}$

34) $\frac{3x\ 1}{x+6} - \frac{2x\ 3}{x-3} = \frac{3x^2}{x^2+3x\ 18}$

Rational Expressions - Dimensional Analysis

Objective: Use dimensional analysis to preform single unit, dual unit, square unit, and cubed unit conversions.

One application of rational expressions deals with converting units. When we convert units of measure we can do so by multiplying several fractions together in a process known as dimensional analysis. The trick will be to decide what fractions to multiply. When multiplying, if we multiply by 1, the value of the expression does not change. One written as a fraction can look like many different things as long as the numerator and denominator are identical in value. Notice the numerator and denominator are not identical in appearance, but rather identical in value. Below are several fractions, each equal to one where numerator and denominator are identical in value.

$$\frac{1}{1} = \frac{4}{4} = \frac{\frac{1}{2}}{\frac{2}{4}} = \frac{100\,\text{cm}}{1\,m} = \frac{1\,\text{lb}}{16\,\text{oz}} = \frac{1\,\text{hr}}{60\,\text{min}} = \frac{60\,\text{min}}{1\,\text{hr}}$$

The last few fractions that include units are called conversion factors. We can make a conversion factor out of any two measurements that represent the same distance. For example, 1 mile = 5280 feet. We could then make a conversion factor $\frac{1\,\text{mi}}{5280\,\text{ft}}$ because both values are the same, the fraction is still equal to one. Similarly we could make a conversion factor $\frac{5280\,\text{ft}}{1\,\text{mi}}$. The trick for conversions will be to use the correct fractions.

The idea behind dimensional analysis is we will multiply by a fraction in such a way that the units we don't want will divide out of the problem. We found out when multiplying rational expressions that if a variable appears in the numerator and denominator we can divide it out of the expression. It is the same with units. Consider the following conversion.

Example 371.

$$17.37 \text{ miles to feet} \quad \text{Write } 17.37 \text{ miles as } a \text{ fraction, put it over } 1$$

$$\left(\frac{17.37 \, \text{mi}}{1}\right) \quad \text{To divide out the miles we need miles in the denominator}$$

$$\left(\frac{17.37 \, \text{mi}}{1}\right)\left(\frac{?? \, \text{ft}}{?? \, \text{mi}}\right) \quad \text{We are converting to feet, so this will go in the numerator}$$

$$\left(\frac{17.37 \, \text{mi}}{1}\right)\left(\frac{5280 \, \text{ft}}{1 \, \text{mi}}\right) \quad \text{Fill in the relationship described above, } 1 \text{ mile} = 5280 \text{ feet}$$

$$\left(\frac{17.37}{1}\right)\left(\frac{5280 \, \text{ft}}{1}\right) \quad \text{Divide out the miles and multiply across}$$

$$91,713.6 \, \text{ft} \quad \text{Our Solution}$$

In the previous example, we had to use the conversion factor $\frac{5280 \text{ft}}{1 \text{mi}}$ so the miles would divide out. If we had used $\frac{1 \text{mi}}{5280 \text{ft}}$ we would not have been able to divide out the miles. This is why when doing dimensional analysis it is very important to use units in the set-up of the problem, so we know how to correctly set up the conversion factor.

Example 372.

If 1 pound = 16 ounces, how many pounds 435 ounces?

$$\left(\frac{435 \, \text{oz}}{1}\right) \quad \text{Write } 435 \text{ as } a \text{ fraction, put it over } 1$$

$$\left(\frac{435 \, \text{oz}}{1}\right)\left(\frac{?? \, \text{lbs}}{?? \, \text{oz}}\right) \quad \text{To divide out oz,}$$

$$\text{put it in the denominator and lbs in numerator}$$

$$\left(\frac{435 \, \text{oz}}{1}\right)\left(\frac{1 \, \text{lbs}}{16 \, \text{oz}}\right) \quad \text{Fill in the given relationship, } 1 \text{ pound} = 16 \text{ ounces}$$

$$\left(\frac{435}{1}\right)\left(\frac{1\,\text{lbs}}{16}\right) = \frac{435\,\text{lbs}}{16} \qquad \text{Divide out oz, multiply across. Divide result}$$

$$27.1875\,\text{lbs} \qquad \text{Our Solution}$$

The same process can be used to convert problems with several units in them. Consider the following example.

Example 373.

A student averaged 45 miles per hour on a trip. What was the student's speed in feet per second?

$$\left(\frac{45\,\text{mi}}{\text{hr}}\right) \qquad \text{"per" is the fraction bar, put hr in denominator}$$

$$\left(\frac{45\,\text{mi}}{\text{hr}}\right)\left(\frac{5280\,\text{ft}}{1\,\text{mi}}\right) \qquad \text{To clear mi they must go in denominator and become ft}$$

$$\left(\frac{45\,\text{mi}}{\text{hr}}\right)\left(\frac{5280\,\text{ft}}{1\,\text{mi}}\right)\left(\frac{1\,\text{hr}}{3600\,\text{sec}}\right) \qquad \text{To clear hr they must go in numerator and become sec}$$

$$\left(\frac{45}{1}\right)\left(\frac{5280\,\text{ft}}{1}\right)\left(\frac{1}{3600\,\text{sec}}\right) \qquad \text{Divide out mi and hr. Multiply across}$$

$$\frac{237600\,\text{ft}}{3600\,\text{sec}} \qquad \text{Divide numbers}$$

$$66\,\text{ft per sec} \qquad \text{Our Solution}$$

If the units are two-dimensional (such as square inches - in^2) or three-dimensional (such as cubic feet - ft^3) we will need to put the same exponent on the conversion factor. So if we are converting square inches (in^2) to square ft (ft^2), the conversion factor would be squared, $\left(\frac{1\,\text{ft}}{12\,\text{in}}\right)^2$. Similarly if the units are cubed, we will cube the convesion factor.

Example 374.

Convert 8 cubic feet to yd^3 Write 8ft^3 as fraction, put it over 1

$$\left(\frac{8\,\text{ft}^3}{1}\right) \qquad \text{To clear ft, put them in denominator, yard in numerator}$$

$$\left(\frac{8\,\text{ft}^3}{1}\right)\left(\frac{??\,\text{yd}}{??\,\text{ft}}\right)^3 \qquad \text{Because the units are cubed,}$$

we cube the conversion factor

$$\left(\frac{8\,\text{ft}^3}{1}\right)\left(\frac{1\,\text{yd}}{3\,\text{ft}}\right)^3 \qquad \text{Evaluate exponent, cubing all numbers and units}$$

$$\left(\frac{8\,\text{ft}^3}{1}\right)\left(\frac{1\,\text{yd}^3}{27\,\text{ft}^3}\right) \qquad \text{Divide out ft}^3$$

$$\left(\frac{8}{1}\right)\left(\frac{1\,\text{yd}^3}{27}\right) = \frac{8\,\text{yd}^3}{27} \qquad \text{Multiply across and divide}$$

$$0.296296\,\text{yd}^3 \qquad \text{Our Solution}$$

When calculating area or volume, be sure to use the units and multiply them as well.

Example 375.

A room is 10 ft by 12 ft. How many square yards are in the room?

$$A = lw = (10\text{ft})(12\,\text{ft}) = 120\,\text{ft}^2 \qquad \text{Multiply length by width, also multiply units}$$

$$\left(\frac{120\text{ft}^2}{1}\right) \qquad \text{Write area as a fraction, put it over 1}$$

$$\left(\frac{120\text{ft}^2}{1}\right)\left(\frac{??\,\text{yd}}{??\,\text{ft}}\right)^2 \qquad \text{Put ft in denominator to clear,}$$

square conversion factor

$$\left(\frac{120\text{ft}^2}{1}\right)\left(\frac{1\,\text{yd}}{3\,\text{ft}}\right)^2 \qquad \text{Evaluate exponent, squaring all numbers and units}$$

$$\left(\frac{120\text{ft}^2}{1}\right)\left(\frac{1\,\text{yd}^2}{9\,\text{ft}^2}\right) \qquad \text{Divide out ft}^2$$

$$\left(\frac{120}{1}\right)\left(\frac{1\,\text{yd}^2}{9}\right) = \frac{120\,\text{yd}^2}{9} \qquad \text{Multiply across and divide}$$

$$13.33\,\text{yd}^2 \qquad \text{Our solution}$$

To focus on the process of conversions, a conversion sheet has been included at the end of this lesson which includes several conversion factors for length, volume, mass and time in both English and Metric units.

The process of dimensional analysis can be used to convert other types of units as well. If we can identify relationships that represent the same value we can make them into a conversion factor.

Example 376.

A child is perscribed a dosage of 12 mg of a certain drug and is allowed to refill his prescription twice. If a there are 60 tablets in a prescription, and each tablet has 4 mg, how many doses are in the 3 prescriptions (original + 2 refills)?

$$\text{Convert 3 Rx to doses} \qquad \text{Identify what the problem is asking}$$

$$1\,Rx = 60\,tab, 1\,tab = 4\,mg, 1\,dose = 12 mg \qquad \text{Identify given conversion factors}$$

$$\left(\frac{3\,Rx}{1}\right) \qquad \text{Write 3 Rx as fraction, put over 1}$$

$$\left(\frac{3\,Rx}{1}\right)\left(\frac{60\,tab}{1\,Rx}\right) \qquad \text{Convert Rx to tab, put Rx in denominator}$$

$$\left(\frac{3\,Rx}{1}\right)\left(\frac{60\,tab}{1\,Rx}\right)\left(\frac{4\,mg}{1\,tab}\right) \qquad \text{Convert tab to mg, put tab in denominator}$$

$$\left(\frac{3\,Rx}{1}\right)\left(\frac{60\,tab}{1\,Rx}\right)\left(\frac{4\,mg}{1\,tab}\right)\left(\frac{1\,dose}{12\,mg}\right) \qquad \text{Convert mg to dose, put mg in denominator}$$

$$\left(\frac{3}{1}\right)\left(\frac{60}{1}\right)\left(\frac{4}{1}\right)\left(\frac{1\,dose}{12}\right) \qquad \text{Divide out Rx, tab, and mg, multiply across}$$

$$\frac{720\,dose}{12} \qquad \text{Divide}$$

$$60\,doses \qquad \text{Our Solution}$$

World View Note: Only three countries in the world still use the English system commercially: Liberia (Western Africa), Myanmar (between India and Vietnam), and the USA.

Conversion Factors

Length

English	Metric
12 in = 1 ft	1000 mm = 1 m
1 yd = 3 ft	10 mm = 1 cm
1 yd = 36 in	100 cm = 1 m
1 mi = 5280 ft	10 dm = 1 m
	1 dam = 10 m
	1 hm = 100 m
	1 km = 1000 m

English/Metric
2.54 cm = 1 in
1 m = 3.28 ft
1.61 km = 1 mi

Area

English	Metric
$1 \text{ ft}^2 = 144 \text{ in}^2$	$1 \text{ a} = 100 \text{ m}^2$
$1 \text{ yd}^2 = 9 \text{ ft}^2$	1 ha = 100 a
$1 \text{ acre} = 43{,}560 \text{ ft}^2$	
$640 \text{ acres} = 1 \text{ mi}^2$	

English/Metric
1 ha = 2.47 acres

Weight (Mass)

English	Metric
1 lb = 16 oz	1 g = 1000 mg
1 T = 2000 lb	1 g = 100 cg
	1000 g = 1 kg
	1000 kg = 1 t

English/Metric
28.3 g = 1 oz
2.2 lb = 1 kg

Volume

English	Metric
1 c = 8 oz	$1 \text{ mL} = 1 \text{ cc} = 1 \text{ cm}^3$
1 pt = 2 c	1 L = 1000 mL
1 qt = 2 pt	1 L = 100 cL
1 gal = 4 qt	1 L = 10 dL
	1000 L = 1 kL

English/Metric
$16.39 \text{ mL} = 1 \text{ in}^3$
1.06 qt = 1 L
3.79 L = 1gal

Time

60 sec = 1 min
60 min = 1 hr
3600 sec = 1 hr
24 hr = 1 day

7.8 Practice - Dimensional Analysis

Use dimensional analysis to convert the following:

1) 7 mi. to yards

2) 234 oz. to tons

3) 11.2 mg to grams

4) 1.35 km to centimeters

5) 9,800,000 mm (milimeters) to miles

6) 4.5 ft^2 to square yards

7) 435,000 m^2 to sqaure kilometers

8) 8 km^2 to square feet

9) 0.0065 km^3 to cubic meters

10) 14.62 in^3 to cubic centimeters

11) 5,500 cm^3 to cubic yards

12) 3.5 mph (miles per hour) to feet per second

13) 185 yd. per min. to miles per hour

14) 153 ft/s (feet per second) to miles per hour

15) 248 mph to meters per second

16) 186,000 mph to kilometers per year

17) 7.50 T/yd^2 (tons per square yard) to pounds per square inch

18) 16 ft/s^2 to kilometers per hour squared

Use dimensional analysis to solve the following:

19) On a recent trip, Jan traveled 260 miles using 8 gallons of gas. How many miles per 1-gallon did she travel? How many yards per 1-ounce?

20) A chair lift at the Divide ski resort in Cold Springs, WY is 4806 feet long and takes 9 minutes. What is the average speed in miles per hour? How many feet per second does the lift travel?

21) A certain laser printer can print 12 pages per minute. Determine this printer's output in pages per day, and reams per month. (1 ream = 5000 pages)

22) An average human heart beats 60 times per minute. If an average person lives to the age of 75, how many times does the average heart beat in a lifetime?

23) Blood sugar levels are measured in miligrams of gluclose per deciliter of blood volume. If a person's blood sugar level measured 128 mg/dL, how much is this in grams per liter?

24) You are buying carpet to cover a room that measures 38 ft by 40 ft. The carpet cost $18 per square yard. How much will the carpet cost?

25) A car travels 14 miles in 15 minutes. How fast is it going in miles per hour? in meters per second?

26) A cargo container is 50 ft long, 10 ft wide, and 8 ft tall. Find its volume in cubic yards and cubic meters.

27) A local zoning ordinance says that a house's "footprint" (area of its ground floor) cannot occupy more than $\frac{1}{4}$ of the lot it is built on. Suppose you own a $\frac{1}{3}$ acre lot, what is the maximum allowed footprint for your house in square feet? in square inches? (1 acre = 43560 ft^2)

28) Computer memory is measured in units of bytes, where one byte is enough memory to store one character (a letter in the alphabet or a number). How many typical pages of text can be stored on a 700-megabyte compact disc? Assume that one typical page of text contains 2000 characters. (1 megabyte = 1,000,000 bytes)

29) In April 1996, the Department of the Interior released a "spike flood" from the Glen Canyon Dam on the Colorado River. Its purpose was to restore the river and the habitants along its bank. The release from the dam lasted a week at a rate of 25,800 cubic feet of water per second. About how much water was released during the 1-week flood?

30) The largest single rough diamond ever found, the Cullinan diamond, weighed 3106 carats; how much does the diamond weigh in miligrams? in pounds? (1 carat - 0.2 grams)

Chapter 8 : Radicals

Radicals - Square Roots

Objective: Simplify expressions with square roots.

Square roots are the most common type of radical used. A square root "un-squares" a number. For example, because $5^2 = 25$ we say the square root of 25 is 5. The square root of 25 is written as $\sqrt{25}$.

World View Note: The radical sign, when first used was an R with a line through the tail, similar to our perscription symbol today. The R came from the latin, "radix", which can be translated as "source" or "foundation". It wasn't until the 1500s that our current symbol was first used in Germany (but even then it was just a check mark with no bar over the numbers!

The following example gives several square roots:

Example 377.

$\sqrt{1} = 1$	$\sqrt{121} = 11$
$\sqrt{4} = 2$	$\sqrt{625} = 25$
$\sqrt{9} = 3$	$\sqrt{-81} =$ Undefined

The final example, $\sqrt{-81}$ is currently undefined as negatives have no square root. This is because if we square a positive or a negative, the answer will be positive. Thus we can only take square roots of positive numbers. In another lesson we will define a method we can use to work with and evaluate negative square roots, but for now we will simply say they are undefined.

Not all numbers have a nice even square root. For example, if we found $\sqrt{8}$ on our calculator, the answer would be 2.82842712474619009760337448419... and even this number is a rounded approximation of the square root. To be as accurate as possible, we will never use the calculator to find decimal approximations of square roots. Instead we will express roots in simplest radical form. We will do this using a property known as the product rule of radicals

$$\text{Product Rule of Square Roots: } \sqrt{a \cdot b} = \sqrt{a} \cdot \sqrt{b}$$

We can use the product rule to simplify an expression such as $\sqrt{36 \cdot 5}$ by splitting it into two roots, $\sqrt{36} \cdot \sqrt{5}$, and simplifying the first root, $6\sqrt{5}$. The trick in this process is being able to translate a problem like $\sqrt{180}$ into $\sqrt{36 \cdot 5}$. There are several ways this can be done. The most common and, with a bit of practice, the fastest method, is to find perfect squares that divide evenly into the radicand, or number under the radical. This is shown in the next example.

Example 378.

$$\sqrt{75} \qquad 75 \text{ is divisible by } 25, a \text{ perfect square}$$
$$\sqrt{25 \cdot 3} \qquad \text{Split into factors}$$
$$\sqrt{25} \cdot \sqrt{3} \qquad \text{Product rule, take the square root of } 25$$
$$5\sqrt{3} \qquad \text{Our Solution}$$

If there is a coefficient in front of the radical to begin with, the problem merely becomes a big multiplication problem.

Example 379.

$$5\sqrt{63} \qquad 63 \text{ is divisible by } 9, a \text{ perfect square}$$
$$5\sqrt{9 \cdot 7} \qquad \text{Split into factors}$$
$$5\sqrt{9} \cdot \sqrt{7} \qquad \text{Product rule, take the square root of } 9$$
$$5 \cdot 3\sqrt{7} \qquad \text{Multiply coefficients}$$
$$15\sqrt{7} \qquad \text{Our Solution}$$

As we simplify radicals using this method it is important to be sure our final answer can be simplified no more.

Example 380.

$$\sqrt{72} \qquad 72 \text{ is divisible by } 9, a \text{ perfect square}$$
$$\sqrt{9 \cdot 8} \qquad \text{Split into factors}$$
$$\sqrt{9} \cdot \sqrt{8} \qquad \text{Product rule, take the square root of } 9$$
$$3\sqrt{8} \qquad \text{But } 8 \text{ is also divisible by } a \text{ perfect square, } 4$$
$$3\sqrt{4 \cdot 2} \qquad \text{Split into factors}$$
$$3\sqrt{4} \cdot \sqrt{2} \qquad \text{Product rule, take the square root of } 4$$
$$3 \cdot 2\sqrt{2} \qquad \text{Multiply}$$

$$6\sqrt{2} \quad \text{Our Solution.}$$

The previous example could have been done in fewer steps if we had noticed that $72 = 36 \cdot 2$, but often the time it takes to discover the larger perfect square is more than it would take to simplify in several steps.

Variables often are part of the radicand as well. When taking the square roots of variables, we can divide the exponent by 2. For example, $\sqrt{x^8} = x^4$, because we divide the exponent of 8 by 2. This follows from the power of a power rule of expoennts, $(x^4)^2 = x^8$. When squaring, we multiply the exponent by two, so when taking a square root we divide the exponent by 2. This is shown in the following example.

Example 381.

$$
\begin{array}{ll}
-5\sqrt{18x^4y^6z^{10}} & 18 \text{ is divisible by } 9, a \text{ perfect square} \\
-5\sqrt{9 \cdot 2x^4y^6z^{10}} & \text{Split into factors} \\
-5\sqrt{9} \cdot \sqrt{2} \cdot \sqrt{x^4} \cdot \sqrt{y^6} \cdot \sqrt{z^{10}} & \text{Product rule, simplify roots, divide exponents by } 2 \\
-5 \cdot 3x^2y^3z^5\sqrt{2} & \text{Multiply coefficients} \\
-15x^2y^3z^5\sqrt{2} & \text{Our Solution}
\end{array}
$$

We can't always evenly divide the exponent on a variable by 2. Sometimes we have a remainder. If there is a remainder, this means the remainder is left inside the radical, and the whole number part is how many are outside the radical. This is shown in the following example.

Example 382.

$$
\begin{array}{ll}
\sqrt{20x^5y^9z^6} & 20 \text{ is divisible by } 4, a \text{ perfect square} \\
\sqrt{4 \cdot 5x^5y^9z^6} & \text{Split into factors} \\
\sqrt{4} \cdot \sqrt{5} \cdot \sqrt{x^5} \cdot \sqrt{y^9} \cdot \sqrt{z^6} & \text{Simplify, divide exponents by } 2, \text{remainder is left inside} \\
2x^2y^4z^3\sqrt{5xy} & \text{Our Solution}
\end{array}
$$

8.1 Practice - Square Roots

Simplify.

1) $\sqrt{245}$

2) $\sqrt{125}$

3) $\sqrt{36}$

4) $\sqrt{196}$

5) $\sqrt{12}$

6) $\sqrt{72}$

7) $3\sqrt{12}$

8) $5\sqrt{32}$

9) $6\sqrt{128}$

10) $7\sqrt{128}$

11) $-8\sqrt{392}$

12) $-7\sqrt{63}$

13) $\sqrt{192n}$

14) $\sqrt{343b}$

15) $\sqrt{196v^2}$

16) $\sqrt{100n^3}$

17) $\sqrt{252x^2}$

18) $\sqrt{200a^3}$

19) $-\sqrt{100k^4}$

20) $-4\sqrt{175p^4}$

21) $-7\sqrt{64x^4}$

22) $-2\sqrt{128n}$

23) $-5\sqrt{36m}$

24) $8\sqrt{112p^2}$

25) $\sqrt{45x^2y^2}$

26) $\sqrt{72a^3b^4}$

27) $\sqrt{16x^3y^3}$

28) $\sqrt{512a^4b^2}$

29) $\sqrt{320x^4y^4}$

30) $\sqrt{512m^4n^3}$

31) $6\sqrt{80xy^2}$

32) $8\sqrt{98mn}$

33) $5\sqrt{245x^2y^3}$

34) $2\sqrt{72x^2y^2}$

35) $-2\sqrt{180u^3v}$

36) $-5\sqrt{72x^3y^4}$

37) $-8\sqrt{180x^4y^2z^4}$

38) $6\sqrt{50a^4bc^2}$

39) $2\sqrt{80hj^4k}$

40) $-\sqrt{32xy^2z^3}$

41) $-4\sqrt{54mnp^2}$

42) $-8\sqrt{32m^2p^4q}$

Radicals - Higher Roots

Objective: Simplify radicals with an index greater than two.

While square roots are the most common type of radical we work with, we can take higher roots of numbers as well: cube roots, fourth roots, fifth roots, etc. Following is a definition of radicals.

$$\sqrt[m]{a} = b \text{ if } b^m = a$$

The small letter m inside the radical is called the index. It tells us which root we are taking, or which power we are "un-doing". For square roots the index is 2. As this is the most common root, the two is not usually written.

World View Note: The word for root comes from the French mathematician Franciscus Vieta in the late 16th century.

The following example includes several higher roots.

Example 383.

$\sqrt[3]{125} = 5$	$\sqrt[3]{-64} = -4$
$\sqrt[4]{81} = 3$	$\sqrt[7]{-128} = -2$
$\sqrt[5]{32} = 2$	$\sqrt[4]{-16} = $ undefined

We must be careful of a few things as we work with higher roots. First its important not to forget to check the index on the root. $\sqrt{81} = 9$ but $\sqrt[4]{81} = 3$. This is because $9^2 = 81$ and $3^4 = 81$. Another thing to watch out for is negatives under roots. We can take an odd root of a negative number, because a negative number raised to an odd power is still negative. However, we cannot take an even root of a negative number, this we will say is undefined. In a later section we will discuss how to work with roots of negative, but for now we will simply say they are undefined.

We can simplify higher roots in much the same way we simplified square roots, using the product property of radicals.

Product Property of Radicals: $\sqrt[m]{ab} = \sqrt[m]{a} \cdot \sqrt[m]{b}$

Often we are not as familiar with higher powers as we are with squares. It is important to remember what index we are working with as we try and work our way to the solution.

Example 384.

$\sqrt[3]{54}$ We are working with a cubed root, want third powers

$2^3 = 8$ Test 2, $2^3 = 8$, 54 is not divisible by 8.

$3^3 = 27$ Test 3, $3^3 = 27$, 54 is divisible by 27!

$\sqrt[3]{27 \cdot 2}$ Write as factors

$\sqrt[3]{27} \cdot \sqrt[3]{2}$ Product rule, take cubed root of 27

$3 \sqrt[3]{2}$ Our Solution

Just as with square roots, if we have a coefficient, we multiply the new coefficients together.

Example 385.

$3 \sqrt[4]{48}$ We are working with a fourth root, want fourth powers

$2^4 = 16$ Test 2, $2^4 = 16$, 48 is divisible by 16!

$3 \sqrt[4]{16 \cdot 3}$ Write as factors

$3 \sqrt[4]{16} \cdot \sqrt[4]{3}$ Product rule, take fourth root of 16

$3 \cdot 2 \sqrt[4]{3}$ Multiply coefficients

$6 \sqrt[4]{3}$ Our Solution

We can also take higher roots of variables. As we do, we will divide the exponent on the variable by the index. Any whole answer is how many of that varible will come out of the square root. Any remainder is how many are left behind inside the square root. This is shown in the following examples.

Example 386.

$\sqrt[5]{x^{25}y^{17}z^3}$ Divide each exponent by 5, whole number outside, remainder inside

$x^5 y^3 \sqrt[5]{y^2 z^3}$ Our Solution

In the previous example, for the x, we divided $\frac{25}{5} = 5\,R\,0$, so x^5 came out, no x's remain inside. For the y, we divided $\frac{17}{5} = 3\,R\,2$, so y^3 came out, and y^2 remains inside. For the z, when we divided $\frac{3}{5} = 0\,R\,3$, all three or z^3 remained inside. The following example includes integers in our problem.

Example 387.

$2 \sqrt[3]{40a^4b^8}$ Looking for cubes that divide into 40. The number 8 works!

$2 \sqrt[3]{8 \cdot 5a^4b^8}$ Take cube root of 8, dividing exponents on variables by 3

$2 \cdot 2ab^2 \sqrt[3]{5ab^2}$ Remainders are left in radical. Multiply coefficients

$4ab^2 \sqrt[3]{5ab^2}$ Our Solution

8.2 Practice - Higher Roots

Simplify.

1) $\sqrt[3]{625}$

2) $\sqrt[3]{375}$

3) $\sqrt[3]{750}$

4) $\sqrt[3]{250}$

5) $\sqrt[3]{875}$

6) $\sqrt[3]{24}$

7) $-4\sqrt[4]{96}$

8) $-8\sqrt[4]{48}$

9) $6\sqrt[4]{112}$

10) $3\sqrt[4]{48}$

11) $-\sqrt[4]{112}$

12) $5\sqrt[4]{243}$

13) $\sqrt[4]{648a^2}$

14) $\sqrt[4]{64n^3}$

15) $\sqrt[5]{224n^3}$

16) $\sqrt[5]{-96x^3}$

17) $\sqrt[5]{224p^5}$

18) $\sqrt[6]{256x^6}$

19) $-3\sqrt[7]{896r}$

20) $-8\sqrt[7]{384b^8}$

21) $-2\sqrt[3]{-48v^7}$

22) $4\sqrt[3]{250a^6}$

23) $-7\sqrt[3]{320n^6}$

24) $-\sqrt[3]{512n^6}$

25) $\sqrt[3]{-135x^5y^3}$

26) $\sqrt[3]{64u^5v^3}$

27) $\sqrt[3]{-32x^4y^4}$

28) $\sqrt[3]{1000a^4b^5}$

29) $\sqrt[3]{256x^4y^6}$

30) $\sqrt[3]{189x^3y^6}$

31) $7\sqrt[3]{-81x^3y^7}$

32) $-4\sqrt[3]{56x^2y^8}$

33) $2\sqrt[3]{375u^2v^8}$

34) $8\sqrt[3]{-750xy}$

35) $-3\sqrt[3]{192ab^2}$

36) $3\sqrt[3]{135xy^3}$

37) $6\sqrt[3]{-54m^8n^3p^7}$

38) $-6\sqrt[4]{80m^4p^7q^4}$

39) $6\sqrt[4]{648x^5y^7z^2}$

40) $-6\sqrt[4]{405a^5b^8c}$

41) $7\sqrt[4]{128h^6j^8k^8}$

42) $-6\sqrt[4]{324x^7yz^7}$

Radicals - Adding Radicals

Objective: Add like radicals by first simplifying each radical.

Adding and subtracting radicals is very similar to adding and subtracting with variables. Consider the following example.

Example 388.

$$5x + 3x - 2x \qquad \text{Combine like terms}$$
$$6x \qquad \text{Our Solution}$$

$$5\sqrt{11} + 3\sqrt{11} - 2\sqrt{11} \qquad \text{Combine like terms}$$
$$6\sqrt{11} \qquad \text{Our Solution}$$

Notice that when we combined the terms with $\sqrt{11}$ it was just like combining terms with x. When adding and subtracting with radicals we can combine like radicals just as like terms. We add and subtract the coefficients in front of the radical, and the radical stays the same. This is shown in the following example.

Example 389.

$$7\sqrt[5]{6} + 4\sqrt[5]{3} - 9\sqrt[5]{3} + \sqrt[5]{6} \qquad \text{Combine like radicals } 7\sqrt[5]{6} + \sqrt[5]{6} \text{ and } 4\sqrt[5]{3} - 8\sqrt[5]{3}$$
$$8\sqrt[5]{6} - 5\sqrt[5]{3} \qquad \text{Our Solution}$$

We cannot simplify this expression any more as the radicals do not match. Often problems we solve have no like radicals, however, if we simplify the radicals first we may find we do in fact have like radicals.

Example 390.

$$5\sqrt{45} + 6\sqrt{18} - 2\sqrt{98} + \sqrt{20} \qquad \text{Simplify radicals, find perfect square factors}$$
$$5\sqrt{9 \cdot 5} + 6\sqrt{9 \cdot 2} - 2\sqrt{49 \cdot 2} + \sqrt{4 \cdot 5} \qquad \text{Take roots where possible}$$
$$5 \cdot 3\sqrt{5} + 6 \cdot 3\sqrt{2} - 2 \cdot 7\sqrt{2} + 2\sqrt{5} \qquad \text{Multiply coefficients}$$
$$15\sqrt{5} + 18\sqrt{2} - 14\sqrt{2} + 2\sqrt{5} \qquad \text{Combine like terms}$$
$$17\sqrt{5} + 4\sqrt{2} \qquad \text{Our Solution}$$

World View Note: The Arab writers of the 16th century used the symbol similar to the greater than symbol with a dot underneath for radicals.

This exact process can be used to add and subtract radicals with higher indices

Example 391.

$$4\sqrt[3]{54} - 9\sqrt[3]{16} + 5\sqrt[3]{9} \qquad \text{Simplify each radical, finding perfect cube factors}$$
$$4\sqrt[3]{27 \cdot 2} - 9\sqrt[3]{8 \cdot 2} + 5\sqrt[3]{9} \qquad \text{Take roots where possible}$$
$$4 \cdot 3\sqrt[3]{2} - 9 \cdot 2\sqrt[3]{2} + 5\sqrt[3]{9} \qquad \text{Multiply coefficients}$$
$$12\sqrt[3]{2} - 18\sqrt[3]{2} + 5\sqrt[3]{9} \qquad \text{Combine like terms } 12\sqrt[3]{2} - 18\sqrt[3]{2}$$
$$-6\sqrt[3]{2} + 5\sqrt[3]{9} \qquad \text{Our Solution}$$

8.3 Practice - Adding Radicals

Simiplify

1) $2\sqrt{5} + 2\sqrt{5} + 2\sqrt{5}$

2) $-3\sqrt{6} - 3\sqrt{3} - 2\sqrt{3}$

3) $-3\sqrt{2} + 3\sqrt{5} + 3\sqrt{5}$

4) $-2\sqrt{6} - \sqrt{3} - 3\sqrt{6}$

5) $-2\sqrt{6} - 2\sqrt{6} - \sqrt{6}$

6) $-3\sqrt{3} + 2\sqrt{3} - 2\sqrt{3}$

7) $3\sqrt{6} + 3\sqrt{5} + 2\sqrt{5}$

8) $-\sqrt{5} + 2\sqrt{3} - 2\sqrt{3}$

9) $2\sqrt{2} - 3\sqrt{18} - \sqrt{2}$

10) $-\sqrt{54} - 3\sqrt{6} + 3\sqrt{27}$

11) $-3\sqrt{6} - \sqrt{12} + 3\sqrt{3}$

12) $-\sqrt{5} - \sqrt{5} - 2\sqrt{54}$

13) $3\sqrt{2} + 2\sqrt{8} - 3\sqrt{18}$

14) $2\sqrt{20} + 2\sqrt{20} - \sqrt{3}$

15) $3\sqrt{18} - \sqrt{2} - 3\sqrt{2}$

16) $-3\sqrt{27} + 2\sqrt{3} - \sqrt{12}$

17) $-3\sqrt{6} - 3\sqrt{6} - \sqrt{3} + 3\sqrt{6}$

18) $-2\sqrt{2} - \sqrt{2} + 3\sqrt{8} + 3\sqrt{6}$

19) $-2\sqrt{18} - 3\sqrt{8} - \sqrt{20} + 2\sqrt{20}$

20) $-3\sqrt{18} - \sqrt{8} + 2\sqrt{8} + 2\sqrt{8}$

21) $-2\sqrt{24} - 2\sqrt{6} + 2\sqrt{6} + 2\sqrt{20}$

22) $-3\sqrt{8} - \sqrt{5} - 3\sqrt{6} + 2\sqrt{18}$

23) $3\sqrt{24} - 3\sqrt{27} + 2\sqrt{6} + 2\sqrt{8}$

24) $2\sqrt{6} - \sqrt{54} - 3\sqrt{27} - \sqrt{3}$

25) $-2\sqrt[3]{16} + 2\sqrt[3]{16} + 2\sqrt[3]{2}$

26) $3\sqrt[3]{135} - \sqrt[3]{81} - \sqrt[3]{135}$

27) $2\sqrt[4]{243} - 2\sqrt[4]{243} - \sqrt[4]{3}$

28) $-3\sqrt[4]{4} + 3\sqrt[4]{324} + 2\sqrt[4]{64}$

29) $3\sqrt[4]{2} - 2\sqrt[4]{2} - \sqrt[4]{243}$

30) $2\sqrt[4]{6} + 2\sqrt[4]{4} + 3\sqrt[4]{6}$

31) $-\sqrt[4]{324} + 3\sqrt[4]{324} - 3\sqrt[4]{4}$

32) $-2\sqrt[4]{243} - \sqrt[4]{96} + 2\sqrt[4]{96}$

33) $2\sqrt[4]{2} + 2\sqrt[4]{3} + 3\sqrt[4]{64} - \sqrt[4]{3}$

34) $2\sqrt[4]{48} - 3\sqrt[4]{405} - 3\sqrt[4]{48} - \sqrt[4]{162}$

35) $-3\sqrt[5]{6} - \sqrt[5]{64} + 2\sqrt[5]{192} - 2\sqrt[5]{64}$

36) $-3\sqrt[7]{3} - 3\sqrt[7]{768} + 2\sqrt[7]{384} + 3\sqrt[7]{5}$

37) $2\sqrt[5]{160} - 2\sqrt[5]{192} - \sqrt[5]{160} - \sqrt[5]{-160}$

38) $-2\sqrt[7]{256} - 2\sqrt[7]{256} - 3\sqrt[7]{2} - \sqrt[7]{640}$

39) $-\sqrt[6]{256} - 2\sqrt[6]{4} - 3\sqrt[6]{320} - 2\sqrt[6]{128}$

Radicals - Multiply and Divide Radicals

Objective: Multiply and divide radicals using the product and quotient rules of radicals.

Multiplying radicals is very simple if the index on all the radicals match. The prodcut rule of radicals which we have already been using can be generalized as follows:

Product Rule of Radicals: $a \sqrt[m]{b} \cdot c \sqrt[m]{d} = ac \sqrt[m]{bd}$

Another way of stating this rule is we are allowed to multiply the factors outside the radical and we are allowed to multiply the factors inside the radicals, as long as the index matches. This is shown in the following example.

Example 392.

$$-5\sqrt{14} \cdot 4\sqrt{6} \qquad \text{Multiply outside and inside the radical}$$
$$-20\sqrt{84} \qquad \text{Simplify the radical, divisible by 4}$$
$$-20\sqrt{4 \cdot 21} \qquad \text{Take the square root where possible}$$
$$-20 \cdot 2\sqrt{21} \qquad \text{Multiply coefficients}$$
$$-40\sqrt{21} \qquad \text{Our Solution}$$

The same process works with higher roots

Example 393.

$$2\sqrt[3]{18} \cdot 6\sqrt[3]{15} \qquad \text{Multiply outside and inside the radical}$$
$$12\sqrt[3]{270} \qquad \text{Simplify the radical, divisible by 27}$$
$$12\sqrt[3]{27 \cdot 10} \qquad \text{Take cube root where possible}$$
$$12 \cdot 3\sqrt[3]{10} \qquad \text{Multiply coefficients}$$
$$36\sqrt[3]{10} \qquad \text{Our Solution}$$

When multiplying with radicals we can still use the distributive property or FOIL just as we could with variables.

Example 394.

$$7\sqrt{6}(3\sqrt{10} - 5\sqrt{15}) \qquad \text{Distribute, following rules for multiplying radicals}$$
$$21\sqrt{60} - 35\sqrt{90} \qquad \text{Simplify each radical, finding perfect square factors}$$
$$21\sqrt{4 \cdot 15} - 35\sqrt{9 \cdot 10} \qquad \text{Take square root where possible}$$
$$21 \cdot 2\sqrt{15} - 35 \cdot 3\sqrt{10} \qquad \text{Multiply coefficients}$$
$$42\sqrt{15} - 105\sqrt{10} \qquad \text{Our Solution}$$

Example 395.

$$(\sqrt{5} - 2\sqrt{3})(4\sqrt{10} + 6\sqrt{6}) \qquad \text{FOIL, following rules for multiplying radicals}$$

$$4\sqrt{50} + 6\sqrt{30} - 8\sqrt{30} - 12\sqrt{18} \quad \text{Simplify radicals, find perfect square factors}$$
$$4\sqrt{25 \cdot 2} + 6\sqrt{30} - 8\sqrt{30} - 12\sqrt{9 \cdot 2} \quad \text{Take square root where possible}$$
$$4 \cdot 5\sqrt{2} + 6\sqrt{30} - 8\sqrt{30} - 12 \cdot 3\sqrt{2} \quad \text{Multiply coefficients}$$
$$20\sqrt{2} + 6\sqrt{30} - 8\sqrt{30} - 36\sqrt{2} \quad \text{Combine like terms}$$
$$-16\sqrt{2} - 2\sqrt{30} \quad \text{Our Solution}$$

World View Note: Clay tablets have been discovered revealing much about Babylonian mathematics dating back from 1800 to 1600 BC. In one of the tables there is an approximation of $\sqrt{2}$ accurate to five decimal places (1.41421)

Example 396.

$$(2\sqrt{5} - 3\sqrt{6})(7\sqrt{2} - 8\sqrt{7}) \quad \text{FOIL, following rules for multiplying radicals}$$
$$14\sqrt{10} - 16\sqrt{35} - 21\sqrt{12} - 24\sqrt{42} \quad \text{Simplify radicals, find perfect square factors}$$
$$14\sqrt{10} - 16\sqrt{35} - 21\sqrt{4 \cdot 3} - 24\sqrt{42} \quad \text{Take square root where possible}$$
$$14\sqrt{10} - 16\sqrt{35} - 21 \cdot 2\sqrt{3} - 24\sqrt{42} \quad \text{Multiply coefficient}$$
$$14\sqrt{10} - 16\sqrt{35} - 42\sqrt{3} - 24\sqrt{42} \quad \text{Our Solution}$$

As we are multiplying we always look at our final solution to check if all the radicals are simplified and all like radicals or like terms have been combined.

Division with radicals is very similar to multiplication, if we think about division as reducing fractions, we can reduce the coefficients outside the radicals and reduce the values inside the radicals to get our final solution.

$$\textbf{Quotient Rule of Radicals:} \quad \frac{a \sqrt[m]{b}}{c \sqrt[m]{d}} = \frac{a}{c} \sqrt[m]{\frac{b}{d}}$$

Example 397.

$$\frac{15\sqrt[3]{108}}{20\sqrt[3]{2}} \quad \text{Reduce } \frac{15}{20} \text{ and } \frac{\sqrt[3]{108}}{\sqrt{2}} \text{ by dividing by 5 and 2 respectively}$$

$$\frac{3\sqrt[3]{54}}{4} \quad \text{Simplify radical, 54 is divisible by 27}$$

$$\frac{3\sqrt[3]{27 \cdot 2}}{4} \quad \text{Take the cube root of 27}$$

$$\frac{3 \cdot 3\sqrt[3]{2}}{4} \quad \text{Multiply coefficients}$$

$$\frac{9\sqrt[3]{2}}{4} \quad \text{Our Solution}$$

There is one catch to dividing with radicals, it is considered bad practice to have a radical in the denominator of our final answer. If there is a radical in the denominator we will rationalize it, or clear out any radicals in the denominator.

We do this by multiplying the numerator and denominator by the same thing. The problems we will consider here will all have a monomial in the denominator. The way we clear a monomial radical in the denominator is to focus on the index. The index tells us how many of each factor we will need to clear the radical. For example, if the index is 4, we will need 4 of each factor to clear the radical. This is shown in the following examples.

Example 398.

$$\frac{\sqrt{6}}{\sqrt{5}}$$ Index is 2, we need two fives in denominator, need 1 more

$$\frac{\sqrt{6}}{\sqrt{5}}\left(\frac{\sqrt{5}}{\sqrt{5}}\right)$$ Multiply numerator and denominator by $\sqrt{5}$

$$\frac{\sqrt{30}}{5}$$ Our Solution

Example 399.

$$\frac{3\sqrt[4]{11}}{\sqrt[4]{2}}$$ Index is 4, we need four twos in denominator, need 3 more

$$\frac{3\sqrt[4]{11}}{\sqrt[4]{2}}\left(\frac{\sqrt[4]{2^3}}{\sqrt[4]{2^3}}\right)$$ Multiply numerator and denominator by $\sqrt[4]{2^3}$

$$\frac{3\sqrt[4]{88}}{2}$$ Our Solution

Example 400.

$$\frac{4\sqrt[3]{2}}{7\sqrt[3]{25}}$$ The 25 can be written as 5^2. This will help us keep the numbers small

$$\frac{4\sqrt[3]{2}}{7\sqrt[3]{5^2}}$$ Index is 3, we need three fives in denominator, need 1 more

$$\frac{4\sqrt[3]{2}}{7\sqrt[3]{5^2}}\left(\frac{\sqrt[3]{5}}{\sqrt[3]{5}}\right)$$ Multiply numerator and denominator by $\sqrt[3]{5}$

$$\frac{4\sqrt[3]{10}}{7\cdot5}$$ Multiply out denominator

$$\frac{4\sqrt[3]{10}}{35}$$ Our Solution

The previous example could have been solved by multiplying numerator and denominator by $\sqrt[3]{25^2}$. However, this would have made the numbers very large and we would have needed to reduce our soultion at the end. This is why re-writing the radical as $\sqrt[3]{5^2}$ and multiplying by just $\sqrt[3]{5}$ was the better way to simplify.

We will also always want to reduce our fractions (inside and out of the radical) before we rationalize.

Example 401.

$$\frac{6\sqrt{14}}{12\sqrt{22}} \qquad \text{Reduce coefficients and inside radical}$$

$$\frac{\sqrt{7}}{2\sqrt{11}} \qquad \text{Index is 2, need two elevens, need 1 more}$$

$$\frac{\sqrt{7}}{2\sqrt{11}}\left(\frac{\sqrt{11}}{\sqrt{11}}\right) \qquad \text{Multiply numerator and denominator by } \sqrt{11}$$

$$\frac{\sqrt{77}}{2\cdot 11} \qquad \text{Multiply denominator}$$

$$\frac{\sqrt{77}}{22} \qquad \text{Our Solution}$$

The same process can be used to rationalize fractions with variables.

Example 402.

$$\frac{18\sqrt[4]{6x^3y^4z}}{8\sqrt[4]{10xy^6z^3}} \qquad \text{Reduce coefficients and inside radical}$$

$$\frac{9\sqrt[4]{3x^2}}{4\sqrt[4]{5y^2z^3}} \qquad \begin{array}{l}\text{Index is 4. We need four of everything to rationalize,}\\ \text{three more fives, two more } y\text{'s and one more } z.\end{array}$$

$$\frac{9\sqrt[4]{3x^2}}{4\sqrt[4]{5y^2z^3}}\left(\frac{\sqrt[4]{5^3y^2z}}{\sqrt[4]{5^3y^2z}}\right) \qquad \text{Multiply numerator and denominator by } \sqrt[4]{5^3y^2z}$$

$$\frac{9\sqrt[4]{375x^2y^2z}}{4\cdot 5yz} \qquad \text{Multiply denominator}$$

$$\frac{9\sqrt[4]{375x^2y^2z}}{20yz} \qquad \text{Our Solution}$$

8.4 Practice - Multiply and Divide Radicals

Multiply or Divide and Simplify.

1) $3\sqrt{5} \cdot -4\sqrt{16}$

2) $-5\sqrt{10} \cdot \sqrt{15}$

3) $\sqrt{12m} \cdot \sqrt{15m}$

4) $\sqrt{5r^3} \cdot -5\sqrt{10r^2}$

5) $\sqrt[3]{4x^3} \cdot \sqrt[3]{2x^4}$

6) $3\sqrt[3]{4a^4} \cdot \sqrt[3]{10a^3}$

7) $\sqrt{6}(\sqrt{2}+2)$

8) $\sqrt{10}(\sqrt{5}+\sqrt{2})$

9) $-5\sqrt{15}(3\sqrt{3}+2)$

10) $5\sqrt{15}(3\sqrt{3}+2)$

11) $5\sqrt{10}(5n+\sqrt{2})$

12) $\sqrt{15}(\sqrt{5}-3\sqrt{3v})$

13) $(2+2\sqrt{2})(-3+\sqrt{2})$

14) $(-2+\sqrt{3})(-5+2\sqrt{3})$

15) $(\sqrt{5}-5)(2\sqrt{5}-1)$

16) $(2\sqrt{3}+\sqrt{5})(5\sqrt{3}+2\sqrt{4})$

17) $(\sqrt{2a}+2\sqrt{3a})(3\sqrt{2a}+\sqrt{5a})$

18) $(-2\sqrt{2p}+5\sqrt{5})(\sqrt{5p}+\sqrt{5p})$

19) $(-5-4\sqrt{3})(-3-4\sqrt{3})$

20) $(5\sqrt{2}-1)(-\sqrt{2m}+5)$

21) $\frac{\sqrt{12}}{5\sqrt{100}}$

22) $\frac{\sqrt{15}}{2\sqrt{4}}$

23) $\frac{\sqrt{5}}{4\sqrt{125}}$

24) $\frac{\sqrt{12}}{\sqrt{3}}$

25) $\frac{\sqrt{10}}{\sqrt{6}}$

26) $\frac{\sqrt{2}}{3\sqrt{5}}$

27) $\frac{2\sqrt{4}}{3\sqrt{3}}$

28) $\frac{4\sqrt{3}}{\sqrt{15}}$

29) $\frac{5x^2}{4\sqrt{3x^3y^3}}$

30) $\frac{4}{5\sqrt{3xy^4}}$

31) $\frac{\sqrt{2p^2}}{\sqrt{3p}}$

32) $\frac{\sqrt{8n^2}}{\sqrt{10n}}$

33) $\frac{3\sqrt[3]{10}}{5\sqrt[3]{27}}$

34) $\frac{\sqrt[3]{15}}{\sqrt[3]{64}}$

35) $\frac{\sqrt[3]{5}}{4\sqrt[3]{4}}$

36) $\frac{\sqrt[4]{2}}{2\sqrt[4]{64}}$

37) $\frac{5\sqrt[4]{5r^4}}{\sqrt[4]{8r^2}}$

38) $\frac{4}{\sqrt[4]{64m^4n^2}}$

Radicals - Rationalize Denominators

Objective: Rationalize the denominators of radical expressions.

It is considered bad practice to have a radical in the denominator of a fraction. When this happens we multiply the numerator and denominator by the same thing in order to clear the radical. In the lesson on dividing radicals we talked about how this was done with monomials. Here we will look at how this is done with binomials.

If the binomial is in the numerator the process to rationalize the denominator is essentially the same as with monomials. The only difference is we will have to distribute in the numerator.

Example 403.

$$\frac{\sqrt{3}-9}{2\sqrt{6}} \qquad \text{Want to clear } \sqrt{6} \text{ in denominator, multiply by } \sqrt{6}$$

$$\frac{(\sqrt{3}-9)}{2\sqrt{6}}\left(\frac{\sqrt{6}}{\sqrt{6}}\right) \qquad \text{We will distribute the } \sqrt{6} \text{ through the numerator}$$

$$\frac{\sqrt{18} - 9\sqrt{6}}{2 \cdot 6}$$ Simplify radicals in numerator, multiply out denominator

$$\frac{\sqrt{9 \cdot 2} - 9\sqrt{6}}{12}$$ Take square root where possible

$$\frac{3\sqrt{2} - 9\sqrt{6}}{12}$$ Reduce by dividing each term by 3

$$\frac{\sqrt{2} - 3\sqrt{6}}{4}$$ Our Solution

It is important to remember that when reducing the fraction we cannot reduce with just the 3 and 12 or just the 9 and 12. When we have addition or subtraction in the numerator or denominator we must divide all terms by the same number.

The problem can often be made easier if we first simplify any radicals in the problem.

$$\frac{2\sqrt{20x^5} - \sqrt{12x^2}}{\sqrt{18x}}$$ Simplify radicals by finding perfect squares

$$\frac{2\sqrt{4 \cdot 5x^3} - \sqrt{4 \cdot 3x^2}}{\sqrt{9 \cdot 2x}}$$ Simplify roots, divide exponents by 2.

$$\frac{2 \cdot 2x^2\sqrt{5x} - 2x\sqrt{3}}{3\sqrt{2x}}$$ Multiply coefficients

$$\frac{4x^2\sqrt{5x} - 2x\sqrt{3}}{3\sqrt{2x}}$$ Multiplying numerator and denominator by $\sqrt{2x}$

$$\frac{(4x^2\sqrt{5x} - 2x\sqrt{3})}{3\sqrt{2x}}\left(\frac{\sqrt{2x}}{\sqrt{2x}}\right)$$ Distribute through numerator

$$\frac{4x^2\sqrt{10x^2} - 2x\sqrt{6x}}{3 \cdot 2x}$$ Simplify roots in numerator, multiply coefficients in denominator

$$\frac{4x^3\sqrt{10} - 2x\sqrt{6x}}{6x}$$ Reduce, dividing each term by $2x$

$$\frac{2x^2\sqrt{10} - \sqrt{6x}}{3x} \quad \text{Our Solution}$$

As we are rationalizing it will always be important to constantly check our problem to see if it can be simplified more. We ask ourselves, can the fraction be reduced? Can the radicals be simplified? These steps may happen several times on our way to the solution.

If the binomial occurs in the denominator we will have to use a different strategy to clear the radical. Consider $\frac{2}{\sqrt{3}-5}$, if we were to multiply the denominator by $\sqrt{3}$ we would have to distribute it and we would end up with $3 - 5\sqrt{3}$. We have not cleared the radical, only moved it to another part of the denominator. So our current method will not work. Instead we will use what is called a conjugate. A **conjugate** is made up of the same terms, with the opposite sign in the middle. So for our example with $\sqrt{3} - 5$ in the denominator, the conjugate would be $\sqrt{3} + 5$. The advantage of a conjugate is when we multiply them together we have $(\sqrt{3} - 5)(\sqrt{3} + 5)$, which is a sum and a difference. We know when we multiply these we get a difference of squares. Squaring $\sqrt{3}$ and 5, with subtraction in the middle gives the product $3 - 25 = -22$. Our answer when multiplying conjugates will no longer have a square root. This is exactly what we want.

Example 404.

$$\frac{2}{\sqrt{3}-5} \qquad \text{Multiply numerator and denominator by conjugate}$$

$$\frac{2}{\sqrt{3}-5}\left(\frac{\sqrt{3}+5}{\sqrt{3}+5}\right) \qquad \text{Distribute numerator, difference of squares in denominator}$$

$$\frac{2\sqrt{3}+10}{3-25} \qquad \text{Simplify denoinator}$$

$$\frac{2\sqrt{3}+10}{-22} \qquad \text{Reduce by dividing all terms by} -2$$

$$\frac{-\sqrt{3}-5}{11} \qquad \text{Our Solution}$$

In the previous example, we could have reduced by dividng by 2, giving the solution $\frac{\sqrt{3}+5}{11}$, both answers are correct.

Example 405.

$$\frac{\sqrt{15}}{\sqrt{5}+\sqrt{3}} \qquad \text{Multiply by conjugate,} \sqrt{5} - \sqrt{3}$$

$$\frac{\sqrt{15}}{\sqrt{5}+\sqrt{3}}\left(\frac{\sqrt{5}-\sqrt{3}}{\sqrt{5}-\sqrt{3}}\right) \qquad \text{Distribute numerator, denominator is difference of squares}$$

$$\frac{\sqrt{75}-\sqrt{45}}{5-3} \qquad \text{Simplify radicals in numerator, subtract in denominator}$$

$$\frac{\sqrt{25\cdot3}-\sqrt{9\cdot5}}{2} \qquad \text{Take square roots where possible}$$

$$\frac{5\sqrt{3}-3\sqrt{5}}{2} \qquad \text{Our Solution}$$

Example 406.

$$\frac{2\sqrt{3x}}{4-\sqrt{5x^3}} \qquad \text{Multiply by conjugate, } 4+\sqrt{5x^3}$$

$$\frac{2\sqrt{3x}}{4-\sqrt{5x^3}}\left(\frac{4+\sqrt{5x^3}}{4+\sqrt{5x^3}}\right) \qquad \text{Distribute numerator, denominator is difference of squares}$$

$$\frac{8\sqrt{3x}+2\sqrt{15x^4}}{16-5x^3} \qquad \text{Simplify radicals where possible}$$

$$\frac{8\sqrt{3x}+2x^2\sqrt{15}}{16-5x^3} \qquad \text{Our Solution}$$

The same process can be used when there is a binomial in the numerator and denominator. We just need to remember to FOIL out the numerator.

Example 407.

$$\frac{3-\sqrt{5}}{2-\sqrt{3}} \qquad \text{Multiply by conjugate, } 2+\sqrt{3}$$

$$\frac{3-\sqrt{5}}{2-\sqrt{3}}\left(\frac{2+\sqrt{3}}{2+\sqrt{3}}\right) \qquad \text{FOIL in numerator, denominator is difference of squares}$$

$$\frac{6+3\sqrt{3}-2\sqrt{5}-\sqrt{15}}{4-3} \qquad \text{Simplify denominator}$$

$$\frac{6+3\sqrt{3}-2\sqrt{5}-\sqrt{15}}{1} \qquad \text{Divide each term by 1}$$

$$6+3\sqrt{3}-2\sqrt{5}-\sqrt{15} \qquad \text{Our Solution}$$

Example 408.

$$\frac{2\sqrt{5} - 3\sqrt{7}}{5\sqrt{6} + 4\sqrt{2}} \qquad \text{Multiply by the conjugate, } 5\sqrt{6} - 4\sqrt{2}$$

$$\frac{2\sqrt{5} - 3\sqrt{7}}{5\sqrt{6} + 4\sqrt{2}}\left(\frac{5\sqrt{6} - 4\sqrt{2}}{5\sqrt{6} - 4\sqrt{2}}\right) \qquad \begin{array}{l}\text{FOIL numerator,} \\ \text{denominator is difference of squares}\end{array}$$

$$\frac{10\sqrt{30} - 8\sqrt{10} - 15\sqrt{42} + 12\sqrt{14}}{25 \cdot 6 - 16 \cdot 2} \qquad \text{Multiply in denominator}$$

$$\frac{10\sqrt{30} - 8\sqrt{10} - 15\sqrt{42} + 12\sqrt{14}}{150 - 32} \qquad \text{Subtract in denominator}$$

$$\frac{10\sqrt{30} - 8\sqrt{10} - 15\sqrt{42} + 12\sqrt{14}}{118} \qquad \text{Our Solution}$$

The same process is used when we have variables

Example 409.

$$\frac{3x\sqrt{2x} + \sqrt{4x^3}}{5x - \sqrt{3x}} \qquad \text{Multiply by the conjugate, } 5x + \sqrt{3x}$$

$$\frac{3x\sqrt{2x} + \sqrt{4x^3}}{5x - \sqrt{3x}}\left(\frac{5x + \sqrt{3x}}{5x + \sqrt{3x}}\right) \qquad \begin{array}{l}\text{FOIL in numerator,} \\ \text{denominator is difference of squares}\end{array}$$

$$\frac{15x^2\sqrt{2x} + 3x\sqrt{6x^2} + 5x\sqrt{4x^3} + \sqrt{12x^4}}{25x^2 - 3x} \qquad \text{Simplify radicals}$$

$$\frac{15x^2\sqrt{2x} + 3x^2\sqrt{6} + 10x^2\sqrt{x} + 2x^2\sqrt{3}}{25x^2 - 3x} \qquad \text{Divide each term by } x$$

$$\frac{15x\sqrt{2x} + 3x\sqrt{6} + 10x\sqrt{x} + 2x\sqrt{3}}{25x - 3} \qquad \text{Our Solution}$$

World View Note: During the 5th century BC in India, Aryabhata published a treatise on astronomy. His work included a method for finding the square root of numbers that have many digits.

8.5 Practice - Rationalize Denominators

Simplify.

1) $\dfrac{4+2\sqrt{3}}{\sqrt{9}}$

2) $\dfrac{4+\sqrt{3}}{4\sqrt{9}}$

3) $\dfrac{4+2\sqrt{3}}{5\sqrt{4}}$

4) $\dfrac{2\sqrt{3} \quad 2}{2\sqrt{16}}$

5) $\dfrac{2 \quad 5\sqrt{5}}{4\sqrt{13}}$

6) $\dfrac{\sqrt{5}+4}{4\sqrt{17}}$

7) $\dfrac{\sqrt{2} \quad 3\sqrt{3}}{\sqrt{3}}$

8) $\dfrac{\sqrt{5}-\sqrt{2}}{3\sqrt{6}}$

9) $\dfrac{5}{3\sqrt{5}+\sqrt{2}}$

10) $\dfrac{5}{\sqrt{3}+4\sqrt{5}}$

11) $\dfrac{2}{5+\sqrt{2}}$

12) $\dfrac{5}{2\sqrt{3} \quad \sqrt{2}}$

13) $\dfrac{3}{4 \quad 3\sqrt{3}}$

14) $\dfrac{4}{\sqrt{2}-2}$

15) $\dfrac{4}{3+\sqrt{5}}$

16) $\dfrac{2}{2\sqrt{5}+2\sqrt{3}}$

17) $-\dfrac{4}{4 \quad 4\sqrt{2}}$

18) $\dfrac{4}{4\sqrt{3} \quad \sqrt{5}}$

19) $\dfrac{1}{1+\sqrt{2}}$

20) $\dfrac{3+\sqrt{3}}{\sqrt{3}-1}$

21) $\dfrac{\sqrt{14} \quad 2}{\sqrt{7}-\sqrt{2}}$

22) $\dfrac{2+\sqrt{10}}{\sqrt{2}+\sqrt{5}}$

23) $\dfrac{\sqrt{ab}-a}{\sqrt{b} \quad \sqrt{a}}$

24) $\dfrac{\sqrt{14}-\sqrt{7}}{\sqrt{14}+\sqrt{7}}$

25) $\dfrac{a+\sqrt{ab}}{\sqrt{a}+\sqrt{b}}$

26) $\dfrac{a+\sqrt{ab}}{\sqrt{a}+\sqrt{b}}$

27) $\dfrac{2+\sqrt{6}}{2+\sqrt{3}}$

28) $\dfrac{2\sqrt{5}+\sqrt{3}}{1 \quad \sqrt{3}}$

29) $\dfrac{a \quad \sqrt{b}}{a+\sqrt{b}}$

30) $\dfrac{a-b}{\sqrt{a}+\sqrt{b}}$

31) $\dfrac{6}{3\sqrt{2} \quad 2\sqrt{3}}$

32) $\dfrac{ab}{a\sqrt{b}-b\sqrt{a}}$

33) $\dfrac{a \quad b}{a\sqrt{b}-b\sqrt{a}}$

34) $\dfrac{4\sqrt{2}+3}{3\sqrt{2}+\sqrt{3}}$

35) $\dfrac{2 \quad \sqrt{5}}{3+\sqrt{5}}$

36) $\dfrac{1+\sqrt{5}}{2\sqrt{5}+5\sqrt{2}}$

37) $\dfrac{5\sqrt{2}+\sqrt{3}}{5+5\sqrt{2}}$ 38) $\dfrac{\sqrt{3}+\sqrt{2}}{2\sqrt{3}\ \ \sqrt{2}}$

Radicals - Rational Exponents

Objective: Convert between radical notation and exponential notation and simplify expressions with rational exponents using the properties of exponents.

When we simplify radicals with exponents, we divide the exponent by the index. Another way to write division is with a fraction bar. This idea is how we will define rational exponents.

Definition of Rational Exponents: $a^{\frac{n}{m}} = (\sqrt[m]{a})^n$

The denominator of a rational exponent becomes the index on our radical, likewise the index on the radical becomes the denominator of the exponent. We can use this property to change any radical expression into an exponential expression.

Example 410.

$(\sqrt[5]{x})^3 = x^{\frac{3}{5}}$	$(\sqrt[6]{3x})^5 = (3x)^{\frac{5}{6}}$
$\dfrac{1}{(\sqrt[7]{a})^3} = a^{-\frac{3}{7}}$	$\dfrac{1}{(\sqrt[3]{xy})^2} = (xy)^{-\frac{2}{3}}$

Index is denominator
Negative exponents from reciprocals

We can also change any rational exponent into a radical expression by using the denominator as the index.

Example 411.

$a^{\frac{5}{3}} = (\sqrt[3]{a})^5$	$(2mn)^{\frac{2}{7}} = (\sqrt[7]{2mn})^2$
$x^{-\frac{4}{5}} = \dfrac{1}{(\sqrt[5]{x})^4}$	$(xy)^{-\frac{2}{9}} = \dfrac{1}{(\sqrt[9]{xy})^2}$

Index is denominator
Negative exponent means reciprocals

World View Note: Nicole Oresme, a Mathematician born in Normandy was the first to use rational exponents. He used the notation $\frac{1}{3} \bullet 9^p$ to represent $9^{\frac{1}{3}}$. However his notation went largely unnoticed.

The ability to change between exponential expressions and radical expressions allows us to evaluate problems we had no means of evaluating before by changing to a radical.

Example 412.

$27^{-\frac{4}{3}}$ Change to radical, denominator is index, negative means reciprocal

$\dfrac{1}{(\sqrt[3]{27})^4}$ Evaluate radical

$\dfrac{1}{(3)^4}$ Evaluate exponent

$\dfrac{1}{81}$ Our solution

The largest advantage of being able to change a radical expression into an exponential expression is we are now allowed to use all our exponent properties to simplify. The following table reviews all of our exponent properties.

Properties of Exponents

$$a^m a^n = a^{m+n} \qquad (ab)^m = a^m b^m \qquad a^{-m} = \dfrac{1}{a^m}$$

$$\dfrac{a^m}{a^n} = a^{m-n} \qquad \left(\dfrac{a}{b}\right)^m = \dfrac{a^m}{b^m} \qquad \dfrac{1}{a^{-m}} = a^m$$

$$(a^m)^n = a^{mn} \qquad a^0 = 1 \qquad \left(\dfrac{a}{b}\right)^{-m} = \dfrac{b^m}{a^m}$$

When adding and subtracting with fractions we need to be sure to have a common denominator. When multiplying we only need to multiply the numerators together and denominators together. The following examples show several different problems, using different properties to simplify the rational exponents.

Example 413.

$$a^{\frac{2}{3}} b^{\frac{1}{2}} a^{\frac{1}{6}} b^{\frac{1}{5}} \qquad \text{Need common denominator on } a's\,(6) \text{ and } b's\,(10)$$
$$a^{\frac{4}{6}} b^{\frac{5}{10}} a^{\frac{1}{6}} b^{\frac{2}{10}} \qquad \text{Add exponents on } a's \text{ and } b's$$
$$a^{\frac{5}{6}} b^{\frac{7}{10}} \qquad \text{Our Solution}$$

Example 414.

$$\left(x^{\frac{1}{3}} y^{\frac{2}{5}} \right)^{\frac{3}{4}} \qquad \text{Multiply } \dfrac{3}{4} \text{ by each exponent}$$

$$x^{\frac{1}{4}} y^{\frac{3}{10}} \qquad \text{Our Solution}$$

Example 415.

$$\dfrac{x^2 y^{\frac{2}{3}} \cdot 2x^{\frac{1}{2}} y^{\frac{5}{6}}}{x^{\frac{7}{2}} y^0} \qquad \text{In numerator, need common denominator to add exponents}$$

$$\frac{x^{\frac{4}{2}} y^{\frac{4}{6}} \cdot 2x^{\frac{1}{2}} y^{\frac{5}{6}}}{x^{\frac{7}{2}} y^0} \qquad \text{Add exponents in numerator, in denominator, } y^0 = 1$$

$$\frac{2x^{\frac{5}{2}} y^{\frac{9}{6}}}{x^{\frac{7}{2}}} \qquad \text{Subtract exponents on } x, \text{ reduce exponent on } y$$

$$2x^{-1} y^{\frac{3}{2}} \qquad \text{Negative exponent moves down to denominator}$$

$$\frac{2y^{\frac{3}{2}}}{x} \qquad \text{Our Solution}$$

Example 416.

$$\left(\frac{25x^{\frac{1}{3}} y^{\frac{2}{5}}}{9x^{\frac{4}{5}} y^{-\frac{3}{2}}} \right)^{\frac{1}{2}} \qquad \begin{array}{l} \text{Using order of operations, simplify inside parenthesis first.} \\ \text{Need common denominators before we can subtract exponents} \end{array}$$

$$\left(\frac{25x^{\frac{5}{15}} y^{\frac{4}{10}}}{9x^{\frac{12}{15}} y^{-\frac{15}{10}}} \right)^{\frac{1}{2}} \qquad \begin{array}{l} \text{Subtract exponents, be careful of the negative:} \\ \dfrac{4}{10} - \left(-\dfrac{15}{10} \right) = \dfrac{4}{10} + \dfrac{15}{10} = \dfrac{19}{10} \end{array}$$

$$\left(\frac{25x^{-\frac{7}{15}} y^{\frac{19}{10}}}{9} \right)^{\frac{1}{2}} \qquad \text{The negative exponent will flip the fraction}$$

$$\left(\frac{9}{25x^{\frac{7}{15}} y^{\frac{19}{10}}} \right)^{\frac{1}{2}} \qquad \text{The exponent } \frac{1}{2} \text{ goes on each factor}$$

$$\frac{9^{\frac{1}{2}}}{25^{\frac{1}{2}} x^{-\frac{7}{30}} y^{\frac{19}{20}}} \qquad \text{Evaluate } 9^{\frac{1}{2}} \text{ and } 25^{\frac{1}{2}} \text{ and move negative exponent}$$

$$\frac{3x^{\frac{7}{30}}}{5y^{\frac{19}{20}}} \qquad \text{Our Solution}$$

It is important to remember that as we simplify with rational exponents we are using the exact same properties we used when simplifying integer exponents. The only difference is we need to follow our rules for fractions as well. It may be worth reviewing your notes on exponent properties to be sure your comfortable with using the properties.

8.6 Practice - Rational Exponents

Write each expression in radical form.

1) $m^{\frac{3}{5}}$

2) $(10r)^{-\frac{3}{4}}$

3) $(7x)^{\frac{3}{2}}$

4) $(6b)^{-\frac{4}{3}}$

Write each expression in exponential form.

5) $\frac{1}{(\sqrt{6x})^3}$

6) $\sqrt{v}$

7) $\frac{1}{(\sqrt[4]{n})^7}$

8) $\sqrt{5a}$

Evaluate.

9) $8^{\frac{2}{3}}$

10) $16^{\frac{1}{4}}$

11) $4^{\frac{3}{2}}$

12) $100^{-\frac{3}{2}}$

Simplify. Your answer should contain only positive exponents.

13) $yx^{\frac{1}{3}} \cdot xy^{\frac{3}{2}}$

14) $4v^{\frac{2}{3}} \cdot v^{-1}$

15) $(a^{\frac{1}{2}}b^{\frac{1}{2}})^{-1}$

16) $(x^{\frac{5}{3}}y^{-2})^0$

17) $\frac{a^2 b^0}{3a^4}$

18) $\frac{2x^{\frac{1}{2}}y^{\frac{1}{3}}}{2x^{\frac{4}{3}}y^{-\frac{7}{4}}}$

19) $uv \cdot u \cdot (v^{\frac{3}{2}})^3$

20) $(x \cdot xy^2)^0$

21) $(x^0 y^{\frac{1}{3}})^{\frac{3}{2}} x^0$

22) $u^{-\frac{5}{4}}v^2 \cdot (u^{\frac{3}{2}})^{-\frac{3}{2}}$

23) $\frac{a^{\frac{3}{4}}b^{-1} \cdot b^{\frac{7}{4}}}{3b^{-1}}$

24) $\frac{2x^{-2}y^{\frac{5}{3}}}{x^{-\frac{5}{4}}y^{-\frac{5}{3}} \cdot xy^{\frac{1}{2}}}$

25) $\frac{3y^{\frac{5}{4}}}{y^{-1} \cdot 2y^{-\frac{1}{3}}}$

26) $\frac{ab^{\frac{1}{3}} \cdot 2b^{\frac{5}{4}}}{4a^{\frac{1}{2}}b^{\frac{2}{3}}}$

27) $\left(\frac{m^{\frac{3}{2}}n^{-2}}{(mn^{\frac{4}{3}})^{-1}}\right)^{\frac{7}{4}}$

28) $\frac{(y^{\frac{1}{2}})^{\frac{3}{2}}}{x^{\frac{3}{2}}y^{\frac{1}{2}}}$

29) $\frac{(m^2 n^{\frac{1}{2}})^0}{n^{\frac{3}{4}}}$

30) $\frac{y^0}{(x^{\frac{3}{4}}y^{-1})^{\frac{1}{3}}}$

31) $\frac{(x^{\frac{4}{3}}y^{\frac{1}{3}} \cdot y)^{-1}}{x^{\frac{1}{3}}y^{-2}}$

32) $\frac{(x^{\frac{1}{2}}y^0)^{\frac{4}{3}}}{y^4 \cdot x^{-2}y^{-\frac{2}{3}}}$

33) $\frac{(uv^2)^{\frac{1}{2}}}{v^{\frac{1}{4}}v^2}$

34) $\left(\frac{y^{\frac{1}{3}}y^{-2}}{(x^{\frac{5}{3}}y^3)^{\frac{3}{2}}}\right)^{\frac{3}{2}}$

Radicals - Radicals of Mixed Index

Objective: Reduce the index on a radical and multiply or divide radicals of different index.

Knowing that a radical has the same properties as exponents (written as a ratio) allows us to manipulate radicals in new ways. One thing we are allowed to do is reduce, not just the radicand, but the index as well. This is shown in the following example.

Example 417.

$$\sqrt[8]{x^6 y^2} \qquad \text{Rewrite as rational exponent}$$
$$(x^6 y^2)^{\frac{1}{8}} \qquad \text{Multiply exponents}$$
$$x^{\frac{6}{8}} y^{\frac{2}{8}} \qquad \text{Reduce each fraction}$$
$$x^{\frac{3}{4}} y^{\frac{1}{4}} \qquad \text{All exponents have denominator of 4, this is our new index}$$
$$\sqrt[4]{x^3 y} \qquad \text{Our Solution}$$

What we have done is reduced our index by dividing the index and all the exponents by the same number (2 in the previous example). If we notice a common factor in the index and all the exponnets on every factor we can reduce by dividing by that common factor. This is shown in the next example

Example 418.

$$\sqrt[24]{a^6 b^9 c^{15}} \qquad \text{Index and all exponents are divisible by 3}$$
$$\sqrt[8]{a^2 b^3 c^5} \qquad \text{Our Solution}$$

We can use the same process when there are coefficients in the problem. We will first write the coefficient as an exponential expression so we can divide the exponet by the common factor as well.

Example 419.

$$\sqrt[9]{8 m^6 n^3} \qquad \text{Write 8 as } 2^3$$
$$\sqrt[9]{2^3 m^6 n^3} \qquad \text{Index and all exponents are divisible by 3}$$
$$\sqrt[3]{2 m^2 n} \qquad \text{Our Solution}$$

We can use a very similar idea to also multiply radicals where the index does not match. First we will consider an example using rational exponents, then identify the pattern we can use.

Example 420.

$$\sqrt[3]{ab^2}\;\sqrt[4]{a^2b} \qquad \text{Rewrite as rational exponents}$$

$$(ab^2)^{\frac{1}{3}}(a^2b)^{\frac{1}{4}} \qquad \text{Multiply exponents}$$

$$a^{\frac{1}{3}}b^{\frac{2}{3}}a^{\frac{2}{4}}b^{\frac{1}{4}} \qquad \text{To have one radical need } a \text{ common denominator, } 12$$

$$a^{\frac{4}{12}}b^{\frac{8}{12}}a^{\frac{6}{12}}b^{\frac{3}{12}} \qquad \text{Write under } a \text{ single radical with common index, } 12$$

$$\sqrt[12]{a^4b^8a^6b^3} \qquad \text{Add exponents}$$

$$\sqrt[12]{a^{10}b^{11}} \qquad \text{Our Solution}$$

To combine the radicals we need a common index (just like the common denominator). We will get a common index by multiplying each index and exponent by an integer that will allow us to build up to that desired index. This process is shown in the next example.

Example 421.

$$\sqrt[4]{a^2b^3}\;\sqrt[6]{a^2b} \qquad \text{Common index is 12.}$$

$$\qquad \text{Multiply first index and exponents by 3, second by 2}$$

$$\sqrt[12]{a^6b^9a^4b^2} \qquad \text{Add exponents}$$

$$\sqrt[12]{a^{10}b^{11}} \qquad \text{Our Solution}$$

Often after combining radicals of mixed index we will need to simplify the resulting radical.

Example 422.

$$\sqrt[5]{x^3y^4}\;\sqrt[3]{x^2y} \qquad \text{Common index: 15.}$$

$$\qquad \text{Multiply first index and exponents by 3, second by 5}$$

$$\sqrt[15]{x^9y^{12}x^{10}y^5} \qquad \text{Add exponents}$$

$$\sqrt[15]{x^{19}y^{17}} \qquad \text{Simplify by dividing exponents by index, remainder is left inside}$$

$$xy\;\sqrt[15]{x^4y^2} \qquad \text{Our Solution}$$

Just as with reducing the index, we will rewrite coefficients as exponential expressions. This will also allow us to use exponent properties to simplify.

Example 423.

$$\sqrt[3]{4x^2y}\;\sqrt[4]{8xy^3} \qquad \text{Rewrite 4 as } 2^2 \text{ and 8 as } 2^3$$

$$\sqrt[3]{2^2x^2y}\;\sqrt[4]{2^3xy^3} \qquad \text{Common index: 12.}$$

$$\qquad \text{Multiply first index and exponents by 4, second by 3}$$

$$\sqrt[12]{2^8x^8y^4 2^9x^3y^9} \qquad \text{Add exponents (even on the 2)}$$

$$\sqrt[12]{2^{17}x^{11}y^{13}} \qquad \text{Simplify by dividing exponents by index, remainder is left inside}$$

$$2y\;\sqrt[12]{2^5x^{11}y} \qquad \text{Simplify } 2^5$$

$$2y\;\sqrt[12]{32x^{11}y} \qquad \text{Our Solution}$$

If there is a binomial in the radical then we need to keep that binomial together through the entire problem.

Example 424.

$$\sqrt{3x(y+z)}\,\sqrt[3]{9x(y+z)^2} \qquad \text{Rewrite 9 as } 3^2$$

$$\sqrt{3x(y+z)}\,\sqrt[3]{3^2x(y+z)^2} \qquad \text{Common index: 6. Multiply first group by 3, second by 2}$$

$$\sqrt[6]{3^3x^3(y+z)^3 3^4 x^2(y+z)^4} \qquad \text{Add exponents, keep } (y+z) \text{ as binomial}$$

$$\sqrt[6]{3^7 x^5(y+z)^7} \qquad \text{Simplify, dividing exponent by index, remainder inside}$$

$$3(y+z)\,\sqrt[6]{3x^5(y+z)} \qquad \text{Our Solution}$$

World View Note: Originally the radical was just a check mark with the rest of the radical expression in parenthesis. In 1637 Rene Descartes was the first to put a line over the entire radical expression.

The same process is used for dividing mixed index as with multiplying mixed index. The only difference is our final answer cannot have a radical over the denominator.

Example 425.

$$\frac{\sqrt[6]{x^4 y^3 z^2}}{\sqrt[8]{x^7 y^2 z}} \qquad \text{Common index is 24. Multiply first group by 4, second by 3}$$

$$\sqrt[24]{\frac{x^{16} y^{12} z^8}{x^{21} y^6 z^3}} \qquad \text{Subtract exponents}$$

$$\sqrt[24]{x^{-5} y^6 z^5} \qquad \text{Negative exponent moves to denominator}$$

$$\sqrt[24]{\frac{y^6 z^5}{x^5}} \qquad \text{Cannot have denominator in radical, need } 12\,x's, \text{ or 7 more}$$

$$\sqrt[24]{\frac{y^6 z^5}{x^5}} \left(\sqrt[24]{\frac{x^{19}}{x^{19}}} \right) \qquad \text{Multiply numerator and denominator by } \sqrt[12]{x^7}$$

$$\frac{\sqrt[24]{x^{19} y^6 z^5}}{x} \qquad \text{Our Solution}$$

8.7 Practice - Radicals of Mixed Index

Reduce the following radicals.

1) $\sqrt[8]{16x^4y^6}$

2) $\sqrt[4]{9x^2y^6}$

3) $\sqrt[12]{64x^4y^6z^8}$

4) $\sqrt[4]{\frac{25x^3}{16x^5}}$

5) $\sqrt[6]{\frac{16x^2}{9y^4}}$

6) $\sqrt[15]{x^9y^{12}z^6}$

7) $\sqrt[12]{x^6y^9}$

8) $\sqrt[10]{64x^8y^4}$

9) $\sqrt[8]{x^6y^4z^2}$

10) $\sqrt[4]{25y^2}$

11) $\sqrt[9]{8x^3y^6}$

12) $\sqrt[16]{81\,x^8y^{12}}$

Combine the following radicals.

13) $\sqrt[3]{5}\sqrt{6}$

14) $\sqrt[3]{7}\sqrt[4]{5}$

15) $\sqrt{x}\sqrt[3]{7y}$

16) $\sqrt[3]{y}\,\sqrt[5]{3z}$

17) $\sqrt{x}\sqrt[3]{x-2}$

18) $\sqrt[4]{3x}\,\sqrt{y+4}$

19) $\sqrt[5]{x^2y}\sqrt{xy}$

20) $\sqrt{ab}\,\sqrt[5]{2a^2b^2}$

21) $\sqrt[4]{xy^2}\sqrt[3]{x^2y}$

22) $\sqrt[5]{a^2b^3}\sqrt[4]{a^2b}$

23) $\sqrt[4]{a^2bc^2}\sqrt[5]{a^2b^3c}$

24) $\sqrt[6]{x^2yz^3}\sqrt[5]{x^2yz^2}$

25) $\sqrt{a}\,\sqrt[4]{a^3}$

26) $\sqrt[3]{x^2}\,\sqrt[6]{x^5}$

27) $\sqrt[5]{b^2}\sqrt{b^3}$

28) $\sqrt[4]{a^3}\,\sqrt[3]{a^2}$

29) $\sqrt{xy^3}\,\sqrt[3]{x^2y}$

30) $\sqrt[5]{a^3b}\,\sqrt{ab}$

31) $\sqrt[4]{9ab^3}\,\sqrt{3a^4b}$

32) $\sqrt{2x^3y^3}\,\sqrt[3]{4xy^2}$

33) $\sqrt[3]{3xy^2z}\,\sqrt[4]{9x^3yz^2}$

34) $\sqrt{a^4b^3c^4}\,\sqrt[3]{ab^2c}$

35) $\sqrt{27a^5(b+1)}\,\sqrt[3]{81a(b+1)^4}$

36) $\sqrt{8x\,(y+z)^5}\,\sqrt[3]{4x^2(y+z)^2}$

37) $\dfrac{\sqrt[3]{a^2}}{\sqrt[4]{a}}$

38) $\dfrac{\sqrt[3]{x^2}}{\sqrt[6]{x}}$

39) $\dfrac{\sqrt[4]{x^2y^3}}{\sqrt[3]{xy}}$

40) $\dfrac{\sqrt[5]{a^4b^2}}{\sqrt[3]{ab^2}}$

41) $\dfrac{\sqrt{ab^3c}}{\sqrt[5]{a^2b^3c^{-1}}}$

42) $\dfrac{\sqrt[5]{x^3y^4z^9}}{\sqrt{xy^{-2}z}}$

43) $\dfrac{\sqrt[4]{(3x-1)^3}}{\sqrt[5]{(3x-1)^3}}$

44) $\dfrac{\sqrt[3]{(2+5x)^2}}{\sqrt[4]{(2+5x)}}$

45) $\dfrac{\sqrt[3]{(2x+1)^2}}{\sqrt[5]{(2x+1)^2}}$

46) $\dfrac{\sqrt[4]{(5-3x)^3}}{\sqrt[3]{(5-3x)^2}}$

Radicals - Complex Numbers

Objective: Add, subtract, multiply, rationalize, and simplify expressions using complex numbers.

World View Note: When mathematics was first used, the primary purpose was for counting. Thus they did not originally use negatives, zero, fractions or irrational numbers. However, the ancient Egyptians quickly developed the need for "a part" and so they made up a new type of number, the ratio or fraction. The Ancient Greeks did not believe in irrational numbers (people were killed for believing otherwise). The Mayans of Central America later made up the number zero when they found use for it as a placeholder. Ancient Chinese Mathematicians made up negative numbers when they found use for them.

In mathematics, when the current number system does not provide the tools to solve the problems the culture is working with, we tend to make up new ways for dealing with the problem that can solve the problem. Throughout history this has been the case with the need for a number that is nothing (0), smaller than zero (negatives), between integers (fractions), and between fractions (irrational numbers). This is also the case for the square roots of negative numbers. To work with the square root of negative numbers mathematicians have defined what are called imaginary and complex numbers.

Definition of Imaginary Numbers: $i^2 = -1$ (thus $i = \sqrt{-1}$)

Examples of imaginary numbers include $3i$, $-6i$, $\frac{3}{5}i$ and $3i\sqrt{5}$. A **complex number** is one that contains both a real and imaginary part, such as $2 + 5i$.

With this definition, the square root of a negative number is no longer undefined. We now are allowed to do basic operations with the square root of negatives. First we will consider exponents on imaginary numbers. We will do this by manipulating our definition of $i^2 = -1$. If we multiply both sides of the definition by i, the equation becomes $i^3 = -i$. Then if we multiply both sides of the equation again by i, the equation becomes $i^4 = -i^2 = -(-1) = 1$, or simply $i^4 = 1$. Multiplying again by i gives $i^5 = i$. One more time gives $i^6 = i^2 = -1$. And if this pattern continues we see a cycle forming, the exponents on i change we cycle through simplified answers of i, -1, $-i$, 1. As there are 4 different possible answers in this cycle, if we divide the exponent by 4 and consider the remainder, we can simplify any exponent on i by learning just the following four values:

Cyclic Property of Powers of i

$$i^0 = 1$$
$$i = i$$
$$i^2 = -1$$
$$i^3 = -i$$

Example 426.

$$i^{35} \quad \text{Divide exponent by 4}$$
$$8\,R\,3 \quad \text{Use remainder as exponent on } i$$
$$i^3 \quad \text{Simplify}$$
$$-i \quad \text{Our Solution}$$

Example 427.

$$i^{124} \quad \text{Divide exponent by 4}$$
$$31\,R\,0 \quad \text{Use remainder as exponent on } i$$
$$i^0 \quad \text{Simplify}$$
$$1 \quad \text{Our Solution}$$

When performing operations (add, subtract, multilpy, divide) we can handle i just like we handle any other variable. This means when adding and subtracting complex numbers we simply add or combine like terms.

Example 428.

$$(2+5i)+(4-7i) \quad \text{Combine like terms } 2+4 \text{ and } 5i-7i$$
$$6-2i \quad \text{Our Solution}$$

It is important to notice what operation we are doing. Students often see the parenthesis and think that means FOIL. We only use FOIL to multiply. This problem is an addition problem so we simply add the terms, or combine like terms.

For subtraction problems the idea is the same, we need to remember to first distribute the negative onto all the terms in the parentheses.

Example 429.

$$(4-8i)-(3-5i) \quad \text{Distribute the negative}$$
$$4-8i-3+5i \quad \text{Combine like terms } 4-3 \text{ and } -8i+5i$$
$$1-3i \quad \text{Our Solution}$$

Addition and subtraction can be combined into one problem.

Example 430.

$$(5i)-(3+8i)+(-4+7i) \quad \text{Distribute the negative}$$
$$5i-3-8i-4+7i \quad \text{Combine like terms } 5i-8i+7i \text{ and } -3-4$$
$$-7+4i \quad \text{Our Solution}$$

Multiplying with complex numbers is the same as multiplying with variables with one exception, we will want to simplify our final answer so there are no exponents on i.

Example 431.

$(3i)(7i)$ Multilpy coefficients and $i's$

$21i^2$ Simplify $i^2 = -1$

$21(-1)$ Multiply

-21 Our Solution

Example 432.

$5i(3i-7)$ Distribute

$15i^2 - 35i$ Simplify $i^2 = -1$

$15(-1) - 35i$ Multiply

$-15 - 35i$ Our Solution

Example 433.

$(2-4i)(3+5i)$ FOIL

$6 + 10i - 12i - 20i^2$ Simplify $i^2 = -1$

$6 + 10i - 12i - 20(-1)$ Multiply

$6 + 10i - 12i + 20$ Combine like terms $6 + 20$ and $10i - 12i$

$26 - 2i$ Our Solution

Example 434.

$(3i)(6i)(2-3i)$ Multiply first two monomials

$18i^2(2-3i)$ Distribute

$36i^2 - 54i^3$ Simplify $i^2 = -1$ and $i^3 = -i$

$36(-1) - 54(-i)$ Multiply

$-36 + 54i$ Our Solution

Remember when squaring a binomial we either have to FOIL or use our shortcut to square the first, twice the product and square the last. The next example uses the shortcut

Example 435.

$(4-5i)^2$ Use perfect square shortcut

$4^2 = 16$ Square the first

$2(4)(-5i) = -40i$ Twice the product

$(5i)^2 = 25i^2 = 25(-1) = -25$ Square the last, simplify $i^2 = -1$

$16 - 40i - 25$ Combine like terms

$-9 - 40i$ Our Solution

Dividing with complex numbers also has one thing we need to be careful of. If i is $\sqrt{-1}$, and it is in the denominator of a fraction, then we have a radical in the denominator! This means we will want to rationalize our denominator so there are no i's. This is done the same way we rationalized denominators with square roots.

Example 436.

$$\frac{7+3i}{-5i} \qquad \text{Just } a \text{ monomial in denominator, multiply by } i$$

$$\frac{7+3i}{-5i}\left(\frac{i}{i}\right) \qquad \text{Distribute } i \text{ in numerator}$$

$$\frac{7i+3i^2}{-5i^2} \qquad \text{Simplify } i^2 = -1$$

$$\frac{7i+3(-1)}{-5(-1)} \qquad \text{Multiply}$$

$$\frac{7i-3}{5} \qquad \text{Our Solution}$$

The solution for these problems can be written several different ways, for example $\frac{-3+7i}{5}$ or $\frac{-3}{5}+\frac{7}{5}i$. The author has elected to use the solution as written, but it is important to express your answer in the form your instructor prefers.

Example 437.

$$\frac{2-6i}{4+8i} \qquad \text{Binomial in denominator, multiply by conjugate, } 4-8i$$

$$\frac{2-6i}{4+8i}\left(\frac{4-8i}{4-8i}\right) \qquad \text{FOIL in numerator, denominator is } a \text{ difference of squares}$$

$$\frac{8-16i-24i+48i^2}{16-64i^2} \qquad \text{Simplify } i^2 = -1$$

$$\frac{8-16i-24i+48(-1)}{16-64(-1)} \qquad \text{Multiply}$$

$$\frac{8-16i-24i-48}{16+64} \qquad \text{Combine like terms } 8-48 \text{ and } -16i-24i \text{ and } 16+64$$

$$\frac{-40-40i}{80} \qquad \text{Reduce, divide each term by } 40$$

$$\frac{-1-i}{2} \qquad \text{Our Solution}$$

321

Using i we can simplify radicals with negatives under the root. We will use the product rule and simplify the negative as a factor of negative one. This is shown in the following examples.

Example 438.

$$\sqrt{-16} \quad \text{Consider the negative as a factor of} -1$$
$$\sqrt{-1 \cdot 16} \quad \text{Take each root, square root of} -1 \text{ is } i$$
$$4i \quad \text{Our Solution}$$

Example 439.

$$\sqrt{-24} \quad \text{Find perfect square factors, including} -1$$
$$\sqrt{-1 \cdot 4 \cdot 6} \quad \text{Square root of} -1 \text{ is } i, \text{square root of 4 is 2}$$
$$2i\sqrt{6} \quad \text{Our Solution}$$

When simplifying complex radicals it is important that we take the -1 out of the radical (as an i) before we combine radicals.

Example 440.

$$\sqrt{-6}\sqrt{-3} \quad \text{Simplify the negatives, bringing } i \text{ out of radicals}$$
$$(i\sqrt{6})(i\sqrt{3}) \quad \text{Multiply } i \text{ by } i \text{ is } i^2 = -1, \text{also multiply radicals}$$
$$-\sqrt{18} \quad \text{Simplify the radical}$$
$$-\sqrt{9 \cdot 2} \quad \text{Take square root of 9}$$
$$-3\sqrt{2} \quad \text{Our Solution}$$

If there are fractions, we need to make sure to reduce each term by the same number. This is shown in the following example.

Example 441.

$$\frac{-15-\sqrt{-200}}{20} \quad \text{Simplify the radical first}$$
$$\sqrt{-200} \quad \text{Find perfect square factors, including} -1$$
$$\sqrt{-1 \cdot 100 \cdot 2} \quad \text{Take square root of} -1 \text{ and 100}$$
$$10i\sqrt{2} \quad \text{Put this back into the expression}$$
$$\frac{-15-10i\sqrt{2}}{20} \quad \text{All the factors are divisible by 5}$$
$$\frac{-3-2i\sqrt{2}}{4} \quad \text{Our Solution}$$

By using $i = \sqrt{-1}$ we will be able to simplify and solve problems that we could not simplify and solve before. This will be explored in more detail in a later section.

8.8 Practice - Complex Numbers

Simplify.

1) $3 - (-8 + 4i)$

2) $(3i) - (7i)$

3) $(7i) - (3 - 2i)$

4) $5 + (-6 - 6i)$

5) $(-6i) - (3 + 7i)$

6) $(-8i) - (7i) - (5 - 3i)$

7) $(3 - 3i) + (-7 - 8i)$

8) $(-4 - i) + (1 - 5i)$

9) $(i) - (2 + 3i) - 6$

10) $(5 - 4i) + (8 - 4i)$

11) $(6i)(-8i)$

12) $(3i)(-8i)$

13) $(-5i)(8i)$

14) $(8i)(-4i)$

15) $(-7i)^2$

16) $(-i)(7i)(4 - 3i)$

17) $(6 + 5i)^2$

18) $(8i)(-2i)(-2 - 8i)$

19) $(-7 - 4i)(-8 + 6i)$

20) $(3i)(-3i)(4 - 4i)$

21) $(-4 + 5i)(2 - 7i)$

22) $-8(4 - 8i) - 2(-2 - 6i)$

23) $(-8 - 6i)(-4 + 2i)$

24) $(-6i)(3 - 2i) - (7i)(4i)$

25) $(1 + 5i)(2 + i)$

26) $(-2 + i)(3 - 5i)$

27) $\frac{-9 + 5i}{i}$

28) $\frac{3 + 2i}{-3i}$

29) $\frac{10 \quad 9i}{6i}$

30) $\frac{-4 + 2i}{3i}$

31) $\frac{3 \quad 6i}{4i}$

32) $\frac{-5 + 9i}{9i}$

33) $\frac{10 - i}{-i}$

34) $\frac{10}{5i}$

35) $\frac{4i}{10 + i}$

36) $\frac{9i}{1 - 5i}$

37) $\frac{8}{7 \quad 6i}$

38) $\frac{4}{4 + 6i}$

39) $\frac{7}{10 \quad 7i}$

40) $\frac{9}{8 \quad 6i}$

41) $\frac{5i}{-6 - i}$

42) $\frac{8i}{6 - 7i}$

43) $\sqrt{-81}$

44) $\sqrt{-45}$

45) $\sqrt{-10}\sqrt{-2}$

46) $\sqrt{-12}\sqrt{-2}$

47 $\dfrac{3+\sqrt{27}}{6}$

48) $\dfrac{-4-\sqrt{-8}}{4}$

49) $\dfrac{8 \quad \sqrt{16}}{4}$

50) $\dfrac{6+\sqrt{-32}}{4}$

51) i^{73}

52) i^{251}

53) i^{48}

54) i^{68}

55) i^{62}

56) i^{181}

57) i^{154}

58) i^{51}

Chapter 9 : Quadratics

Quadratics - Solving with Radicals

Objective: Solve equations with radicals and check for extraneous solutions.

Here we look at equations that have roots in the problem. As you might expect, to clear a root we can raise both sides to an exponent. So to clear a square root we can rise both sides to the second power. To clear a cubed root we can raise both sides to a third power. There is one catch to solving a problem with roots in it, sometimes we end up with solutions that do not actually work in the equation. This will only happen if the index on the root is even, and it will not happen all the time. So for these problems it will be required that we check our answer in the original problem. If a value does not work it is called an extraneous solution and not included in the final solution.

When solving a radical problem with an even index: check answers!

Example 442.

$$\sqrt{7x+2} = 4 \qquad \text{Even index! We will have to check answers}$$
$$(\sqrt{7x+2})^2 = 4^2 \qquad \text{Square both sides, simplify exponents}$$
$$7x+2 = 16 \qquad \text{Solve}$$
$$\underline{-2 \quad -2} \qquad \text{Subtract 2 from both sides}$$
$$\frac{7x = 14}{7 \quad 7} \qquad \text{Divide both sides by 7}$$
$$x = 2 \qquad \text{Need to check answer in original problem}$$
$$\sqrt{7(2)+2} = 4 \qquad \text{Multiply}$$
$$\sqrt{14+2} = 4 \qquad \text{Add}$$
$$\sqrt{16} = 4 \qquad \text{Square root}$$
$$4 = 4 \qquad \text{True! It works!}$$
$$x = 2 \qquad \text{Our Solution}$$

Example 443.

$$\sqrt[3]{x-1} = -4 \qquad \text{Odd index, we don't need to check answer}$$
$$(\sqrt[3]{x-1})^3 = (-4)^3 \qquad \text{Cube both sides, simplify exponents}$$
$$x-1 = -64 \qquad \text{Solve}$$

$$\begin{array}{ll} \underline{+1 \quad +1} & \text{Add 1 to both sides} \\ x = -63 & \text{Our Solution} \end{array}$$

Example 444.

$$\begin{array}{ll} \sqrt[4]{3x+6} = -3 & \text{Even index! We will have to check answers} \\ (\sqrt[4]{3x+6}) = (-3)^4 & \text{Rise both sides to fourth power} \\ 3x + 6 = 81 & \text{Solve} \\ \underline{-6 \quad -6} & \text{Subtract 6 from both sides} \\ \dfrac{3x = 75}{3 \quad 3} & \text{Divide both sides by 3} \\ \\ x = 25 & \text{Need to check answer in original problem} \\ \sqrt[4]{3(25)+6} = -3 & \text{Multiply} \\ \sqrt[4]{75+6} = -3 & \text{Add} \\ \sqrt[4]{81} = -3 & \text{Take root} \\ 3 = -3 & \text{False, extraneous solution} \\ \text{No Solution} & \text{Our Solution} \end{array}$$

If the radical is not alone on one side of the equation we will have to solve for the radical before we raise it to an exponent

Example 445.

$$\begin{array}{ll} x + \sqrt{4x+1} = 5 & \text{Even index! We will have to check solutions} \\ \underline{-x \qquad\qquad -x} & \text{Isolate radical by subtracting } x \text{ from both sides} \\ \sqrt{4x+1} = 5 - x & \text{Square both sides} \\ (\sqrt{4x+1})^2 = (5-x)^2 & \text{Evaluate exponents, recal } (a-b)^2 = a^2 - 2ab + b^2 \\ 4x + 1 = 25 - 10x + x^2 & \text{Re} - \text{order terms} \\ 4x + 1 = x^2 - 10x + 25 & \text{Make equation equal zero} \\ \underline{-4x - 1 \qquad -4x \ -1} & \text{Subtract } 4x \text{ and 1 from both sides} \\ 0 = x^2 - 14x + 24 & \text{Factor} \\ 0 = (x-12)(x-2) & \text{Set each factor equal to zero} \\ x - 12 = 0 \ \text{ or } \ x - 2 = 0 & \text{Solve each equation} \\ \underline{+12 + 12 \qquad +2 + 2} & \\ x = 12 \ \text{ or } \ x = 2 & \text{Need to check answers in original problem} \\ \\ (12) + \sqrt{4(12)+1} = 5 & \text{Check } x = 5 \text{ first} \end{array}$$

327

$$12 + \sqrt{48 + 1} = 5 \qquad \text{Add}$$
$$12 + \sqrt{49} = 5 \qquad \text{Take root}$$
$$12 + 7 = 5 \qquad \text{Add}$$
$$19 = 5 \qquad \text{False, extraneous root}$$

$$(2) + \sqrt{4(2) + 1} = 5 \qquad \text{Check } x = 2$$
$$2 + \sqrt{8 + 1} = 5 \qquad \text{Add}$$
$$2 + \sqrt{9} = 5 \qquad \text{Take root}$$
$$2 + 3 = 5 \qquad \text{Add}$$
$$5 = 5 \qquad \text{True! It works}$$

$$x = 2 \qquad \text{Our Solution}$$

The above example illustrates that as we solve we could end up with an x^2 term or a quadratic. In this case we remember to set the equation to zero and solve by factoring. We will have to check both solutions if the index in the problem was even. Sometimes both values work, sometimes only one, and sometimes neither works.

World View Note: The babylonians were the first known culture to solve quadratics in radicals - as early as 2000 BC!

If there is more than one square root in a problem we will clear the roots one at a time. This means we must first isolate one of them before we square both sides.

Example 446.

$$\sqrt{3x - 8} - \sqrt{x} = 0 \qquad \text{Even index! We will have to check answers}$$
$$+ \sqrt{x} + \sqrt{x} \qquad \text{Isolate first root by adding } \sqrt{x} \text{ to both sides}$$
$$\sqrt{3x - 8} = \sqrt{x} \qquad \text{Square both sides}$$
$$(\sqrt{3x - 8})^2 = (\sqrt{x})^2 \qquad \text{Evaluate exponents}$$
$$3x - 8 = x \qquad \text{Solve}$$
$$\underline{-3x \qquad -3x} \qquad \text{Subtract } 3x \text{ from both sides}$$
$$-8 = -2x \qquad \text{Divide both sides by } -2$$
$$\overline{-2} \quad \overline{-2}$$
$$4 = x \qquad \text{Need to check answer in original}$$
$$\sqrt{3(4) - 8} - \sqrt{4} = 0 \qquad \text{Multiply}$$
$$\sqrt{12 - 8} - \sqrt{4} = 0 \qquad \text{Subtract}$$
$$\sqrt{4} - \sqrt{4} = 0 \qquad \text{Take roots}$$

$$2 - 2 = 0 \quad \text{Subtract}$$
$$0 = 0 \quad \text{True! It works}$$
$$x = 4 \quad \text{Our Solution}$$

When there is more than one square root in the problem, after isolating one root and squaring both sides we may still have a root remaining in the problem. In this case we will again isolate the term with the second root and square both sides. When isolating, we will isolate the *term* with the square root. This means the square root can be multiplied by a number after isolating.

Example 447.

$$\sqrt{2x+1} - \sqrt{x} = 1 \qquad \text{Even index! We will have to check answers}$$
$$\underline{\quad + \sqrt{x} + \sqrt{x}} \qquad \text{Isolate first root by adding } \sqrt{x} \text{ to both sides}$$
$$\sqrt{2x+1} = \sqrt{x} + 1 \qquad \text{Square both sides}$$
$$(\sqrt{2x+1})^2 = (\sqrt{x}+1)^2 \qquad \text{Evaluate exponents, recall } (a+b)^2 = a^2 + 2ab + b^2$$
$$2x+1 = x + 2\sqrt{x} + 1 \qquad \text{Isolate the term with the root}$$
$$\underline{-x - 1 - x \qquad\qquad -1} \qquad \text{Subtract } x \text{ and } 1 \text{ from both sides}$$
$$x = 2\sqrt{x} \qquad \text{Square both sides}$$
$$(x)^2 = (2\sqrt{x})^2 \qquad \text{Evaluate exponents}$$
$$x^2 = 4x \qquad \text{Make equation equal zero}$$
$$\underline{-4x - 4x} \qquad \text{Subtract } x \text{ from both sides}$$
$$x^2 - 4x = 0 \qquad \text{Factor}$$
$$x(x-4) = 0 \qquad \text{Set each factor equal to zero}$$
$$x = 0 \text{ or } x - 4 = 0 \qquad \text{Solve}$$
$$\underline{\qquad\qquad +4 + 4} \qquad \text{Add } 4 \text{ to both sides of second equation}$$
$$x = 0 \text{ or } x = 4 \qquad \text{Need to check answers in original}$$

$$\sqrt{2(0)+1} - \sqrt{(0)} = 1 \qquad \text{Check } x = 0 \text{ first}$$
$$\sqrt{1} - \sqrt{0} = 1 \qquad \text{Take roots}$$
$$1 - 0 = 1 \qquad \text{Subtract}$$
$$1 = 1 \qquad \text{True! It works}$$

$$\sqrt{2(4)+1} - \sqrt{(4)} = 1 \qquad \text{Check } x = 4$$
$$\sqrt{8+1} - \sqrt{4} = 1 \qquad \text{Add}$$
$$\sqrt{9} - \sqrt{4} = 1 \qquad \text{Take roots}$$
$$3 - 2 = 1 \qquad \text{Subtract}$$
$$1 = 1 \qquad \text{True! It works}$$

$$x = 0 \text{ or } 4 \quad \text{Our Solution}$$

Example 448.

$\sqrt{3x+9} - \sqrt{x+4} = -1$	Even index! We will have to check answers
$\underline{+\sqrt{x+4} + \sqrt{x+4}}$	Isolate the first root by adding $\sqrt{x+4}$
$\sqrt{3x+9} = \sqrt{x+4} - 1$	Square both sides
$(\sqrt{3x+9})^2 = (\sqrt{x+4} - 1)^2$	Evaluate exponents
$3x+9 = x+4 - 2\sqrt{x+4} + 1$	Combine like terms
$3x+9 = x+5 - 2\sqrt{x+4}$	Isolate the term with radical
$\underline{-x-5 \quad -x-5}$	Subtract x and 5 from both sides
$2x+4 = -2\sqrt{x+4}$	Square both sides
$(2x+4)^2 = (-2\sqrt{x+4})^2$	Evaluate exponents
$4x^2 + 16x + 16 = 4(x+4)$	Distribute
$4x^2 + 16x + 16 = 4x + 16$	Make equation equal zero
$\underline{-4x-16 \quad -4x-16}$	Subtract $4x$ and 16 from both sides
$4x^2 + 12x = 0$	Factor
$4x(x+3) = 0$	Set each factor equal to zero
$4x = 0 \text{ or } x+3 = 0$	Solve
$\overline{4} \ \ \overline{4} \qquad -3-3$	
$x = 0 \text{ or } x = -3$	Check solutions in original

$\sqrt{3(0)+9} - \sqrt{(0)+4} = -1$	Check $x = 0$ first
$\sqrt{9} - \sqrt{4} = -1$	Take roots
$3 - 2 = -1$	Subtract
$1 = -1$	False, extraneous solution

$\sqrt{3(-3)+9} - \sqrt{(-3)+4} = -1$	Check $x = -3$
$\sqrt{-9+9} - \sqrt{(-3)+4} = -1$	Add
$\sqrt{0} - \sqrt{1} = -1$	Take roots
$0 - 1 = -1$	Subtract
$-1 = -1$	True! It works

$$x = -3 \quad \text{Our Solution}$$

9.1 Practice - Solving with Radicals

Solve.

1) $\sqrt{2x+3} - 3 = 0$

2) $\sqrt{5x+1} - 4 = 0$

3) $\sqrt{6x-5} - x = 0$

4) $\sqrt{x+2} - \sqrt{x} = 2$

5) $3 + x = \sqrt{6x+13}$

6) $x - 1 = \sqrt{7-x}$

7) $\sqrt{3-3x} - 1 = 2x$

8) $\sqrt{2x+2} = 3 + \sqrt{2x-1}$

9) $\sqrt{4x+5} - \sqrt{x+4} = 2$

10) $\sqrt{3x+4} - \sqrt{x+2} = 2$

11) $\sqrt{2x+4} - \sqrt{x+3} = 1$

12) $\sqrt{7x+2} - \sqrt{3x+6} = 6$

13) $\sqrt{2x+6} - \sqrt{x+4} = 1$

14) $\sqrt{4x-3} - \sqrt{3x+1} = 1$

15) $\sqrt{6-2x} - \sqrt{2x+3} = 3$

16) $\sqrt{2-3x} - \sqrt{3x+7} = 3$

Quadratics - Solving with Exponents

Objective: Solve equations with exponents using the odd root property and the even root property.

Another type of equation we can solve is one with exponents. As you might expect we can clear exponents by using roots. This is done with very few unexpected results when the exponent is odd. We solve these problems very straight forward using the odd root property

Odd Root Property: if $a^n = b$, then $a = \sqrt[n]{b}$ when n is odd

Example 449.

$$x^5 = 32 \qquad \text{Use odd root property}$$
$$\sqrt[5]{x^5} = \sqrt[5]{32} \qquad \text{Simplify roots}$$
$$x = 2 \qquad \text{Our Solution}$$

However, when the exponent is even we will have two results from taking an even root of both sides. One will be positive and one will be negative. This is because both $3^2 = 9$ and $(-3)^2 = 9$. so when solving $x^2 = 9$ we will have two solutions, one positive and one negative: $x = 3$ and -3

Even Root Property: if $a^n = b$, then $a = \pm \sqrt[n]{b}$ when n is even

Example 450.

$$x^4 = 16 \qquad \text{Use even root property} (\pm)$$

$$\sqrt[4]{x^4} = \pm\sqrt[4]{16} \qquad \text{Simplify roots}$$
$$x = \pm 2 \qquad \text{Our Solution}$$

World View Note: In 1545, French Mathematicain Gerolamo Cardano published his book *The Great Art, or the Rules of Algebra* which included the solution of an equation with a fourth power, but it was considered absurd by many to take a quantity to the fourth power because there are only three dimensions!

Example 451.

$(2x+4)^2 = 36$	Use even root property ($\pm$)
$\sqrt{(2x+4)^2} = \pm\sqrt{36}$	Simplify roots
$2x+4 = \pm 6$	To avoid sign errors we need two equations
$2x+4 = 6$ or $2x+4 = -6$	One equation for $+$, one equation for $-$
$\underline{-4-4} \qquad \underline{-4 \quad -4}$	Subtract 4 from both sides
$2x = 2$ or $2x = -10$	Divide both sides by 2
$\overline{2} \;\; \overline{2} \qquad \overline{2} \quad \overline{2}$	
$x = 1$ or $x = -5$	Our Solutions

In the previous example we needed two equations to simplify because when we took the root, our solutions were two rational numbers, 6 and -6. If the roots did not simplify to rational numbers we can keep the $\pm$ in the equation.

Example 452.

$(6x-9)^2 = 45$	Use even root property ($\pm$)
$\sqrt{(6x-9)^2} = \pm\sqrt{45}$	Simplify roots
$6x-9 = \pm 3\sqrt{5}$	Use one equation because root did not simplify to rational
$\underline{+9+9}$	Add 9 to both sides
$6x = 9 \pm 3\sqrt{5}$	Divide both sides by 6
$\overline{6} \qquad \overline{6}$	
$x = \dfrac{9 \pm 3\sqrt{5}}{6}$	Simplify, divide each term by 3
$x = \dfrac{3 \pm \sqrt{5}}{2}$	Our Solution

When solving with exponents, it is important to first isolate the part with the exponent before taking any roots.

Example 453.

$$(x+4)^3 - 6 = 119 \qquad \text{Isolate part with exponent}$$
$$\underline{\ +6\ \ +6}$$
$$(x+4)^3 = 125 \qquad \text{Use odd root property}$$
$$\sqrt[3]{(x+4)^3} = \sqrt[3]{125} \qquad \text{Simplify roots}$$
$$x+4 = 5 \qquad \text{Solve}$$
$$\underline{-4 - 4} \qquad \text{Subtract 4 from both sides}$$
$$x = 1 \qquad \text{Our Solution}$$

Example 454.

$$(6x+1)^2 + 6 = 10 \qquad \text{Isolate part with exponent}$$
$$\underline{\ -6 - 6} \qquad \text{Subtract 6 from both sides}$$
$$(6x+1)^2 = 4 \qquad \text{Use even root property } (\pm)$$
$$\sqrt{(6x+1)^2} = \pm\sqrt{4} \qquad \text{Simplify roots}$$
$$6x+1 = \pm 2 \qquad \text{To avoid sign errors, we need two equations}$$
$$6x+1 = 2 \ \text{ or } \ 6x+1 = -2 \qquad \text{Solve each equation}$$
$$\underline{-1 - 1 -1\ \ -1} \qquad \text{Subtract 1 from both sides}$$
$$6x = 1 \ \text{ or } \ 6x = -3 \qquad \text{Divide both sides by 6}$$
$$\overline{6}\ \ \overline{6} \ \overline{6}\ \ \overline{6}$$
$$x = \frac{1}{6} \ \text{ or } \ x = -\frac{1}{2} \qquad \text{Our Solution}$$

When our exponents are a fraction we will need to first convert the fractional exponent into a radical expression to solve. Recall that $a^{\frac{m}{n}} = \left(\sqrt[n]{a}\right)^m$. Once we have done this we can clear the exponent using either the even ($\pm$) or odd root property. Then we can clear the radical by raising both sides to an exponent (remember to check answers if the index is even).

Example 455.

$$(4x+1)^{\frac{2}{5}} = 9 \qquad \text{Rewrite as a radical expression}$$
$$\left(\sqrt[5]{4x+1}\right)^2 = 9 \qquad \text{Clear exponent first with even root property } (\pm)$$
$$\sqrt{\left(\sqrt[5]{4x+1}\right)^2} = \pm\sqrt{9} \qquad \text{Simplify roots}$$

$$\sqrt[5]{4x+1} = \pm 3 \qquad \text{Clear radical by raising both sides to 5th power}$$
$$(\sqrt[5]{4x+1})^5 = (\pm 3)^5 \qquad \text{Simplify exponents}$$
$$4x + 1 = \pm 243 \qquad \text{Solve, need 2 equations!}$$
$$4x + 1 = 243 \text{ or } 4x + 1 = -243$$
$$\underline{-1 \quad -1 \qquad\quad -1 \quad\; -1} \qquad \text{Subtract 1 from both sides}$$
$$4x = 242 \text{ or } 4x = -244 \qquad \text{Divide both sides by 4}$$
$$\overline{\;4\;} \quad\; \overline{\;4\;} \qquad \overline{\;4\;} \qquad \overline{\;4\;}$$
$$x = \frac{121}{2}, -61 \qquad \text{Our Solution}$$

Example 456.

$$(3x-2)^{\frac{3}{4}} = 64 \qquad \text{Rewrite as radical expression}$$
$$(\sqrt[4]{3x-2})^3 = 64 \qquad \text{Clear exponent first with odd root property}$$
$$\sqrt[3]{(\sqrt[4]{3x-2})^3} = \sqrt[3]{64} \qquad \text{Simplify roots}$$
$$\sqrt[4]{3x-2} = 4 \qquad \text{Even Index! Check answers.}$$
$$(\sqrt[4]{3x-2})^4 = 4^4 \qquad \text{Raise both sides to 4th power}$$
$$3x - 2 = 256 \qquad \text{Solve}$$
$$\underline{+2 \quad +2} \qquad \text{Add 2 to both sides}$$
$$3x = 258 \qquad \text{Divide both sides by 3}$$
$$\overline{\;3\;} \quad\; \overline{\;3\;}$$
$$x = 86 \qquad \text{Need to check answer in radical form of problem}$$
$$(\sqrt[4]{3(86)-2})^3 = 64 \qquad \text{Multiply}$$
$$(\sqrt[4]{258-2})^3 = 64 \qquad \text{Subtract}$$
$$(\sqrt[4]{256})^3 = 64 \qquad \text{Evaluate root}$$
$$4^3 = 64 \qquad \text{Evaluate exponent}$$
$$64 = 64 \qquad \text{True! It works}$$
$$x = 86 \qquad \text{Our Solution}$$

With rational exponents it is very helpful to convert to radical form to be able to see if we need a $\pm$ because we used the even root property, or to see if we need to check our answer because there was an even root in the problem. When checking we will usually want to check in the radical form as it will be easier to evaluate.

9.2 Practice - Solving with Exponents

Solve.

1) $x^2 = 75$

2) $x^3 = -8$

3) $x^2 + 5 = 13$

4) $4x^3 - 2 = 106$

5) $3x^2 + 1 = 73$

6) $(x - 4)^2 = 49$

7) $(x + 2)^5 = -243$

8) $(5x + 1)^4 = 16$

9) $(2x + 5)^3 - 6 = 21$

10) $(2x + 1)^2 + 3 = 21$

11) $(x - 1)^{\frac{2}{3}} = 16$

12) $(x - 1)^{\frac{3}{2}} = 8$

13) $(2 - x)^{\frac{3}{2}} = 27$

14) $(2x + 3)^{\frac{4}{3}} = 16$

15) $(2x - 3)^{\frac{2}{3}} = 4$

16) $(x + 3)^{\frac{1}{3}} = 4$

17) $(x + \frac{1}{2})^{-\frac{2}{3}} = 4$

18) $(x - 1)^{\frac{5}{3}} = 32$

19) $(x - 1)^{-\frac{5}{2}} = 32$

20) $(x + 3)^{\frac{3}{2}} = -8$

21) $(3x - 2)^{\frac{4}{5}} = 16$

22) $(2x + 3)^{\frac{3}{2}} = 27$

23) $(4x + 2)^{\frac{3}{5}} = -8$

24) $(3 - 2x)^{\frac{4}{3}} = -81$

Quadratics - Complete the Square

Objective: Solve quadratic equations by completing the square.

When solving quadratic equations in the past we have used factoring to solve for our variable. This is exactly what is done in the next example.

Example 457.

$$\begin{aligned}
x^2 + 5x + 6 &= 0 \quad && \text{Factor} \\
(x+3)(x+2) &= 0 \quad && \text{Set each factor equal to zero} \\
x + 3 = 0 \quad &\text{or} \quad x + 2 = 0 \quad && \text{Solve each equation} \\
\underline{-3 - 3} \quad & \quad \underline{-2 - 2} \\
x = -3 \quad &\text{or} \quad x = -2 \quad && \text{Our Solutions}
\end{aligned}$$

However, the problem with factoring is all equations cannot be factored. Consider the following equation: $x^2 - 2x - 7 = 0$. The equation cannot be factored, however there are two solutions to this equation, $1 + 2\sqrt{2}$ and $1 - 2\sqrt{2}$. To find these two solutions we will use a method known as completing the square. When completing the square we will change the quadratic into a perfect square which can easily be solved with the square root property. The next example reviews the square root property.

Example 458.

$$\begin{aligned}
(x+5)^2 &= 18 \quad && \text{Square root of both sides} \\
\sqrt{(x+5)^2} &= \pm\sqrt{18} \quad && \text{Simplify each radical} \\
x + 5 &= \pm 3\sqrt{2} \quad && \text{Subtract 5 from both sides} \\
\underline{-5 \quad -5} \\
x &= -5 \pm 3\sqrt{2} \quad && \text{Our Solution}
\end{aligned}$$

To complete the square, or make our problem into the form of the previous example, we will be searching for the third term in a trinomial. If a quadratic is of the form $x^2 + bx + c$, and a perfect square, the third term, c, can be easily found by the formula $\left(\frac{1}{2} \cdot b\right)^2$. This is shown in the following examples, where we find the number that completes the square and then factor the perfect square.

Example 459.

$$x^2 + 8x + c \quad c = \left(\frac{1}{2} \cdot b\right)^2 \text{ and our } b = 8$$

$$\left(\frac{1}{2} \cdot 8\right)^2 = 4^2 = 16 \quad \text{The third term to complete the square is } 16$$

$$x^2 + 8x + 16 \quad \text{Our equation as } a \text{ perfect square, factor}$$

$$(x + 4)^2 \quad \text{Our Solution}$$

Example 460.

$$x^2 - 7x + c \quad c = \left(\frac{1}{2} \cdot b\right)^2 \text{ and our } b = 7$$

$$\left(\frac{1}{2} \cdot 7\right)^2 = \left(\frac{7}{2}\right)^2 = \frac{49}{4} \quad \text{The third term to complete the square is } \frac{49}{4}$$

$$x^2 - 11x + \frac{49}{4} \quad \text{Our equation as } a \text{ perfect square, factor}$$

$$\left(x - \frac{7}{2}\right)^2 \quad \text{Our Solution}$$

Example 461.

$$x^2 + \frac{5}{3}x + c \quad c = \left(\frac{1}{2} \cdot b\right)^2 \text{ and our } b = 8$$

$$\left(\frac{1}{2} \cdot \frac{5}{3}\right)^2 = \left(\frac{5}{6}\right)^2 = \frac{25}{36} \quad \text{The third term to complete the square is } \frac{25}{36}$$

$$x^2 + \frac{5}{3}x + \frac{25}{36} \quad \text{Our equation as a perfect square, factor}$$

$$\left(x + \frac{5}{6}\right)^2 \quad \text{Our Solution}$$

The process in the previous examples, combined with the even root property, is used to solve quadratic equations by completing the square. The following five steps describe the process used to complete the square, along with an example to demonstrate each step.

Problem	$3x^2 + 18x - 6 = 0$
1. Separate constant term from variables	$\begin{array}{r} +6 +6 \\ \hline 3x^2 + 18x \quad\ = 6 \end{array}$
2. Divide each term by a	$\begin{array}{l} \frac{3}{3}x^2 + \frac{18}{3}x \quad = \frac{6}{3} \\ x^2 + 6x \qquad\ = 2 \end{array}$
3. Find value to complete the square: $\left(\frac{1}{2} \cdot b\right)^2$	$\left(\frac{1}{2} \cdot 6\right)^2 = 3^2 = 9$
4. Add to both sides of equation	$\begin{array}{r} x^2 + 6x \qquad = 2 \\ +9 \ +9 \\ \hline x^2 + 6x + 9 = 11 \end{array}$
5. Factor	$(x+3)^2 = 11$
Solve by even root property	$\begin{array}{r} \sqrt{(x+3)^2} = \pm\sqrt{11} \\ x + 3 = \pm\sqrt{11} \\ -3 \ \ -3 \\ \hline x = -3 \pm \sqrt{11} \end{array}$

World View Note: The Chinese in 200 BC were the first known culture group to use a method similar to completing the square, but their method was only used to calculate positive roots.

The advantage of this method is it can be used to solve any quadratic equation. The following examples show how completing the square can give us rational solutions, irrational solutions, and even complex solutions.

Example 462.

$$2x^2 + 20x + 48 = 0 \quad \text{Separate constant term from varaibles}$$

$$\underline{-48 \ -48} \qquad \text{Subtract 24}$$

$$\frac{2x^2 + 20x}{2 \quad 2} \ \ = \frac{-48}{2} \qquad \text{Divide by } a \text{ or } 2$$

$$x^2 + 10x \ \ = -24 \qquad \text{Find number to complete the square: } \left(\frac{1}{2} \cdot b\right)^2$$

$$\left(\frac{1}{2} \cdot 10\right)^2 = 5^2 = 25 \qquad \text{Add 25 to both sides of the equation}$$

$$x^2 + 10x \ \ = -24$$

$$\underline{+25 \ \ +25}$$

$$x^2 + 10x + 25 = 1 \qquad \text{Factor}$$

$$(x+5)^2 = 1 \qquad \text{Solve with even root property}$$

$$\sqrt{(x+5)^2} = \pm\sqrt{1} \qquad \text{Simplify roots}$$

$$x + 5 = \pm 1 \qquad \text{Subtract 5 from both sides}$$

$$\underline{-5 \ -5}$$

$$x = -5 \pm 1 \qquad \text{Evaluate}$$

$$x = -4 \text{ or } -6 \qquad \text{Our Solution}$$

Example 463.

$$x^2 - 3x - 2 = 0 \qquad \text{Separate constant from variables}$$

$$\underline{+2+2} \qquad \text{Add 2 to both sides}$$

$$x^2 - 3x \ \ = 2 \qquad \text{No } a, \text{ find number to complete the square } \left(\frac{1}{2} \cdot b\right)^2$$

$$\left(\frac{1}{2} \cdot 3\right)^2 = \left(\frac{3}{2}\right)^2 = \frac{9}{4} \qquad \text{Add } \frac{9}{4} \text{ to both sides,}$$

$$\frac{2}{1}\left(\frac{4}{4}\right) + \frac{9}{4} = \frac{8}{4} + \frac{9}{4} = \frac{17}{4} \qquad \text{Need common denominator (4) on right}$$

$$x^2 - 3x + \frac{9}{4} = \frac{8}{4} + \frac{9}{4} = \frac{17}{4} \qquad \text{Factor}$$

$$\left(x - \frac{3}{2}\right)^2 = \frac{17}{4} \qquad \text{Solve using the even root property}$$

$$\sqrt{\left(x - \frac{3}{2}\right)^2} = \pm\sqrt{\frac{17}{4}} \qquad \text{Simplify roots}$$

$$x - \frac{3}{2} = \frac{\pm\sqrt{17}}{2} \qquad \text{Add } \frac{3}{2} \text{ to both sides,}$$

$$+\frac{3}{2}+\frac{3}{2} \qquad \text{we already have } a \text{ common denominator}$$

$$x=\frac{3\pm\sqrt{17}}{2} \qquad \text{Our Solution}$$

Example 464.

$$3x^2 = 2x - 7 \qquad \text{Separate the constant from the variables}$$

$$-2x - 2x \qquad \text{Subtract } 2x \text{ from both sides}$$

$$\frac{3x^2 - 2x = -7}{3 \quad 3 \quad 3} \qquad \text{Divide each term by } a \text{ or } 3$$

$$x^2 - \frac{2}{3}x \quad = -\frac{7}{3} \qquad \text{Find the number to complete the square } \left(\frac{1}{2}\cdot b\right)^2$$

$$\left(\frac{1}{2}\cdot\frac{2}{3}\right)^2 = \left(\frac{1}{3}\right)^2 = \frac{1}{9} \qquad \text{Add to both sides,}$$

$$-\frac{7}{3}\left(\frac{3}{3}\right)+\frac{1}{9}=\frac{-21}{3}+\frac{1}{9}=\frac{-20}{9} \qquad \text{get common denominator on right}$$

$$x^2 - \frac{2}{3}x + \frac{1}{3} = -\frac{20}{9} \qquad \text{Factor}$$

$$\left(x-\frac{1}{3}\right)^2 = -\frac{20}{9} \qquad \text{Solve using the even root property}$$

$$\sqrt{\left(x-\frac{1}{3}\right)^2} = \pm\sqrt{\frac{-20}{9}} \qquad \text{Simplify roots}$$

$$x - \frac{1}{3} = \frac{\pm 2i\sqrt{5}}{3} \qquad \text{Add } \frac{1}{3} \text{ to both sides,}$$

$$+\frac{1}{3}+\frac{1}{3} \qquad \text{Already have common denominator}$$

$$x=\frac{1\pm 2i\sqrt{5}}{3} \qquad \text{Our Solution}$$

As several of the examples have shown, when solving by completing the square we will often need to use fractions and be comfortable finding common denominators and adding fractions together. Once we get comfortable solving by completing the square and using the five steps, any quadratic equation can be easily solved.

9.3 Practice - Complete the Square

Find the value that completes the square and then rewrite as a perfect square.

1) $x^2 - 30x + \_\_$

2) $a^2 - 24a + \_\_$

3) $m^2 - 36m + \_\_$

4) $x^2 - 34x + \_\_$

5) $x^2 - 15x + \_\_$

6) $r^2 - \frac{1}{9}r + \_\_$

7) $y^2 - y + \_\_$

8) $p^2 - 17p + \_\_$

Solve each equation by completing the square.

9) $x^2 - 16x + 55 = 0$

10) $n^2 - 8n - 12 = 0$

11) $v^2 - 8v + 45 = 0$

12) $b^2 + 2b + 43 = 0$

13) $6x^2 + 12x + 63 = 0$

14) $3x^2 - 6x + 47 = 0$

15) $5k^2 - 10k + 48 = 0$

16) $8a^2 + 16a - 1 = 0$

17) $x^2 + 10x - 57 = 4$

18) $p^2 - 16p - 52 = 0$

19) $n^2 - 16n + 67 = 4$

20) $m^2 - 8m - 3 = 6$

21) $2x^2 + 4x + 38 = -6$

22) $6r^2 + 12r - 24 = -6$

23) $8b^2 + 16b - 37 = 5$

24) $6n^2 - 12n - 14 = 4$

25) $x^2 = -10x - 29$

26) $v^2 = 14v + 36$

27) $n^2 = -21 + 10n$

28) $a^2 - 56 = -10a$

29) $3k^2 + 9 = 6k$

30) $5x^2 = -26 + 10x$

31) $2x^2 + 63 = 8x$

32) $5n^2 = -10n + 15$

33) $p^2 - 8p = -55$

34) $x^2 + 8x + 15 = 8$

35) $7n^2 - n + 7 = 7n + 6n^2$

36) $n^2 + 4n = 12$

37) $13b^2 + 15b + 44 = -5 + 7b^2 + 3b$

38) $-3r^2 + 12r + 49 = -6r^2$

39) $5x^2 + 5x = -31 - 5x$

40) $8n^2 + 16n = 64$

41) $v^2 + 5v + 28 = 0$

42) $b^2 + 7b - 33 = 0$

43) $7x^2 - 6x + 40 = 0$

44) $4x^2 + 4x + 25 = 0$

45) $k^2 - 7k + 50 = 3$

46) $a^2 - 5a + 25 = 3$

47) $5x^2 + 8x - 40 = 8$

48) $2p^2 - p + 56 = -8$

49) $m^2 = -15 + 9m$

50) $n^2 - n = -41$

51) $8r^2 + 10r = -55$

52) $3x^2 - 11x = -18$

53) $5n^2 - 8n + 60 = -3n + 6 + 4n^2$

54) $4b^2 - 15b + 56 = 3b^2$

55) $-2x^2 + 3x - 5 = -4x^2$

56) $10v^2 - 15v = 27 + 4v^2 - 6v$

Quadratics - Quadratic Formula

Objective: Solve quadratic equations by using the quadratic formula.

The general from of a quadratic is $ax^2 + bx + c = 0$. We will now solve this formula for x by completing the square

Example 465.

$$ax^2 + bc + c = 0 \qquad \text{Separate constant from variables}$$

$$\underline{\quad -c - c} \qquad \text{Subtract } c \text{ from both sides}$$

$$\frac{ax^2 + bx}{a} \quad \frac{}{a} = \frac{-c}{a} \qquad \text{Divide each term by } a$$

$$x^2 + \frac{b}{a}x = \frac{-c}{a} \qquad \text{Find the number that completes the square}$$

$$\left(\frac{1}{2} \cdot \frac{b}{a}\right)^2 = \left(\frac{b}{2a}\right)^2 = \frac{b^2}{4a^2} \qquad \text{Add to both sides,}$$

$$\frac{b^2}{4a^2} - \frac{c}{a}\left(\frac{4a}{4a}\right) = \frac{b^2}{4a^2} - \frac{4ac}{4a^2} = \frac{b^2 - 4ac}{4a^2} \qquad \text{Get common denominator on right}$$

$$x^2 + \frac{b}{a}x + \frac{b^2}{4a^2} = \frac{b^2}{4a^2} - \frac{4ac}{4a^2} = \frac{b^2 - 4ac}{4a^2} \qquad \text{Factor}$$

$$\left(x + \frac{b}{2a}\right)^2 = \frac{b^2 - 4ac}{4a^2} \qquad \text{Solve using the even root property}$$

$$\sqrt{\left(x + \frac{b}{2a}\right)^2} = \pm\sqrt{\frac{b^2 - 4ac}{4a^2}} \qquad \text{Simplify roots}$$

$$x + \frac{b}{2a} = \frac{\pm\sqrt{b^2 - 4ac}}{2a} \qquad \text{Subtract } \frac{b}{2a} \text{ from both sides}$$

$$x = \frac{-b \pm \sqrt{b^2 - 4ac}}{2a} \qquad \text{Our Solution}$$

This solution is a very important one to us. As we solved a general equation by completing the square, we can use this formula to solve any quadratic equation. Once we identify what a, b, and c are in the quadratic, we can substitute those values into $x = \frac{b = \sqrt{b^2 \ 4ac}}{2a}$ and we will get our two solutions. This formula is known as the quadratic fromula

Quadratic Formula: if $ax^2 + bx + c = 0$ then $x = \dfrac{-b \pm \sqrt{b^2 - 4ac}}{2a}$

World View Note: Indian mathematician Brahmagupta gave the first explicit formula for solving quadratics in 628. However, at that time mathematics was not done with variables and symbols, so the formula he gave was, "To the absolute number multiplied by four times the square, add the square of the middle term; the square root of the same, less the middle term, being divided by twice the square is the value." This would translate to $\frac{\sqrt{4ac + b^2} - b}{2a}$ as the solution to the equation $ax^2 + bx = c$.

We can use the quadratic formula to solve any quadratic, this is shown in the following examples.

Example 466.

$$x^2 + 3x + 2 = 0 \qquad a = 1, b = 3, c = 2, \text{use quadratic formula}$$
$$x = \frac{-3 \pm \sqrt{3^2 - 4(1)(2)}}{2(1)} \qquad \text{Evaluate exponent and multiplication}$$
$$x = \frac{-3 \pm \sqrt{9 - 8}}{2} \qquad \text{Evaluate subtraction under root}$$
$$x = \frac{-3 \pm \sqrt{1}}{2} \qquad \text{Evaluate root}$$
$$x = \frac{-3 \pm 1}{2} \qquad \text{Evaluate} \pm \text{to get two answers}$$
$$x = \frac{-2}{2} \text{ or } \frac{-4}{2} \qquad \text{Simplify fractions}$$
$$x = -1 \text{ or } -2 \qquad \text{Our Solution}$$

As we are solving using the quadratic formula, it is important to remember the equation must fist be equal to zero.

Example 467.

$$25x^2 = 30x + 11 \qquad \text{First set equal to zero}$$
$$\underline{-30x - 11 \quad -30x - 11} \qquad \text{Subtract } 30x \text{ and } 11 \text{ from both sides}$$
$$25x^2 - 30x - 11 = 0 \qquad a = 25, b = -30, c = -11, \text{use quadratic formula}$$
$$x = \frac{30 \pm \sqrt{(-30)^2 - 4(25)(-11)}}{2(25)} \qquad \text{Evaluate exponent and multiplication}$$

344

$$x = \frac{30 \pm \sqrt{900 + 1100}}{50} \qquad \text{Evaluate addition inside root}$$

$$x = \frac{30 \pm \sqrt{2000}}{50} \qquad \text{Simplify root}$$

$$x = \frac{30 \pm 20\sqrt{5}}{50} \qquad \text{Reduce fraction by dividing each term by 10}$$

$$x = \frac{3 \pm 2\sqrt{5}}{5} \qquad \text{Our Solution}$$

Example 468.

$$3x^2 + 4x + 8 = 2x^2 + 6x - 5 \qquad \text{First set equation equal to zero}$$
$$\underline{-2x^2 - 6x + 5 \quad -2x^2 - 6x + 5} \qquad \text{Subtract } 2x^2 \text{ and } 6x \text{ and add } 5$$
$$x^2 - 2x + 13 = 0 \qquad a = 1, b = -2, c = 13, \text{ use quadratic formula}$$

$$x = \frac{2 \pm \sqrt{(-2)^2 - 4(1)(13)}}{2(1)} \qquad \text{Evaluate exponent and multiplication}$$

$$x = \frac{2 \pm \sqrt{4 - 52}}{2} \qquad \text{Evaluate subtraction inside root}$$

$$x = \frac{2 \pm \sqrt{-48}}{2} \qquad \text{Simplify root}$$

$$x = \frac{2 \pm 4i\sqrt{3}}{2} \qquad \text{Reduce fraction by dividing each term by 2}$$

$$x = 1 \pm 2i\sqrt{3} \qquad \text{Our Solution}$$

When we use the quadratic formula we don't necessarily get two unique answers. We can end up with only one solution if the square root simplifies to zero.

Example 469.

$$4x^2 - 12x + 9 = 0 \qquad a = 4, b = -12, c = 9, \text{ use quadratic formula}$$

$$x = \frac{12 \pm \sqrt{(-12)^2 - 4(4)(9)}}{2(4)} \qquad \text{Evaluate exponents and multiplication}$$

$$x = \frac{12 \pm \sqrt{144 - 144}}{8} \qquad \text{Evaluate subtraction inside root}$$

$$x = \frac{12 \pm \sqrt{0}}{8} \qquad \text{Evaluate root}$$

$$x = \frac{12 \pm 0}{8} \qquad \text{Evaluate } \pm$$

$$x = \frac{12}{8} \qquad \text{Reduce fraction}$$

$$x = \frac{3}{2} \qquad \text{Our Solution}$$

If a term is missing from the quadratic, we can still solve with the quadratic formula, we simply use zero for that term. The order is important, so if the term with x is missing, we have $b=0$, if the constant term is missing, we have $c=0$.

Example 470.

$$3x^2 + 7 = 0 \qquad a = 3, b = 0 \text{(missing term)}, c = 7$$

$$x = \frac{-0 \pm \sqrt{0^2 - 4(3)(7)}}{2(3)} \qquad \text{Evaluate exponnets and multiplication, zeros not needed}$$

$$x = \frac{\pm\sqrt{-84}}{6} \qquad \text{Simplify root}$$

$$x = \frac{\pm 2i\sqrt{21}}{6} \qquad \text{Reduce, dividing by 2}$$

$$x = \frac{\pm i\sqrt{21}}{3} \qquad \text{Our Solution}$$

We have covered three different methods to use to solve a quadratic: factoring, complete the square, and the quadratic formula. It is important to be familiar with all three as each has its advantage to solving quadratics. The following table walks through a suggested process to decide which method would be best to use for solving a problem.

1. If it can easily factor, solve by factoring	$x^2 - 5x + 6 = 0$ $(x-2)(x-3) = 0$ $x = 2$ or $x = 3$
2. If $a = 1$ and b is even, complete the square	$x^2 + 2x = 4$ $\left(\frac{1}{2} \cdot 2\right)^2 = 1^2 = 1$ $x^2 + 2x + 1 = 5$ $(x+1)^2 = 5$ $x + 1 = \pm\sqrt{5}$ $x = -1 \pm \sqrt{5}$
3. Otherwise, solve by the quadratic formula	$x^2 - 3x + 4 = 0$ $x = \frac{3 = \sqrt{(3)^2 \quad 4(1)(4)}}{2(1)}$ $x = \frac{3 \quad i\sqrt{7}}{2}$

The above table is mearly a suggestion for deciding how to solve a quadtratic. Remember completing the square and quadratic formula will always work to solve any quadratic. Factoring only woks if the equation can be factored.

9.4 Practice - Quadratic Formula

Solve each equation with the quadratic formula.

1) $4a^2 + 6 = 0$

2) $3k^2 + 2 = 0$

3) $2x^2 - 8x - 2 = 0$

4) $6n^2 - 1 = 0$

5) $2m^2 - 3 = 0$

6) $5p^2 + 2p + 6 = 0$

7) $3r^2 - 2r - 1 = 0$

8) $2x^2 - 2x - 15 = 0$

9) $4n^2 - 36 = 0$

10) $3b^2 + 6 = 0$

11) $v^2 - 4v - 5 = -8$

12) $2x^2 + 4x + 12 = 8$

13) $2a^2 + 3a + 14 = 6$

14) $6n^2 - 3n + 3 = -4$

15) $3k^2 + 3k - 4 = 7$

16) $4x^2 - 14 = -2$

17) $7x^2 + 3x - 16 = -2$

18) $4n^2 + 5n = 7$

19) $2p^2 + 6p - 16 = 4$

20) $m^2 + 4m - 48 = -3$

21) $3n^2 + 3n = -3$

22) $3b^2 - 3 = 8b$

23) $2x^2 = -7x + 49$

24) $3r^2 + 4 = -6r$

25) $5x^2 = 7x + 7$

26) $6a^2 = -5a + 13$

27) $8n^2 = -3n - 8$

28) $6v^2 = 4 + 6v$

29) $2x^2 + 5x = -3$

30) $x^2 = 8$

31) $4a^2 - 64 = 0$

32) $2k^2 + 6k - 16 = 2k$

33) $4p^2 + 5p - 36 = 3p^2$

34) $12x^2 + x + 7 = 5x^2 + 5x$

35) $-5n^2 - 3n - 52 = 2 - 7n^2$

36) $7m^2 - 6m + 6 = -m$

37) $7r^2 - 12 = -3r$

38) $3x^2 - 3 = x^2$

39) $2n^2 - 9 = 4$

40) $6b^2 = b^2 + 7 - b$

Quadratics - Build Quadratics From Roots

Objective: Find a quadratic equation that has given roots using reverse factoring and reverse completing the square.

Up to this point we have found the solutions to quadratics by a method such as factoring or completing the square. Here we will take our solutions and work backwards to find what quadratic goes with the solutions.

We will start with rational solutions. If we have rational solutions we can use factoring in reverse, we will set each solution equal to x and then make the equation equal to zero by adding or subtracting. Once we have done this our expressions will become the factors of the quadratic.

Example 471.

$$
\begin{array}{ll}
\text{The solutions are } 4 \text{ and} -2 & \text{Set each solution equal to } x \\
x = 4 \ \text{ or } \ x = -2 & \text{Make each equation equal zero} \\
\underline{-4 - 4 \quad +2 \quad +2} & \text{Subtract 4 from first, add 2 to second} \\
x - 4 = 0 \ \text{ or } \ x + 2 = 0 & \text{These expressions are the factors} \\
(x - 4)(x + 2) = 0 & \text{FOIL} \\
x^2 + 2x - 4x - 8 & \text{Combine like terms} \\
x^2 - 2x - 8 = 0 & \text{Our Solution}
\end{array}
$$

If one or both of the solutions are fractions we will clear the fractions by multiplying by the denominators.

Example 472.

$$
\begin{array}{ll}
\text{The solution are } \dfrac{2}{3} \text{ and } \dfrac{3}{4} & \text{Set each solution equal to } x \\
x = \dfrac{2}{3} \ \text{ or } \ x = \dfrac{3}{4} & \text{Clear fractions by multiplying by denominators} \\
3x = 2 \ \text{ or } \ 4x = 3 & \text{Make each equation equal zero} \\
\underline{-2 - 2 \quad -3 - 3} & \text{Subtract 2 from the first, subtract 3 from the second} \\
3x - 2 = 0 \ \text{ or } \ 4x - 3 = 0 & \text{These expressions are the factors} \\
(3x - 2)(4x - 3) = 0 & \text{FOIL} \\
12x^2 - 9x - 8x + 6 = 0 & \text{Combine like terms}
\end{array}
$$

$$12x^2 - 17x + 6 = 0 \quad \text{Our Solution}$$

If the solutions have radicals (or complex numbers) then we cannot use reverse factoring. In these cases we will use reverse completing the square. When there are radicals the solutions will always come in pairs, one with a plus, one with a minus, that can be combined into "one" solution using $\pm$. We will then set this solution equal to x and square both sides. This will clear the radical from our problem.

Example 473.

$$
\begin{array}{ll}
\text{The solutions are } \sqrt{3} \text{ and } -\sqrt{3} & \text{Write as "one" expression equal to } x \\
x = \pm\sqrt{3} & \text{Square both sides} \\
x^2 = 3 & \text{Make equal to zero} \\
\underline{-3 \; -3} & \text{Subtract 3 from both sides} \\
x^2 - 3 = 0 & \text{Our Solution}
\end{array}
$$

We may have to isolate the term with the square root (with plus or minus) by adding or subtracting. With these problems, remember to square a binomial we use the formula $(a + b)^2 = a^2 + 2ab + b^2$

Example 474.

$$
\begin{array}{ll}
\text{The solutions are } 2 - 5\sqrt{2} \text{ and } 2 + 5\sqrt{2} & \text{Write as "one" expression equal to } x \\
x = 2 \pm 5\sqrt{2} & \text{Isolate the square root term} \\
\underline{-2 \; -2} & \text{Subtract 2 from both sides} \\
x - 2 = \pm 5\sqrt{2} & \text{Square both sides} \\
x^2 - 4x + 4 = 25 \cdot 2 & \\
x^2 - 4x + 4 = 50 & \text{Make equal to zero} \\
\underline{-50 \; -50} & \text{Subtract 50} \\
x^2 - 4x - 46 = 0 & \text{Our Solution}
\end{array}
$$

World View Note: Before the quadratic formula, before completing the square, before factoring, quadratics were solved geometrically by the Greeks as early as 300 BC! In 1079 Omar Khayyam, a Persian mathematician solved cubic equations geometrically!

If the solution is a fraction we will clear it just as before by multiplying by the denominator.

Example 475.

The solutions are $\dfrac{2+\sqrt{3}}{4}$ and $\dfrac{2-\sqrt{3}}{4}$ Write as "one" expresion equal to x

$$x = \frac{2 \pm \sqrt{3}}{4}$$ Clear fraction by multiplying by 4

$$4x = 2 \pm \sqrt{3}$$ Isolate the square root term

$$\underline{-2\ -2}$$ Subtract 2 from both sides

$$4x - 2 = \pm\sqrt{3}$$ Square both sides

$$16x^2 - 16x + 4 = 3$$ Make equal to zero

$$\underline{-3\ -3}$$ Subtract 3

$$16x^2 - 16x + 1 = 0$$ Our Solution

The process used for complex solutions is identical to the process used for radicals.

Example 476.

The solutions are $4 - 5i$ and $4 + 5i$ Write as "one" expression equal to x

$$x = 4 \pm 5i$$ Isolate the i term

$$\underline{-4\ -4}$$ Subtract 4 from both sides

$$x - 4 = \pm 5i$$ Square both sides

$$x^2 - 8x + 16 = 25i^2$$ $i^2 = -1$

$$x^2 - 8x + 16 = -25$$ Make equal to zero

$$\underline{+25\ \ +25}$$ Add 25 to both sides

$$x^2 - 8x + 41 = 0$$ Our Solution

Example 477.

The solutions are $\dfrac{3-5i}{2}$ and $\dfrac{3+5i}{2}$ Write as "one" expression equal to x

$$x = \frac{3 \pm 5i}{2}$$ Clear fraction by multiplying by denominator

$$2x = 3 \pm 5i$$ Isolate the i term

$$\underline{-3\ -3}$$ Subtract 3 from both sides

$$2x - 3 = \pm 5i$$ Square both sides

$$4x^2 - 12x + 9 = 5i^2$$ $i^2 = -1$

$$4x^2 - 12x + 9 = -25$$ Make equal to zero

$$\underline{+25\ \ +25}$$ Add 25 to both sides

$$4x^2 - 12x + 34 = 0$$ Our Solution

9.5 Practice - Build Quadratics from Roots

From each problem, find a quadratic equation with those numbers as
its solutions.

1) 2, 5

2) 3, 6

3) 20, 2

4) 13, 1

5) 4, 4

6) 0, 9

7) 0, 0

8) $-2, -5$

9) $-4, 11$

10) 3, -1

11) $\frac{3}{4}, \frac{1}{4}$

12) $\frac{5}{8}, \frac{5}{7}$

13) $\frac{1}{2}, \frac{1}{3}$

14) $\frac{1}{2}, \frac{2}{3}$

15) $\frac{3}{7}, 4$

16) $2, \frac{2}{9}$

17) $-\frac{1}{3}, \frac{5}{6}$

18) $\frac{5}{3}, -\frac{1}{2}$

19) $-6, \frac{1}{9}$

20) $-\frac{2}{5}, 0$

21) ± 5

22) ± 1

23) $\pm \frac{1}{5}$

24) $\pm \sqrt{7}$

25) $\pm \sqrt{11}$

26) $\pm 2\sqrt{3}$

27) $\pm \frac{\sqrt{3}}{4}$

28) $\pm 11i$

29) $\pm i\sqrt{13}$

30) $\pm 5i\sqrt{2}$

31) $2 \pm \sqrt{6}$

32) $-3 \pm \sqrt{2}$

33) $1 \pm 3i$

34) $-2 \pm 4i$

35) $6 \pm i\sqrt{3}$

36) $-9 \pm i\sqrt{5}$

37) $\frac{1 \pm \sqrt{6}}{2}$

38) $\frac{2 = 5i}{3}$

39) $\frac{6 \pm i\sqrt{2}}{8}$

40) $\frac{2 \pm i\sqrt{15}}{2}$

Quadratics - Quadratic in Form

Objective: Solve equations that are quadratic in form by substitution to create a quadratic equation.

We have seen three different ways to solve quadratics: factoring, completing the square, and the quadratic formula. A quadratic is any equation of the form $0 = ax^2 + bx + c$, however, we can use the skills learned to solve quadratics to solve problems with higher (or sometimes lower) powers if the equation is in what is called quadratic form.

Quadratic Form: $0 = ax^m + bx^n + c$ where $m = 2n$

An equation is in quadratic form if one of the exponents on a variable is double the exponent on the same variable somewhere else in the equation. If this is the case we can create a new variable, set it equal to the variable with smallest exponent. When we substitute this into the equation we will have a quadratic equation we can solve.

World View Note: Arab mathematicians around the year 1000 were the first to use this method!

Example 478.

$$x^4 - 13x^2 + 36 = 0 \quad \text{Quadratic form, one exponent, 4, double the other, 2}$$
$$y = x^2 \quad \text{New variable equal to the variable with smaller exponent}$$
$$y^2 = x^4 \quad \text{Square both sides}$$
$$y^2 - 13y + 36 = 0 \quad \text{Substitute } y \text{ for } x^2 \text{ and } y^2 \text{ for } x^4$$
$$(y - 9)(y - 4) = 0 \quad \text{Solve. We can solve this equation by factoring}$$
$$y - 9 = 0 \text{ or } y - 4 = 0 \quad \text{Set each factor equal to zero}$$
$$\underline{+9 + 9 \qquad +4 + 4} \quad \text{Solve each equation}$$
$$y = 9 \text{ or } y = 4 \quad \text{Solutions for } y, \text{ need } x. \text{ We will use } y = x^2 \text{ equation}$$
$$9 = x^2 \text{ or } 4 = x^2 \quad \text{Substitute values for } y$$
$$\pm\sqrt{9} = \sqrt{x^2} \text{ or } \pm\sqrt{4} = \sqrt{x^2} \quad \text{Solve using the even root property, simplify roots}$$
$$x = \pm 3, \pm 2 \quad \text{Our Solutions}$$

When we have higher powers of our variable, we could end up with many more solutions. The previous equation had four unique solutions.

Example 479.

$$a^2 - a^1 - 6 = 0 \qquad \text{Quadratic form, one exponent, } -2, \text{ is double the other, } -1$$

$$b = a^{-1} \qquad \text{Make } a \text{ new variable equal to the variable with lowest exponent}$$

$$b^2 = a^{-2} \qquad \text{Square both sides}$$

$$b^2 - b - 6 = 0 \qquad \text{Substitute } b^2 \text{ for } a^2 \text{ and } b \text{ for } a^1$$

$$(b-3)(b+2) = 0 \qquad \text{Solve. We will solve by factoring}$$

$$b - 3 = 0 \ \text{ or } \ b + 2 = 0 \qquad \text{Set each factor equal to zero}$$

$$\underline{+3 + 3 \qquad\quad -2 - 2} \qquad \text{Solve each equation}$$

$$b = 3 \ \text{ or } \ b = -2 \qquad \text{Solutions for } b, \text{ still need } a, \text{ substitute into } b = a^1$$

$$3 = a^{-1} \ \text{ or } \ -2 = a^{-1} \qquad \text{Raise both sides to } -1 \text{ power}$$

$$3^1 = a \ \text{ or } \ (-2)^1 = a \qquad \text{Simplify negative exponents}$$

$$a = \frac{1}{3}, -\frac{1}{2} \qquad \text{Our Solution}$$

Just as with regular quadratics, these problems will not always have rational solutions. We also can have irrational or complex solutions to our equations.

Example 480.

$$2x^4 + x^2 = 6 \qquad \text{Make equation equal to zero}$$

$$\underline{\quad -6 - 6\quad} \qquad \text{Subtract 6 from both sides}$$

$$2x^4 + x^2 - 6 = 0 \qquad \text{Quadratic form, one exponent, 4, double the other, 2}$$

$$y = x^2 \qquad \text{New variable equal variable with smallest exponent}$$

$$y^2 = x^4 \qquad \text{Square both sides}$$

$$2y^2 + y - 6 = 0 \qquad \text{Solve. We will factor this equation}$$

$$(2y - 3)(y + 2) = 0 \qquad \text{Set each factor equal to zero}$$

$$2y - 3 = 0 \ \text{ or } \ y + 2 = 0 \qquad \text{Solve each equation}$$

$$\underline{+3 + 3 \qquad\quad -2 - 2}$$

$$2y = 3 \ \text{ or } \ y = -2$$

$$\overline{2} \ \ \overline{2}$$

$$y = \frac{3}{2} \ \text{ or } \ y = -2 \qquad \text{We have } y, \text{ still need } x. \text{ Substitute into } y = x^2$$

$$\frac{3}{2} = x^2 \ \text{ or } \ -2 = x^2 \qquad \text{Square root of each side}$$

$$\pm\sqrt{\frac{3}{2}} = \sqrt{x^2} \ \text{ or } \ \pm\sqrt{-2} = \sqrt{x^2} \qquad \text{Simplify each root, rationalize denominator}$$

$$x = \frac{\pm\sqrt{6}}{2}, \pm i\sqrt{2} \qquad \text{Our Solution}$$

353

When we create a new variable for our substitution, it won't always be equal to just another variable. We can make our substitution variable equal to an expression as shown in the next example.

Example 481.

$$3(x-7)^2 - 2(x-7) + 5 = 0 \quad \text{Quadratic form}$$
$$y = x - 7 \quad \text{Define new variable}$$
$$y^2 = (x-7)^2 \quad \text{Square both sides}$$
$$3y^2 - 2y + 5 = 0 \quad \text{Substitute values into original}$$
$$(3y-5)(y+1) = 0 \quad \text{Factor}$$
$$3y - 5 = 0 \text{ or } y + 1 = 0 \quad \text{Set each factor equal to zero}$$
$$\underline{+5+5} \qquad \underline{-1-1} \quad \text{Solve each equation}$$
$$3y = 5 \text{ or } y = -1$$
$$\overline{3} \quad \overline{3}$$
$$y = \frac{5}{3} \text{ or } y = -1 \quad \text{We have } y, \text{ we still need } x.$$
$$\frac{5}{3} = x - 7 \text{ or } -1 = x - 7 \quad \text{Substitute into } y = x - 7$$
$$+\frac{21}{3} \quad +7 \qquad \underline{+7} \quad \underline{+7} \quad \text{Add 7. Use common denominator as needed}$$
$$x = \frac{26}{3}, 6 \quad \text{Our Solution}$$

Example 482.

$$(x^2 - 6x)^2 = 7(x^2 - 6x) - 12 \quad \text{Make equation equal zero}$$
$$-7(x^2 - 6x) + 12 - 7(x^2 - 6x) + 12 \quad \text{Move all terms to left}$$
$$(x^2 - 6x)^2 - 7(x^2 - 6x) + 12 = 0 \quad \text{Quadratic form}$$
$$y = x^2 - 6x \quad \text{Make new variable}$$
$$y^2 = (x^2 - 6x)^2 \quad \text{Square both sides}$$
$$y^2 - 7y + 12 = 0 \quad \text{Substitute into original equation}$$
$$(y-3)(y-4) = 0 \quad \text{Solve by factoring}$$
$$y - 3 = 0 \text{ or } y - 4 = 0 \quad \text{Set each factor equal to zero}$$
$$\underline{+3+3} \qquad \underline{+4+4} \quad \text{Solve each equation}$$
$$y = 3 \text{ or } y = 4 \quad \text{We have } y, \text{ still need } x.$$
$$3 = x^2 - 6x \text{ or } 4 = x^3 - 6x \quad \text{Solve each equation, complete the square}$$
$$\left(\frac{1}{2} \cdot 6\right)^2 = 3^2 = 9 \quad \text{Add 9 to both sides of each equation}$$
$$12 = x^2 - 6x + 9 \text{ or } 13 = x^2 - 6x + 9 \quad \text{Factor}$$

$$12 = (x-3)^2 \text{ or } 13 = (x-3)^2 \qquad \text{Use even root property}$$
$$\pm\sqrt{12} = \sqrt{(x-3)^2} \text{ or } \pm\sqrt{13} = \sqrt{(x-3)^2} \qquad \text{Simplify roots}$$
$$\pm 2\sqrt{3} = x-3 \text{ or } \pm\sqrt{13} = x-3 \qquad \text{Add 3 to both sides}$$
$$\underline{+3 \qquad +3 \qquad\qquad +3 \qquad +3}$$
$$x = 3 \pm 2\sqrt{3}, 3 \pm \sqrt{13} \qquad \text{Our Solution}$$

The higher the exponent, the more solution we could have. This is illustrated in the following example, one with six solutions.

Example 483.

$$x^6 - 9x^3 + 8 = 0 \qquad \text{Quadratic form, one exponent, 6, double the other, 3}$$
$$y = x^3 \qquad \text{New variable equal to variable with lowest exponent}$$
$$y^2 = x^6 \qquad \text{Square both sides}$$
$$y^2 - 9y + 8 = 0 \qquad \text{Substitute } y^2 \text{ for } x^6 \text{ and } y \text{ for } x^3$$
$$(y-1)(y-8) = 0 \qquad \text{Solve. We will solve by factoring.}$$
$$y - 1 = 0 \text{ or } y - 8 = 0 \qquad \text{Set each factor equal to zero}$$
$$\underline{+1+1 \qquad\qquad +8+8} \qquad \text{Solve each equation}$$
$$y = 1 \text{ or } y = 8 \qquad \text{Solutions for } y, \text{ we need } x. \text{ Substitute into } y = x^3$$
$$x^3 = 1 \text{ or } x^3 = 8 \qquad \text{Set each equation equal to zero}$$
$$\underline{-1-1 \qquad -8-8}$$
$$x^3 - 1 = 0 \text{ or } x^3 - 8 = 0 \qquad \text{Factor each equation, difference of cubes}$$
$$(x-1)(x^2+x+1) = 0 \qquad \text{First equation factored. Set each factor equal to zero}$$
$$x - 1 = 0 \text{ or } x^2 + x + 1 = 0 \qquad \text{First equation is easy to solve}$$
$$\underline{+1+1}$$
$$x = 1 \qquad \text{First solution}$$
$$\frac{-1 \pm \sqrt{1^2 - 4(1)(1)}}{2(1)} = \frac{1 \pm i\sqrt{3}}{2} \qquad \text{Quadratic formula on second factor}$$
$$(x-2)(x^2+2x+4) = 0 \qquad \text{Factor the second difference of cubes}$$
$$x - 2 = 0 \text{ or } x^2 + 2x + 4 = 0 \qquad \text{Set each factor equal to zero.}$$
$$\underline{+2+2} \qquad \text{First equation is easy to solve}$$
$$x = 2 \qquad \text{Our fourth solution}$$
$$\frac{-2 \pm \sqrt{2^2 - 4(1)(4)}}{2(1)} = -1 \pm i\sqrt{3} \qquad \text{Quadratic formula on second factor}$$
$$x = 1, 2, \frac{1 \pm i\sqrt{3}}{2}, -1 \pm i\sqrt{3} \qquad \text{Our final six solutions}$$

9.6 Practice - Quadratic in Form

Solve each of the following equations. Some equations will have complex roots.

1) $x^4 - 5x^2 + 4 = 0$

2) $y^4 - 9y^2 + 20 = 0$

3) $m^4 - 7m^2 - 8 = 0$

4) $y^4 - 29y^2 + 100 = 0$

5) $a^4 - 50a^2 + 49 = 0$

6) $b^4 - 10b^2 + 9 = 0$

7) $x^4 - 25x^2 + 144 = 0$

8) $y^4 - 40y^2 + 144 = 0$

9) $m^4 - 20m^2 + 64 = 0$

10) $x^6 - 35x^3 + 216 = 0$

11) $z^6 - 216 = 19z^3$

12) $y^4 - 2y^2 = 24$

13) $6z^4 - z^2 = 12$

14) $x^{\,2} - x^{\,1} - 12 = 0$

15) $x^{\frac{2}{3}} - 35 = 2x^{\frac{1}{3}}$

16) $5y^{-2} - 20 = 21y^{-1}$

17) $y^{-6} + 7y^{-3} = 8$

18) $x^4 - 7x^2 + 12 = 0$

19) $x^4 - 2x^2 - 3 = 0$

20) $x^4 + 7x^2 + 10 = 0$

21) $2x^4 - 5x^2 + 2 = 0$

22) $2x^4 - x^2 - 3 = 0$

23) $x^4 - 9x^2 + 8 = 0$

24) $x^6 - 10x^3 + 16 = 0$

25) $8x^6 - 9x^3 + 1 = 0$

26) $8x^6 + 7x^3 - 1 = 0$

27) $x^8 - 17x^4 + 16 = 0$

28) $(x-1)^2 - 4(x-1) = 5$

29) $(y+b)^2 - 4(y+b) = 21$

30) $(x+1)^2 + 6(x+1) + 9 = 0$

31) $(y+2)^2 - 6(y+2) = 16$

32) $(m-1)^2 - 5(m-1) = 14$

33) $(x-3)^2 - 2(x-3) = 35$

34) $(a+1)^2 + 2(a-1) = 15$

35) $(r-1)^2 - 8(r-1) = 20$

36) $2(x-1)^2 - (x-1) = 3$

37) $3(y+1)^2 - 14(y+1) = 5$

38) $(x^2-3)^2 - 2(x^2-3) = 3$

39) $(3x^2 - 2x)^2 + 5 = 6(3x^2 - 2x)$

40) $(x^2 + x + 3)^2 + 15 = 8(x^2 + x + 3)$

41) $2(3x+1)^{\frac{2}{3}} - 5(3x+1)^{\frac{1}{3}} = 88$

42) $(x^2 + x)^2 - 8(x^2 + x) + 12 = 0$

43) $(x^2 + 2x)^2 - 2(x^2 + 2x) = 3$

44) $(2x^2 + 3x)^2 = 8(2x^2 + 3x) + 9$

45) $(2x^2 - x)^2 - 4(2x^2 - x) + 3 = 0$

46) $(3x^2 - 4x)^2 = 3(3x^2 - 4x) + 4$

Quadratics - Rectangles

Objective: Solve applications of quadratic equations using rectangles.

An application of solving quadratic equations comes from the formula for the area of a rectangle. The area of a rectangle can be calculated by multiplying the width by the length. To solve problems with rectangles we will first draw a picture to represent the problem and use the picture to set up our equation.

Example 484.

The length of a rectangle is 3 more than the width. If the area is 40 square inches, what are the dimensions?

$$\boxed{40}\,x \qquad \text{We do not know the width, } x.$$
$$x+3 \qquad\qquad \text{Length is 4 more, or } x+4, \text{ and area is 40.}$$

$$x(x+3)=40 \qquad \text{Multiply length by width to get area}$$
$$x^2+3x=40 \qquad \text{Distribute}$$
$$\underline{-40-40} \qquad \text{Make equation equal zero}$$
$$x^2+3x-40=0 \qquad \text{Factor}$$
$$(x-5)(x+8)=0 \qquad \text{Set each factor equal to zero}$$
$$x-5=0 \text{ or } x+8=0 \qquad \text{Solve each equation}$$
$$\underline{+5+5 \qquad\quad -8-8}$$
$$x=5 \text{ or } x=-8 \qquad \text{Our } x \text{ is a width, cannot be negative.}$$
$$(5)+3=8 \qquad \text{Length is } x+3, \text{ substitute 5 for } x \text{ to find length}$$
$$5 \text{ in by } 8 \text{ in} \qquad \text{Our Solution}$$

The above rectangle problem is very simple as there is only one rectangle involved. When we compare two rectangles, we may have to get a bit more creative.

Example 485.

If each side of a square is increased by 6, the area is multiplied by 16. Find the side of the original square.

$\boxed{x^2}\,x$ Square has all sides the same length

x Area is found by multiplying length by width

$\boxed{16x^2}\,x+6$ Each side is increased by 6,

$x+6$ Area is 16 times original area

$$(x+6)(x+6)=16x^2 \qquad \text{Multiply length by width to get area}$$
$$x^2+12x+36=16x^2 \qquad \text{FOIL}$$
$$\underline{-16x^2 \qquad\qquad -16x^2} \qquad \text{Make equation equal zero}$$
$$-15x^2+12x+36=0 \qquad \text{Divide each term by } -1, \text{ changes the signs}$$
$$15x^2-12x-36=0 \qquad \text{Solve using the quadratic formula}$$
$$x=\frac{12\pm\sqrt{(-12)^2-4(15)(-36)}}{2(15)} \qquad \text{Evaluate}$$
$$x=\frac{16\pm\sqrt{2304}}{30}$$
$$x=\frac{16\pm48}{30} \qquad \text{Can't have a negative solution, we will only add}$$
$$x=\frac{60}{30}=2 \qquad \text{Our } x \text{ is the original square}$$

358

Example 486.

The length of a rectangle is 4 ft greater than the width. If each dimension is increased by 3, the new area will be 33 square feet larger. Find the dimensions of the original rectangle.

$$\boxed{x(x+4)}\ x \qquad \text{We don't know width, } x, \text{ length is 4 more, } x+4$$

$$x+4 \qquad\qquad \text{Area is found by multiplying length by width}$$

$$\boxed{x(x+4)+33}\ x+3 \qquad \begin{array}{l}\text{Increase each side by 3.}\\ \text{width becomes } x+3, \text{ length } x+4+3=x+7\end{array}$$

$$x+7 \qquad\qquad \text{Area is 33 more than original, } x(x+4)+33$$

$$(x+3)(x+7)=x(x+4)+33 \qquad \text{Set up equation, length times width is area}$$

$$x^2+10x+21=x^2+4x+33 \qquad \text{Subtract } x^2 \text{ from both sides}$$

$$\underline{-x^2 \qquad\qquad -x^2}$$

$$10x+21=4x+33 \qquad \text{Move variables to one side}$$

$$\underline{-4x \qquad -4x} \qquad \text{Subtract } 4x \text{ from each side}$$

$$6x+21=33 \qquad \text{Subtract 21 from both sides}$$

$$\underline{-21\ -21}$$

$$\dfrac{6x}{6}=\dfrac{12}{6} \qquad \text{Divide both sides by 6}$$

$$x=2 \qquad x \text{ is the width of the original}$$

$$(2)+4=6 \qquad x+4 \text{ is the length. Substitute 2 to find}$$

$$2\,\text{ft by }6\,\text{ft} \qquad \text{Our Solution}$$

From one rectangle we can find two equations. Perimeter is found by adding all the sides of a polygon together. A rectangle has two widths and two lengths, both the same size. So we can use the equation $P = 2l + 2w$ (twice the length plus twice the width).

Example 487.

The area of a rectangle is 168 cm^2. The perimeter of the same rectangle is 52 cm. What are the dimensions of the rectangle?

$$\boxed{}\ x \qquad \text{We don't know anything about length or width}$$

$$y \qquad\qquad \text{Use two variables, } x \text{ and } y$$

$$xy=168 \qquad \text{Length times width gives the area.}$$

$$2x+2y=52 \qquad \text{Also use perimeter formula.}$$

$$\underline{-2x \qquad -2x} \qquad \text{Solve by substitution, isolate } y$$

$$2y=-2x+52 \qquad \text{Divide each term by 2}$$

359

$$\overline{2} \quad \overline{2} \quad \overline{2}$$

$y = -x + 26$	Substitute into area equation
$x(-x + 26) = 168$	Distribute
$-x^2 + 26x = 168$	Divide each term by -1, changing all the signs
$x^2 - 26x = -168$	Solve by completing the square.
$\left(\dfrac{1}{2} \cdot 26\right)^2 = 13^2 = 169$	Find number to complete the square: $\left(\dfrac{1}{2} \cdot b\right)^2$
$x^2 - 26x + 324 = 1$	Add 169 to both sides
$(x - 13)^2 = 1$	Factor
$x - 13 = \pm 1$	Square root both sides
$\underline{+13 \quad +13}$	
$x = 13 \pm 1$	Evaluate
$x = 14 \text{ or } 12$	Two options for first side.
$y = -(14) + 26 = 12$	Substitute 14 into $y = -x + 26$
$y = -(12) + 26 = 14$	Substitute 12 into $y = -x + 26$
	Both are the same rectangle, variables switched!
12 cm by 14 cm	Our Solution

World View Note: Indian mathematical records from the 9th century demonstrate that their civilization had worked extensivly in geometry creating religious alters of various shapes including rectangles.

Another type of rectangle problem is what we will call a "frame problem". The idea behind a frame problem is that a rectangle, such as a photograph, is centered inside another rectangle, such as a frame. In these cases it will be important to rememember that the frame extends on all sides of the rectangle. This is shown in the following example.

Example 488.

An 8 in by 12 in picture has a frame of uniform width around it. The area of the frame is equal to the area of the picture. What is the width of the frame?

	Draw picture, picture if 8 by 10
	If frame has width x, on both sides, we add $2x$
$8 \cdot 12 = 96$	Area of the picture, length times width
$2 \cdot 96 = 192$	Frame is the same as the picture. Total area is double this.
$(12 + 2x)(8 + 2x) = 192$	Area of everything, length times width
$96 + 24x + 16x + 4x^2 = 192$	FOIL
$4x^2 + 40x + 96 = 192$	Combine like terms
$\underline{-192 \quad -192}$	Make equation equal to zero by subtracting 192
$4x^2 + 40x - 96 = 0$	Factor out GCF of 4

360

$$4(x^2 + 10x - 24) = 0 \quad \text{Factor trinomial}$$
$$4(x - 2)(x + 12) = 0 \quad \text{Set each factor equal to zero}$$
$$x - 2 = 0 \quad \text{or} \quad x + 12 = 0 \quad \text{Solve each equation}$$
$$\underline{+2 + 2} \qquad \underline{-12 - 12}$$
$$x = 2 \quad \text{or} \quad -12 \quad \text{Can't have negative frame width.}$$
$$2 \text{ inches} \quad \text{Our Solution}$$

Example 489.

A farmer has a field that is 400 rods by 200 rods. He is mowing the field in a spiral pattern, starting from the outside and working in towards the center. After an hour of work, 72% of the field is left uncut. What is the size of the ring cut around the outside?

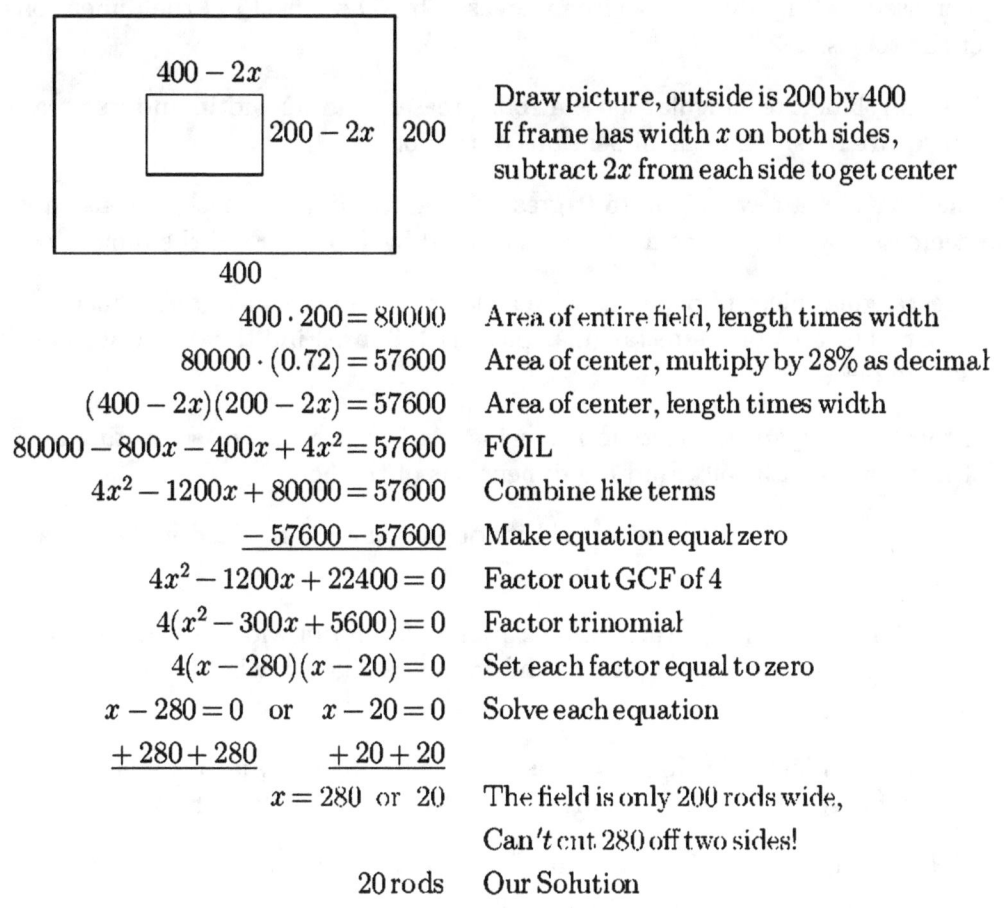

Draw picture, outside is 200 by 400
If frame has width x on both sides,
subtract $2x$ from each side to get center

$$400 \cdot 200 = 80000 \quad \text{Area of entire field, length times width}$$
$$80000 \cdot (0.72) = 57600 \quad \text{Area of center, multiply by 28\% as decimal}$$
$$(400 - 2x)(200 - 2x) = 57600 \quad \text{Area of center, length times width}$$
$$80000 - 800x - 400x + 4x^2 = 57600 \quad \text{FOIL}$$
$$4x^2 - 1200x + 80000 = 57600 \quad \text{Combine like terms}$$
$$\underline{-57600 - 57600} \quad \text{Make equation equal zero}$$
$$4x^2 - 1200x + 22400 = 0 \quad \text{Factor out GCF of 4}$$
$$4(x^2 - 300x + 5600) = 0 \quad \text{Factor trinomial}$$
$$4(x - 280)(x - 20) = 0 \quad \text{Set each factor equal to zero}$$
$$x - 280 = 0 \quad \text{or} \quad x - 20 = 0 \quad \text{Solve each equation}$$
$$\underline{+280 + 280} \qquad \underline{+20 + 20}$$
$$x = 280 \quad \text{or} \quad 20 \quad \text{The field is only 200 rods wide,}$$
$$\text{Can't cut 280 off two sides!}$$
$$20 \text{ rods} \quad \text{Our Solution}$$

For each of the frame problems above we could have also completed the square or use the quadratic formula to solve the trinomials. Remember that completing the square or the quadratic formula always will work when solving, however, factoring only works if we can factor the trinomial.

9.7 Practice - Rectangles

1) In a landscape plan, a rectangular flowerbed is designed to be 4 meters longer than it is wide. If 60 square meters are needed for the plants in the bed, what should the dimensions of the rectangular bed be?

2) If the side of a square is increased by 5 the area is multiplied by 4. Find the side of the original square.

3) A rectangular lot is 20 yards longer than it is wide and its area is 2400 square yards. Find the dimensions of the lot.

4) The length of a room is 8 ft greater than it is width. If each dimension is increased by 2 ft, the area will be increased by 60 sq. ft. Find the dimensions of the rooms.

5) The length of a rectangular lot is 4 rods greater than its width, and its area is 60 square rods. Find the dimensions of the lot.

6) The length of a rectangle is 15 ft greater than its width. If each dimension is decreased by 2 ft, the area will be decreased by 106 ft^2. Find the dimensions.

7) A rectangular piece of paper is twice as long as a square piece and 3 inches wider. The area of the rectangular piece is 108 in^2. Find the dimensions of the square piece.

8) A room is one yard longer than it is wide. At 75¢ per sq. yd. a covering for the floor costs $31.50. Find the dimensions of the floor.

9) The area of a rectangle is 48 ft^2 and its perimeter is 32 ft. Find its length and width.

10) The dimensions of a picture inside a frame of uniform width are 12 by 16 inches. If the whole area (picture and frame) is 288 in^2, what is the width of the frame?

11) A mirror 14 inches by 15 inches has a frame of uniform width. If the area of the frame equals that of the mirror, what is the width of the frame.

12) A lawn is 60 ft by 80 ft. How wide a strip must be cut around it when mowing the grass to have cut half of it.

13) A grass plot 9 yards long and 6 yards wide has a path of uniform width around it. If the area of the path is equal to the area of the plot, determine the width of the path.

14) A landscape architect is designing a rectangular flowerbed to be border with 28 plants that are placed 1 meter apart. He needs an inner rectangular space in the center for plants that must be 1 meter from the border of the bed and

that require 24 square meters for planting. What should the overall dimensions of the flowerbed be?

15) A page is to have a margin of 1 inch, and is to contain 35 in^2 of painting. How large must the page be if the length is to exceed the width by 2 inches?

16) A picture 10 inches long by 8 inches wide has a frame whose area is one half the area of the picture. What are the outside dimensions of the frame?

17) A rectangular wheat field is 80 rods long by 60 rods wide. A strip of uniform width is cut around the field, so that half the grain is left standing in the form of a rectangular plot. How wide is the strip that is cut?

18) A picture 8 inches by 12 inches is placed in a frame of uniform width. If the area of the frame equals the area of the picture find the width of the frame.

19) A rectangular field 225 ft by 120 ft has a ring of uniform width cut around the outside edge. The ring leaves 65% of the field uncut in the center. What is the width of the ring?

20) One Saturday morning George goes out to cut his lot that is 100 ft by 120 ft. He starts cutting around the outside boundary spiraling around towards the center. By noon he has cut 60% of the lawn. What is the width of the ring that he has cut?

21) A frame is 15 in by 25 in and is of uniform width. The inside of the frame leaves 75% of the total area available for the picture. What is the width of the frame?

22) A farmer has a field 180 ft by 240 ft. He wants to increase the area of the field by 50% by cultivating a band of uniform width around the outside. How wide a band should he cultivate?

23) The farmer in the previous problem has a neighbor who has a field 325 ft by 420 ft. His neighbor wants to increase the size of his field by 20% by cultivating a band of uniform width around the outside of his lot. How wide a band should his neighbor cultivate?

24) A third farmer has a field that is 500 ft by 550 ft. He wants to increase his field by 20%. How wide a ring should he cultivate around the outside of his field?

25) Donna has a garden that is 30 ft by 36 ft. She wants to increase the size of the garden by 40%. How wide a ring around the outside should she cultivate?

26) A picture is 12 in by 25 in and is surrounded by a frame of uniform width. The area of the frame is 30% of the area of the picture. How wide is the frame?

Quadratics - Teamwork

Objective: Solve teamwork problems by creating a rational equation to model the problem.

If it takes one person 4 hours to paint a room and another person 12 hours to paint the same room, working together they could paint the room even quicker, it turns out they would paint the room in 3 hours together. This can be reasoned by the following logic, if the first person paints the room in 4 hours, she paints $\frac{1}{4}$ of the room each hour. If the second person takes 12 hours to paint the room, he paints $\frac{1}{12}$ of the room each hour. So together, each hour they paint $\frac{1}{4} + \frac{1}{12}$ of the room. Using a common denominator of 12 gives: $\frac{3}{12} + \frac{1}{12} = \frac{4}{12} = \frac{1}{3}$. This means each hour, working together they complete $\frac{1}{3}$ of the room. If $\frac{1}{3}$ is completed each hour, it follows that it will take 3 hours to complete the entire room.

This pattern is used to solve teamwork problems. If the first person does a job in A, a second person does a job in B, and together they can do a job in T (total). We can use the team work equation.

$$\text{Teamwork Equation: } \frac{1}{A} + \frac{1}{B} = \frac{1}{T}$$

Often these problems will involve fractions. Rather than thinking of the first fraction as $\frac{1}{A}$, it may be better to think of it as the reciprocal of A's time.

World View Note: When the Egyptians, who were the first to work with fractions, wrote fractions, they were all unit fractions (numerator of one). They only used these type of fractions for about 2000 years! Some believe that this cumbersome style of using fractions was used for so long out of tradition, others believe the Egyptians had a way of thinking about and working with fractions that has been completely lost in history.

Example 490.

Adam can clean a room in 3 hours. If his sister Maria helps, they can clean it in $2\frac{2}{5}$ hours. How long will it take Maria to do the job alone?

$$2\frac{2}{5} = \frac{12}{5} \quad \text{Together time, } 2\frac{2}{5}, \text{ needs to be converted to fraction}$$

$$\text{Adam: } 3, \text{Maria: } x, \text{Total: } \frac{5}{12} \quad \text{Clearly state times for each and total, using } x \text{ for Maria}$$

$$\frac{1}{3} + \frac{1}{x} = \frac{5}{12} \qquad \text{Using reciprocals, add the individual times gives total}$$

$$\frac{1(12x)}{3} + \frac{1(12x)}{x} = \frac{5(12x)}{12} \qquad \text{Multiply each term by LCD of } 12x$$

$$4x + 12 = 5x \qquad \text{Reduce each fraction}$$
$$\underline{-4x \qquad\qquad -4x} \qquad \text{Move variables to one side, subtracting } 4x$$
$$12 = x \qquad \text{Our solution for } x$$

It takes Maria 12 hours   Our Solution

Somtimes we only know how two people's times are related to eachother as in the next example.

Example 491.

Mike takes twice as long as Rachel to complete a project. Together they can complete the project in 10 hours. How long will it take each of them to complete the project alone?

Mike: $2x$, Rachel: x, Total: 10   Clearly define variables. If Rachel is x, Mike is $2x$

$$\frac{1}{2x} + \frac{1}{x} = \frac{1}{10} \qquad \text{Using reciprocals, add individal times equaling total}$$

$$\frac{1(10x)}{2x} + \frac{1(10x)}{x} = \frac{1(10x)}{10} \qquad \text{Multiply each term by LCD, } 10x$$

$$5 + 10 = x \qquad \text{Combine like terms}$$
$$15 = x \qquad \text{We have our } x, \text{ we said } x \text{ was Rachel}'s \text{ time}$$
$$2(15) = 30 \qquad \text{Mike is double Rachel, this gives Mike}'s \text{ time.}$$

Mike: 30 hr, Rachel: 15 hr   Our Solution

With problems such as these we will often end up with a quadratic to solve.

Example 492.

Brittney can build a large shed in 10 days less than Cosmo can. If they built it together it would take them 12 days. How long would it take each of them working alone?

Britney: $x - 10$, Cosmo: x, Total: 12   If Cosmo is x, Britney is $x - 10$

$$\frac{1}{x - 10} + \frac{1}{x} = \frac{1}{12} \qquad \text{Using reciprocals, make equation}$$

$$\frac{1(12x(x-10))}{x-10} + \frac{1(12x(x-10))}{x} = \frac{1(12x(x-10))}{12} \qquad \text{Multiply by LCD: } 12x(x-10)$$

$$
\begin{aligned}
12x + 12(x-10) &= x(x-10) & &\text{Reduce fraction}\\
12x + 12x - 120 &= x^2 - 10x & &\text{Distribute}\\
24x - 120 &= x^2 - 10x & &\text{Combine like terms}\\
\underline{-24x + 120 \quad -24x + 120} & & &\text{Move all terms to one side}\\
0 &= x^2 - 34x + 120 & &\text{Factor}\\
0 &= (x-30)(x-4) & &\text{Set each factor equal to zero}\\
x - 30 = 0 \ \ &\text{or} \ \ x - 4 = 0 & &\text{Solve each equation}\\
\underline{+30+30 \qquad +4+4} & & &\\
x = 30 \ \ &\text{or} \ \ x = 4 & &\text{This, } x \text{, was defined as Cosmo.}\\
30 - 10 = 20 \ \ &\text{or} \ \ 4 - 10 = -6 & &\text{Find Britney, can}'t \text{ have negative time}\\
\text{Britney: 20 days, } &\text{Cosmo: 30 days} & &\text{Our Solution}
\end{aligned}
$$

In the previous example, when solving, one of the possible times ended up negative. We can't have a negative amount of time to build a shed, so this possibility is ignored for this problem. Also, as we were solving, we had to factor $x^2 - 34x + 120$. This may have been difficult to factor. We could have also chosen to complete the square or use the quadratic formula to find our solutions.

It is important that units match as we solve problems. This means we may have to convert minutes into hours to match the other units given in the problem.

Example 493.

An electrician can complete a job in one hour less than his apprentice. Together they do the job in 1 hour and 12 minutes. How long would it take each of them working alone?

$$1 \text{ hr } 12 \text{ min} = 1\frac{12}{60} \text{ hr} \qquad \text{Change 1 hour 12 minutes to mixed number}$$

$$1\frac{12}{60} = 1\frac{1}{5} = \frac{6}{5} \qquad \text{Reduce and convert to fraction}$$

$$\text{Electrician: } x - 1, \text{ Apprentice: } x, \text{ Total: } \frac{6}{5} \qquad \text{Clearly define variables}$$

$$\frac{1}{x-1} + \frac{1}{x} = \frac{5}{6} \qquad \text{Using reciprocals, make equation}$$

$$\frac{1(6x(x-1))}{x-1} + \frac{1(6x(x-1))}{x} = \frac{5(6x(x-1))}{6} \qquad \text{Multiply each term by LCD } 6x(x-1)$$

$$6x + 6(x - 1) = 5x(x - 1) \quad \text{Reduce each fraction}$$
$$6x + 6x - 6 = 5x^2 - 5x \quad \text{Distribute}$$
$$12x - 6 = 5x^2 - 5x \quad \text{Combine like terms}$$
$$\underline{-12x + 6 \quad -12x + 6} \quad \text{Move all terms to one side of equation}$$
$$0 = 5x^2 - 17x + 6 \quad \text{Factor}$$
$$0 = (5x - 2)(x - 3) \quad \text{Set each factor equal to zero}$$
$$5x - 2 = 0 \text{ or } x - 3 = 0 \quad \text{Solve each equation}$$
$$\underline{+2 + 2 \qquad +3 + 3}$$
$$5x = 2 \text{ or } x = 3$$
$$\overline{5} \quad \overline{5}$$
$$x = \frac{2}{5} \text{ or } x = 3 \quad \text{Subtract 1 from each to find electrician}$$
$$\frac{2}{5} - 1 = \frac{-3}{5} \text{ or } 3 - 1 = 2 \quad \text{Ignore negative.}$$

Electrician: 2 hr, Apprentice: 3 hours   Our Solution

Very similar to a teamwork problem is when the two involved parts are working against each other. A common example of this is a sink that is filled by a pipe and emptied by a drain. If they are working against eachother we need to make one of the values negative to show they oppose eachother. This is shown in the next example..

Example 494.

A sink can be filled by a pipe in 5 minutes but it takes 7 minutes to drain a full sink. If both the pipe and the drain are open, how long will it take to fill the sink?

$$\text{Sink: 5, Drain: 7, Total: } x \quad \text{Define variables, drain is negative}$$
$$\frac{1}{5} - \frac{1}{7} = \frac{1}{x} \quad \text{Using reciprocals to make equation,}$$
$$\text{Subtract because they are opposite}$$
$$\frac{1(35x)}{5} - \frac{1(35x)}{7} = \frac{1(35x)}{x} \quad \text{Multiply each term by LCD: } 35x$$

$$7x - 5x = 35 \quad \text{Reduce fractions}$$
$$2x = 35 \quad \text{Combine like terms}$$
$$\overline{2} \quad \overline{2} \quad \text{Divide each term by 2}$$
$$x = 17.5 \quad \text{Our answer for } x$$

17.5 min or 17 min 30 sec   Our Solution

367

9.8 Practice - Teamwork

1) Bills father can paint a room in two hours less than Bill can paint it. Working together they can complete the job in two hours and 24 minutes. How much time would each require working alone?

2) Of two inlet pipes, the smaller pipe takes four hours longer than the larger pipe to fill a pool. When both pipes are open, the pool is filled in three hours and forty-five minutes. If only the larger pipe is open, how many hours are required to fill the pool?

3) Jack can wash and wax the family car in one hour less than Bob can. The two working together can complete the job in $1\frac{1}{5}$ hours. How much time would each require if they worked alone?

4) If A can do a piece of work alone in 6 days and B can do it alone in 4 days, how long will it take the two working together to complete the job?

5) Working alone it takes John 8 hours longer than Carlos to do a job. Working together they can do the job in 3 hours. How long will it take each to do the job working alone?

6) A can do a piece of work in 3 days, B in 4 days, and C in 5 days each working alone. How long will it take them to do it working together?

7) A can do a piece of work in 4 days and B can do it in half the time. How long will it take them to do the work together?

8) A cistern can be filled by one pipe in 20 minutes and by another in 30 minutes. How long will it take both pipes together to fill the tank?

9) If A can do a piece of work in 24 days and A and B together can do it in 6 days, how long would it take B to do the work alone?

10) A carpenter and his assistant can do a piece of work in $3\frac{3}{4}$ days. If the carpenter himself could do the work alone in 5 days, how long would the assistant take to do the work alone?

11) If Sam can do a certain job in 3 days, while it takes Fred 6 days to do the same job, how long will it take them, working together, to complete the job?

12) Tim can finish a certain job in 10 hours. It take his wife JoAnn only 8 hours to do the same job. If they work together, how long will it take them to complete the job?

13) Two people working together can complete a job in 6 hours. If one of them works twice as fast as the other, how long would it take the faster person, working alone, to do the job?

14) If two people working together can do a job in 3 hours, how long will it take the slower person to do the same job if one of them is 3 times as fast as the other?

15) A water tank can be filled by an inlet pipe in 8 hours. It takes twice that long for the outlet pipe to empty the tank. How long will it take to fill the tank if both pipes are open?

16) A sink can be filled from the faucet in 5 minutes. It takes only 3 minutes to empty the sink when the drain is open. If the sink is full and both the faucet and the drain are open, how long will it take to empty the sink?

17) It takes 10 hours to fill a pool with the inlet pipe. It can be emptied in 15 hrs with the outlet pipe. If the pool is half full to begin with, how long will it take to fill it from there if both pipes are open?

18) A sink is $\frac{1}{4}$ full when both the faucet and the drain are opened. The faucet alone can fill the sink in 6 minutes, while it takes 8 minutes to empty it with the drain. How long will it take to fill the remaining $\frac{3}{4}$ of the sink?

19) A sink has two faucets, one for hot water and one for cold water. The sink can be filled by a cold-water faucet in 3.5 minutes. If both faucets are open, the sink is filled in 2.1 minutes. How long does it take to fill the sink with just the hot-water faucet open?

20) A water tank is being filled by two inlet pipes. Pipe A can fill the tank in $4\frac{1}{2}$ hrs, while both pipes together can fill the tank in 2 hours. How long does it take to fill the tank using only pipe B?

21) A tank can be emptied by any one of three caps. The first can empty the tank in 20 minutes while the second takes 32 minutes. If all three working together could empty the tank in $8\frac{8}{59}$ minutes, how long would the third take to empty the tank?

22) One pipe can fill a cistern in $1\frac{1}{2}$ hours while a second pipe can fill it in $2\frac{1}{3}$ hrs. Three pipes working together fill the cistern in 42 minutes. How long would it take the third pipe alone to fill the tank?

23) Sam takes 6 hours longer than Susan to wax a floor. Working together they can wax the floor in 4 hours. How long will it take each of them working alone to wax the floor?

24) It takes Robert 9 hours longer than Paul to repair a transmission. If it takes them $2\frac{2}{5}$ hours to do the job if they work together, how long will it take each of them working alone?

25) It takes Sally $10\frac{1}{2}$ minutes longer than Patricia to clean up their dorm room. If they work together they can clean it in 5 minutes. How long will it take each of them if they work alone?

26) A takes $7\frac{1}{2}$ minutes longer than B to do a job. Working together they can do the job in 9 minutes. How long does it take each working alone?

27) Secretary A takes 6 minutes longer than Secretary B to type 10 pages of manuscript. If they divide the job and work together it will take them $8\frac{3}{4}$ minutes to type 10 pages. How long will it take each working alone to type the 10 pages?

28) It takes John 24 minutes longer than Sally to mow the lawn. If they work together they can mow the lawn in 9 minutes. How long will it take each to mow the lawn if they work alone?

Quadratics - Simultaneous Products

Objective: Solve simultaneous product equations using substitution to create a rational equation.

When solving a system of equations where the variables are multiplied together we can use the same idea of substitution that we used with linear equations. When we do so we may end up with a quadratic equation to solve. When we used substitution we solved for a variable and substitute this expression into the other equation. If we have two products we will choose a variable to solve for first and divide both sides of the equations by that variable or the factor containing the variable. This will create a situation where substitution can easily be done.

Example 495.

$$xy = 48$$
$$(x+3)(y-2) = 54$$

To solve for x, divide first equation by x, second by $x+3$

$$y = \frac{48}{x} \text{ and } y - 2 = \frac{54}{x+3}$$

Substitute $\frac{48}{x}$ for y in the second equation

$$\frac{48}{x} - 2 = \frac{54}{x+3}$$

Multiply each term by LCD: $x(x+3)$

$$\frac{48x(x+3)}{x} - 2x(x+3) = \frac{54x(x+3)}{x+3}$$

Reduce each fraction

$$48(x+3) - 2x(x+3) = 54x$$ Distribute

$$48x + 144 - 2x^2 - 6x = 54x$$ Combine like terms

$$-2x^2 + 42x + 144 = 54x$$ Make equation equal zero

$$\underline{\quad -54x \qquad\quad -54x}$$ Subtract $54x$ from both sides

$$-2x^2 - 12x + 144 = 0$$ Divide each term by GCF of -2

$$x^2 + 6x - 72 = 0$$ Factor

$$(x-6)(x+12) = 0$$ Set each factor equal to zero

$$x - 6 = 0 \text{ or } x + 12 = 0$$ Solve each equation

$$\underline{+6+6 \qquad\quad -12-12}$$

$$x = 6 \text{ or } x = -12$$ Substitute each solution into $xy = 48$

$$6y = 48 \text{ or } -12y = 48$$ Solve each equation

$$\overline{6 \quad 6} \qquad\quad \overline{-12 \ -12}$$

$$y = 8 \text{ or } y = -4$$ Our solutions for y,

$$(6, 8) \text{ or } (-12, -4)$$ Our Solutions as ordered pairs

These simultaneous product equations will also solve by the exact same pattern. We pick a variable to solve for, divide each side by that variable, or factor containing the variable. This will allow us to use substitution to create a rational expression we can use to solve. Quite often these problems will have two solutions.

Example 496.

$$xy = -35$$
$$(x+6)(y-2) = 5$$

To solve for x, divide the first equation by x, second by $x+6$

$$y = \frac{-35}{x} \quad \text{and} \quad y - 2 = \frac{5}{x+6}$$

Substitute $\frac{-35}{x}$ for y in the second equation

$$\frac{-35}{x} - 2 = \frac{5}{x+6}$$

Multiply each term by LCD: $x(x+6)$

$$\frac{-35x(x+6)}{x} - 2x(x+6) = \frac{5x(x+6)}{x+6}$$

Reduce fractions

$$-35(x+6) - 2x(x+6) = 5x$$ Distribute
$$-35x - 210 - 2x^2 - 12x = 5x$$ Combine like terms
$$-2x^2 - 47x - 210 = 5x$$ Make equation equal zero
$$\underline{\; -5x -5x}$$
$$-2x^2 - 52x - 210 = 0$$ Divide each term by -2
$$x^2 + 26x + 105 = 0$$ Factor
$$(x+5)(x+21) = 0$$ Set each factor equal to zero
$$x + 5 = 0 \quad \text{or} \quad x + 21 = 0$$ Solve each equation
$$\underline{-5-5} \qquad \underline{-21-21}$$
$$x = -5 \quad \text{or} \quad x = -21$$ Substitute each solution into $xy = -35$
$$-5y = -35 \quad \text{or} \quad -21y = -35$$ Solve each equation
$$\overline{-5} \quad \overline{-5} \qquad \overline{-21} \quad \overline{-21}$$
$$y = 7 \quad \text{or} \quad y = \frac{5}{3}$$ Our solutions for y
$$(-5, 7) \quad \text{or} \quad \left(-21, \frac{5}{3}\right)$$ Our Solutions as ordered pairs

The processes used here will be used as we solve applications of quadratics including distance problems and revenue problems. These will be covered in another section.

World View Note: William Horner, a British mathematician from the late 18th century/early 19th century is credited with a method for solving simultaneous equations, however, Chinese mathematician Chu Shih-chieh in 1303 solved these equations with exponents as high as 14!

9.9 Practice - Simultaneous Product

Solve.

1) $xy = 72$
      $(x+2)(y-4) = 128$

2) $xy = 180$
      $(x-1)(y-\frac{1}{2}) = 205$

3) $xy = 150$
      $(x-6)(y+1) = 64$

4) $xy = 120$
      $(x+2)(y-3) = 120$

5) $xy = 45$
      $(x+2)(y+1) = 70$

6) $xy = 65$
      $(x-8)(y+2) = 35$

7) $xy = 90$
      $(x-5)(y+1) = 120$

8) $xy = 48$
      $(x-6)(y+3) = 60$

9) $xy = 12$
      $(x+1)(y-4) = 16$

10) $xy = 60$
      $(x+5)(y+3) = 150$

11) $xy = 45$
      $(x-5)(y+3) = 160$

12) $xy = 80$
      $(x-5)(y+5) = 45$

Quadratics - Revenue and Distance

Objective: Solve revenue and distance applications of quadratic equations.

A common application of quadratics comes from revenue and distance problems. Both are set up almost identical to each other so they are both included together. Once they are set up, we will solve them in exactly the same way we solved the simultaneous product equations.

Revenue problems are problems where a person buys a certain number of items for a certain price per item. If we multiply the number of items by the price per item we will get the total paid. To help us organize our information we will use the following table for revenue problems

	Number	Price	Total
First			
Second			

The price column will be used for the individual prices, the total column is used for the total paid, which is calculated by multiplying the number by the price. Once we have the table filled out we will have our equations which we can solve. This is shown in the following examples.

Example 497.

A man buys several fish for $56. After three fish die, he decides to sell the rest at a profit of $5 per fish. His total profit was $4. How many fish did he buy to begin with?

	Number	Price	Total
Buy	n	p	56
Sell			

Using our table, we don't know the number he bought, or at what price, so we use varibles n and p. Total price was $56.

	Number	Price	Total
Buy	n	p	56
Sell	$n-3$	$p+5$	60

When he sold, he sold 3 less $(n-3)$, for $5 more $(p+5)$. Total profit was $4, combined with $56 spent is $60

$$np = 56$$
$$(n-3)(p+5) = 60$$

Find equatinos by multiplying number by price

These are a simultaneous product

$$p = \frac{56}{n} \quad \text{and} \quad p+5 = \frac{60}{n-3}$$

Solving for number, divide by n or $(n-3)$

$$\frac{56}{n} + 5 = \frac{60}{n-3}$$ Substitute $\frac{56}{n}$ for p in second equation

$$\frac{56n(n-3)}{n} + 5n(n-3) = \frac{60n(n-3)}{n-3}$$ Multiply each term by LCD: $n(n-3)$

$$56(n-3) + 5n(n-3) = 60n$$ Reduce fractions
$$56n - 168 + 5n^2 - 15n = 60n$$ Combine like terms
$$5n^2 + 41n - 168 = 60n$$ Move all terms to one side
$$\underline{ -60n \qquad\qquad -60n}$$
$$5n^2 - 19n - 168 = 0$$ Solve with quadratic formula

$$n = \frac{19 \pm \sqrt{(-19)^2 - 4(5)(-168)}}{2(5)}$$ Simplify

$$n = \frac{19 \pm \sqrt{3721}}{10} = \frac{19 \pm 61}{10}$$ We don't want negative solutions, only do $+$

$$n = \frac{80}{10} = 8$$ This is our n

$$8 \text{ fish} \qquad \text{Our Solution}$$

Example 498.

A group of students together bought a couch for their dorm that cost \$96. However, 2 students failed to pay their share, so each student had to pay \$4 more. How many students were in the original group?

	Number	Price	Total
Deal	n	p	96
Paid			

\$96 was paid, but we don't know the number or the price agreed upon by each student.

	Number	Price	Total
Deal	n	p	96
Paid	$n-2$	$p+4$	96

There were 2 less that actually paid $(n-2)$ and they had to pay \$4 more $(p+4)$. The total here is still \$96.

$$np = 96$$ Equations are product of number and price
$$(n-2)(p+4) = 96$$ This is a simultaneous product

$$p = \frac{96}{n} \text{ and } p + 4 = \frac{96}{n-2}$$ Solving for number, divide by n and $n-2$

$$\frac{96}{n} + 4 = \frac{96}{n-2}$$ Substitute $\frac{96}{n}$ for p in the second equation

374

$$\frac{96n(n-2)}{n} + 4n(n-2) = \frac{96n(n-2)}{n-2} \qquad \text{Multiply each term by LCD: } n(n-2)$$

$$96(n-2) + 4n(n-2) = 96n \qquad \text{Reduce fractions}$$

$$96n - 192 + 4n^2 - 8n = 96n \qquad \text{Distribute}$$

$$4n^2 + 88n - 192 = 96n \qquad \text{Combine like terms}$$

$$\underline{-96n \qquad\qquad -96n} \qquad \text{Set equation equal to zero}$$

$$4n^2 - 8n - 192 = 0 \qquad \text{Solve by completing the square,}$$

$$\underline{+192 + 192} \qquad \text{Separate variables and constant}$$

$$\frac{4n^2 - 8n}{4} = \frac{192}{4} \qquad \text{Divide each term by } a \text{ or } 4$$

$$n^2 - 2n = 48 \qquad \text{Complete the square: } \left(b \cdot \frac{1}{2}\right)^2$$

$$\left(2 \cdot \frac{1}{2}\right)^2 = 1^2 = 1 \qquad \text{Add to both sides of equation}$$

$$n^2 - 2n + 1 = 49 \qquad \text{Factor}$$

$$(n-1)^2 = 49 \qquad \text{Square root of both sides}$$

$$n - 1 = \pm 7 \qquad \text{Add 1 to both sides}$$

$$n = 1 \pm 7 \qquad \text{We don't want a negative solution}$$

$$n = 1 + 7 = 8$$

$$8 \text{ students} \qquad \text{Our Solution}$$

The above examples were solved by the quadratic formula and completing the square. For either of these we could have used either method or even factoring. Remember we have several options for solving quadratics. Use the one that seems easiest for the problem.

Distance problems work with the same ideas that the revenue problems work. The only difference is the variables are r and t (for rate and time), instead of n and p (for number and price). We already know that distance is calculated by multiplying rate by time. So for our distance problems our table becomes the following:

	rate	time	distance
First			
Second			

Using this table and the exact same patterns as the revenue problems is shown in the following example.

Example 499.

375

Greg went to a conference in a city 120 miles away. On the way back, due to road construction he had to drive 10 mph slower which resulted in the return trip taking 2 hours longer. How fast did he drive on the way to the conference?

	rate	time	distance
There	r	t	120
Back			

We do not know rate, r, or time, t he traveled on the way to the conference. But we do know the distance was 120 miles.

	rate	time	distance
There	r	t	120
Back	$r - 10$	$t+2$	120

Coming back he drove 10 mph slower $(r - 10)$ and took 2 hours longer $(t + 2)$. The distance was still 120 miles.

$$rt = 120$$ Equations are product of rate and time

$$(r - 10)(t + 2) = 120$$ We have simultaneous product equations

$$t = \frac{120}{r} \text{ and } t + 2 = \frac{120}{r - 10}$$ Solving for rate, divide by r and $r - 10$

$$\frac{120}{r} + 2 = \frac{120}{r - 10}$$ Substitute $\frac{120}{r}$ for t in the second equation

$$\frac{120r(r - 10)}{r} + 2r(r - 10) = \frac{120r(r - 10)}{r - 10}$$ Multiply each term by LCD: $r(r - 10)$

$$120(r - 10) + 2r^2 - 20r = 120r$$ Reduce each fraction

$$120r - 1200 + 2r^2 - 20r = 120r$$ Distribute

$$2r^2 + 100r - 1200 = 120r$$ Combine like terms

$$\underline{ -120r -120r}$$ Make equation equal to zero

$$2r^2 - 20r - 1200 = 0$$ Divide each term by 2

$$r^2 - 10r - 600 = 0$$ Factor

$$(r - 30)(r + 20) = 0$$ Set each factor equal to zero

$$r - 30 = 0 \text{ and } r + 20 = 0$$ Solve each equation

$$\underline{+30 + 30 -20 - 20}$$

$$r = 30 \text{ and } r = -20$$ Can't have a negative rate

$$30 \text{ mph}$$ Our Solution

World View Note: The world's fastest man (at the time of printing) is Jamaican Usain Bolt who set the record of running 100 m in 9.58 seconds on August 16, 2009 in Berlin. That is a speed of over 23 miles per hour!

Another type of simultaneous product distance problem is where a boat is traveling in a river with the current or against the current (or an airplane flying with the wind or against the wind). If a boat is traveling downstream, the current will push it or increase the rate by the speed of the current. If a boat is traveling upstream, the current will pull against it or decrease the rate by the speed of the current. This is demonstrated in the following example.

Example 500.

A man rows down stream for 30 miles then turns around and returns to his original location, the total trip took 8 hours. If the current flows at 2 miles per hour, how fast would the man row in still water?

8   Write total time above time column

	rate	time	distance
down		t	30
up		$8-t$	30

We know the distance up and down is 30. Put t for time downstream. Subtracting $8-t$ becomes time upstream

	rate	time	distance
down	$r+2$	t	30
up	$r-2$	$8-t$	30

Downstream the current of 2 mph pushes the boat $(r+2)$ and upstream the current pulls the boat $(r-2)$

$$(r+2)t=30$$ Multiply rate by time to get equations
$$(r-2)(8-t)=30$$ We have a simultaneous product

$$t=\frac{30}{r+2} \text{ and } 8-t=\frac{30}{r-2}$$ Solving for rate, divide by $r+2$ or $r-2$

$$8-\frac{30}{r+2}=\frac{30}{r-2}$$ Substitute $\frac{30}{r+2}$ for t in second equation

$$8(r+2)(r-2)-\frac{30(r+2)(r-2)}{r+2}=\frac{30(r+2)(r-2)}{r-2}$$ Multiply each term by LCD: $(r+2)(r-2)$

$$8(r+2)(r-2)-30(r-2)=30(r+2)$$ Reduce fractions
$$8r^2-32-30r+60=30r+60$$ Multiply and distribute
$$8r^2-30r+28=30r+60$$ Make equation equal zero
$$\underline{-30r-60-30r-60}$$

$$8r^2-60r-32=0$$ Divide each term by 4
$$2r^2-15r-8=0$$ Factor
$$(2r+1)(r-8)=0$$ Set each factor equal to zero
$$2r+1=0 \text{ or } r-8=0$$ Solve each equation
$$\underline{-1-1 \qquad +8+8}$$
$$2r=-1 \text{ or } r=8$$
$$\overline{2} \quad \overline{2}$$

$$r=-\frac{1}{2} \text{ or } r=8$$ Can't have a negative rate
8 mph   Our Solution

9.10 Practice - Revenue and Distance

1) A merchant bought some pieces of silk for $900. Had he bought 3 pieces more for the same money, he would have paid $15 less for each piece. Find the number of pieces purchased.

2) A number of men subscribed a certain amount to make up a deficit of $100 but 5 men failed to pay and thus increased the share of the others by $1 each. Find the amount that each man paid.

3) A merchant bought a number of barrels of apples for $120. He kept two barrels and sold the remainder at a profit of $2 per barrel making a total profit of $34. How many barrels did he originally buy?

4) A dealer bought a number of sheep for $440. After 5 had died he sold the remainder at a profit of $2 each making a profit of $60 for the sheep. How many sheep did he originally purchase?

5) A man bought a number of articles at equal cost for $500. He sold all but two for $540 at a profit of $5 for each item. How many articles did he buy?

6) A clothier bought a lot of suits for $750. He sold all but 3 of them for $864 making a profit of $7 on each suit sold. How many suits did he buy?

7) A group of boys bought a boat for $450. Five boys failed to pay their share, hence each remaining boys were compelled to pay $4.50 more. How many boys were in the original group and how much had each agreed to pay?

8) The total expenses of a camping party were $72. If there had been 3 fewer persons in the party, it would have cost each person $2 more than it did. How many people were in the party and how much did it cost each one?

9) A factory tests the road performance of new model cars by driving them at two different rates of speed for at least 100 kilometers at each rate. The speed rates range from 50 to 70 km/hr in the lower range and from 70 to 90 km/hr in the higher range. A driver plans to test a car on an available speedway by driving it for 120 kilometers at a speed in the lower range and then driving 120 kilometers at a rate that is 20 km/hr faster. At what rates should he drive if he plans to complete the test in $3\frac{1}{2}$ hours?

10) A train traveled 240 kilometers at a certain speed. When the engine was replaced by an improved model, the speed was increased by 20 km/hr and the travel time for the trip was decreased by 1 hour. What was the rate of each engine?

11) The rate of the current in a stream is 3 km/hr. A man rowed upstream for 3 kilometers and then returned. The round trip required 1 hour and 20 minutes. How fast was he rowing?

12) A pilot flying at a constant rate against a headwind of 50 km/hr flew for 750 kilometers, then reversed direction and returned to his starting point. He completed the round trip in 8 hours. What was the speed of the plane?

13) Two drivers are testing the same model car at speeds that differ by 20 km/hr. The one driving at the slower rate drives 70 kilometers down a speedway and returns by the same route. The one driving at the faster rate drives 76 kilometers down the speedway and returns by the same route. Both drivers leave at the same time, and the faster car returns $\frac{1}{2}$ hour earlier than the slower car. At what rates were the cars driven?

14) An athlete plans to row upstream a distance of 2 kilometers and then return to his starting point in a total time of 2 hours and 20 minutes. If the rate of the current is 2 km/hr, how fast should he row?

15) An automobile goes to a place 72 miles away and then returns, the round trip occupying 9 hours. His speed in returning is 12 miles per hour faster than his speed in going. Find the rate of speed in both going and returning.

16) An automobile made a trip of 120 miles and then returned, the round trip occupying 7 hours. Returning, the rate was increased 10 miles an hour. Find the rate of each.

17) The rate of a stream is 3 miles an hour. If a crew rows downstream for a distance of 8 miles and then back again, the round trip occupying 5 hours, what is the rate of the crew in still water?

18) The railroad distance between two towns is 240 miles. If the speed of a train were increased 4 miles an hour, the trip would take 40 minutes less. What is the usual rate of the train?

19) By going 15 miles per hour faster, a train would have required 1 hour less to travel 180 miles. How fast did it travel?

20) Mr. Jones visits his grandmother who lives 100 miles away on a regular basis. Recently a new freeway has opend up and, although the freeway route is 120 miles, he can drive 20 mph faster on average and takes 30 minutes less time to make the trip. What is Mr. Jones rate on both the old route and on the freeway?

21) If a train had traveled 5 miles an hour faster, it would have needed $1\frac{1}{2}$ hours less time to travel 150 miles. Find the rate of the train.

22) A traveler having 18 miles to go, calculates that his usual rate would make him one-half hour late for an appointment; he finds that in order to arrive on time he must travel at a rate one-half mile an hour faster. What is his usual rate?

Quadratics - Graphs of Quadratics

Objective: Graph quadratic equations using the vertex, x-intercepts, and y-intercept.

Just as we drew pictures of the solutions for lines or linear equations, we can draw a picture of solution to quadratics as well. One way we can do that is to make a table of values.

Example 501.

$$y = x^2 - 4x + 3$$ Make a table of values

x	y
0	
1	
2	
3	
4	

We will test 5 values to get an idea of shape

$y = (0)^2 + 4(0) + 3 = 0 - 0 + 3 = 3$ Plug 0 in for x and evaluate

$y = (1)^2 - 4(1) + 3 = 1 - 4 + 3 = 0$ Plug 1 in for x and evaluate

$y = (2)^2 - 4(2) + 3 = 4 - 8 + 3 = -1$ Plug 2 in for x and evaluate

$y = (3)^2 - 4(3) + 3 = 9 - 12 + 3 = 0$ Plug 3 in for x and evaluate

$y = (4)^2 - 4(4) + 3 = 16 - 16 + 3 = 3$ Plug 4 in for x and evaluate

x	y
0	3
1	0
2	-1
3	0
4	3

Our completed table. Plot points on graph

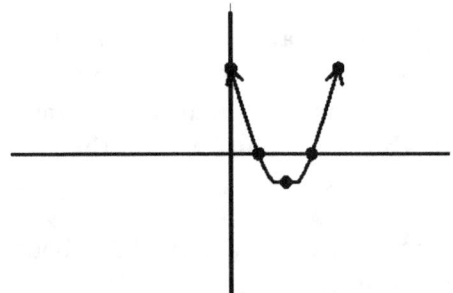

Plot the points $(0, 3)$, $(1, 0)$, $(2, -1)$, $(3, 0)$, and $(4, 3)$.

Connect the dots with a smooth curve.

Our Solution

When we have x^2 in our equations, the graph will no longer be a straight line. Quadratics have a graph that looks like a U shape that is called a parabola.

World View Note: The first major female mathematician was Hypatia of Egypt who was born around 370 AD. She studied conic sections. The parabola is one type of conic section.

The above method to graph a parabola works for any equation, however, it can be very tedious to find all the correct points to get the correct bend and shape. For this reason we identify several key points on a graph and in the equation to help us graph parabolas more efficiently. These key points are described below.

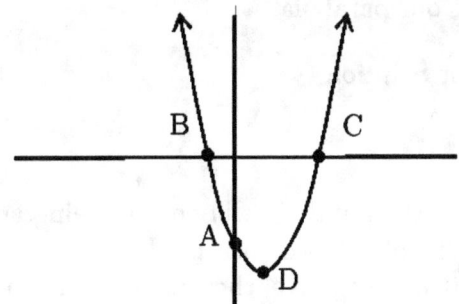

Point A: y-intercept: Where the graph crosses the vertical y-axis.

Points B and C: x-intercepts: Where the graph crosses the horizontal x-axis

Point D: Vertex: The point where the graph curves and changes directions.

We will use the following method to find each of the points on our parabola.

To graph the equation $y = ax^2 + bx + c$, find the following points

1. y-intercept: Found by making $x = 0$, this simplifies down to $y = c$

2. x-intercepts: Found by making $y = 0$, this means solving $0 = ax^2 + bx + c$

3. Vertex: Let $x = \frac{-b}{2a}$ to find x. Then plug this value into the equation to find y.

After finding these points we can connect the dots with a smooth curve to find our graph!

Example 502.

$$y = x^2 + 4x + 3 \quad \text{Find the key points}$$

$$y = 3 \quad y = c \text{ is the } y - \text{intercept}$$

$$0 = x^2 + 4x + 3 \quad \text{To find } x - \text{intercept we solve the equation}$$
$$0 = (x+3)(x+1) \quad \text{Factor}$$
$$x + 3 = 0 \text{ and } x + 1 = 0 \quad \text{Set each factor equal to zero}$$
$$\underline{-3 \quad -3} \qquad \underline{-1 \quad -1} \quad \text{Solve each equation}$$
$$x = -3 \text{ and } x = -1 \quad \text{Our } x - \text{intercepts}$$

$$x = \frac{-4}{2(1)} = \frac{-4}{2} = -2 \quad \text{To find the vertex, first use } x = \frac{-b}{2a}$$
$$y = (-2)^2 + 4(-2) + 3 \quad \text{Plug this answer into equation to find } y - \text{coordinate}$$
$$y = 4 - 8 + 3 \quad \text{Evaluate}$$
$$y = -1 \quad \text{The } y - \text{coordinate}$$
$$(-2, -1) \quad \text{Vertex as a point}$$

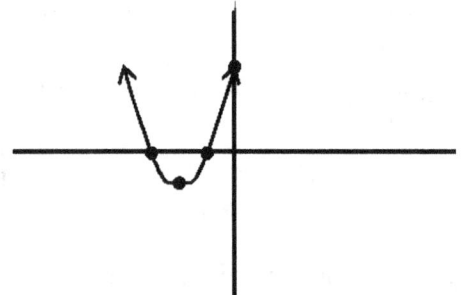

Graph the y-intercept at 3, the x-intercepts at -3 and -1, and the vertex at $(-2, -1)$. Connect the dots with a smooth curve in a U shape to get our parabola.

Our Solution

If the a in $y = ax^2 + bx + c$ is a negative value, the parabola will end up being an upside-down U. The process to graph it is identical, we just need to be very careful of how our signs operate. Remember, if a is negative, then ax^2 will also be negative because we only square the x, not the a.

Example 503.

$$y = -3x^2 + 12x - 9 \qquad \text{Find key points}$$

$$y = -9 \qquad y - \text{intercept is } y = c$$

$$0 = -3x^2 + 12x - 9 \qquad \text{To find } x - \text{intercept solve this equation}$$
$$0 = -3(x^2 - 4x + 3) \qquad \text{Factor out GCF first, then factor rest}$$
$$0 = -3(x - 3)(x - 1) \qquad \text{Set each factor with } a \text{ varaible equal to zero}$$
$$x - 3 = 0 \text{ and } x - 1 = 0 \qquad \text{Solve each equation}$$
$$\underline{+3 + 3 \qquad\qquad +1 + 1}$$
$$x = 3 \text{ and } x = 1 \qquad \text{Our } x - \text{intercepts}$$

$$x = \frac{-12}{2(-3)} = \frac{-12}{-6} = 2 \qquad \text{To find the vertex, first use } x = \frac{-b}{2a}$$
$$y = -3(2)^2 + 12(2) - 9 \qquad \text{Plug this value into equation to find } y - \text{coordinate}$$
$$y = -3(4) + 24 - 9 \qquad \text{Evaluate}$$
$$y = -12 + 24 - 9$$
$$y = 3 \qquad y - \text{value of vertex}$$
$$(2, 3) \qquad \text{Vertex as } a \text{ point}$$

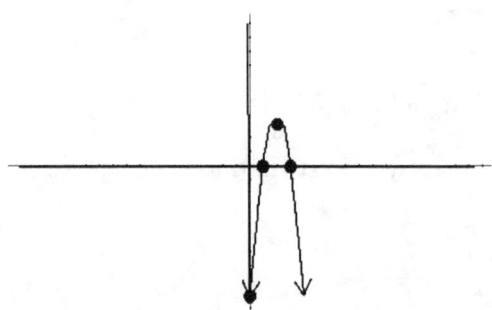

Graph the y-intercept at -9, the x-intercepts at 3 and 1, and the vertex at $(2, 3)$. Connect the dots with smooth curve in an upside-down U shape to get our parabola.

Our Solution

It is important to remember the graph of all quadratics is a parabola with the same U shape (they could be upside-down). If you plot your points and we cannot connect them in the correct U shape then one of your points must be wrong. Go back and check your work to be sure they are correct!

Just as all quadratics (equation with $y = x^2$) all have the same U-shape to them and all linear equations (equations such as $y = x$) have the same line shape when graphed, different equations have different shapes to them. Below are some common equations (some we have yet to cover!) with their graph shape drawn.

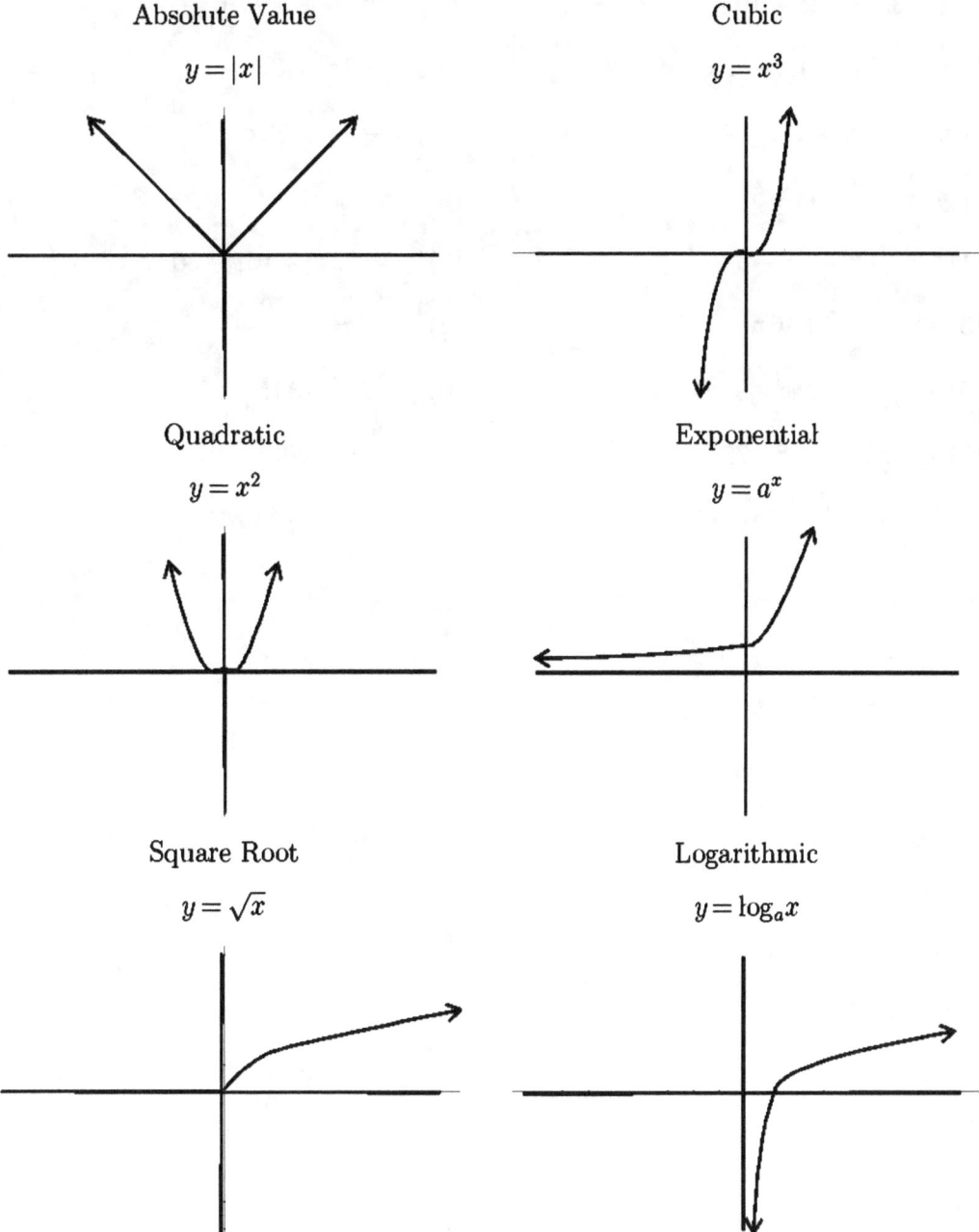

Absolute Value

$$y = |x|$$

Cubic

$$y = x^3$$

Quadratic

$$y = x^2$$

Exponential

$$y = a^x$$

Square Root

$$y = \sqrt{x}$$

Logarithmic

$$y = \log_a x$$

9.11 Practice - Graphs of Quadratics

Find the vertex and intercepts of the following quadratics. Use this information to graph the quadratic.

1) $y = x^2 - 2x - 8$

2) $y = x^2 - 2x - 3$

3) $y = 2x^2 - 12x + 10$

4) $y = 2x^2 - 12x + 16$

5) $y = -2x^2 + 12x - 18$

6) $y = -2x^2 + 12x - 10$

7) $y = -3x^2 + 24x - 45$

8) $y = -3x^2 + 12x - 9$

9) $y = -x^2 + 4x + 5$

10) $y = -x^2 + 4x - 3$

11) $y = -x^2 + 6x - 5$

12) $y = -2x^2 + 16x - 30$

13) $y = -2x^2 + 16x - 24$

14) $y = 2x^2 + 4x - 6$

15) $y = 3x^2 + 12x + 9$

16) $y = 5x^2 + 30x + 45$

17) $y = 5x^2 - 40x + 75$

18) $y = 5x^2 + 20x + 15$

19) $y = -5x^2 - 60x - 175$

20) $y = -5x^2 + 20x - 15$

Chapter 10 : Functions

Functions - Function Notation

Objective: Idenfity functions and use correct notation to evaluate functions at numerical and variable values.

There are many different types of equations that we can work with in algebra. An equation gives the relationship between variables and numbers. Examples of several relationships are below:

$$\frac{(x-3)^2}{9} - \frac{(y+2)^2}{4} = 1 \quad \text{and} \quad y = x^2 - 2x + 7 \quad \text{and} \quad \sqrt{y+x} - 7 = xy$$

There is a spcical classification of relationships known as functions. **Functions** have at most one output for any input. Generally x is the variable that we plug into an equation and evaluate to find y. For this reason x is considered an input variable and y is considered an output variable. This means the definition of a function, in terms of equations in x and y could be said, for any x value there is at most one y value that corresponds with it.

A great way to visualize this definition is to look at the graphs of a few relationships. Because x values are vertical lines we will draw a vertical line through the graph. If the vertical line crosses the graph more than once, that means we have too many possible y values. If the graph crosses the graph only once, then we say the relationship is a function.

Example 504.

Which of the following graphs are graphs of functions?

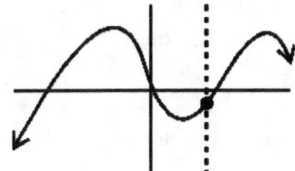 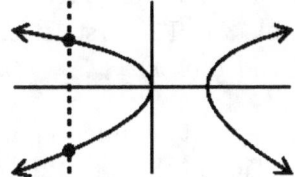 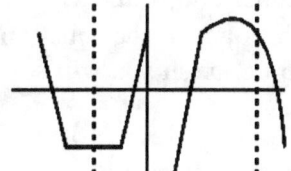

Drawing a vertical line through this graph will only cross the graph once, it is a function.

Drawing a vertical line through this graph will cross the graph twice, once at top and once at bottom. This is not a function.

Drawing a vertical line through this graph will cross the graph only once, it is a function.

We can look at the above idea in an algebraic method by taking a relationship and solving it for y. If we have only one solution then it is a function.

Example 505.

$$\text{Is } 3x^2 - y = 5 \text{ a function?} \qquad \text{Solve the relation for } y$$
$$\underline{-3x^2 \quad -3x^2} \qquad\qquad \text{Subtract } 3x^2 \text{ from both sides}$$
$$-y = -3x^2 + 5 \qquad\qquad \text{Divide each term by} -1$$
$$\overline{-1} \quad \overline{-1} \ \overline{-1}$$
$$y = 3x^2 - 5 \qquad\qquad \text{Only one solution for } y.$$
$$\qquad\qquad\qquad \text{Yes!} \quad \text{It is } a \text{ function}$$

Example 506.

$$\text{Is } y^2 - x = 5 \text{ a function?} \qquad \text{Solve the relation for } y$$
$$+x + x \qquad\qquad\qquad \text{Add } x \text{ to both sides}$$
$$y^2 = x + 5 \qquad\qquad\qquad \text{Square root of both sdies}$$
$$\sqrt{y^2} = \pm \sqrt{x + 5} \qquad\qquad \text{Simplify}$$
$$y = \pm \sqrt{x + 5} \qquad\qquad \text{Two solutions for } y \, (\text{one} +, \text{one} -)$$
$$\qquad\qquad\qquad \text{No!} \quad \text{Not } a \text{ function}$$

Once we know we have a function, often we will change the notation used to emphasis the fact that it is a function. Instead of writing $y =$, we will use function notation which can be written $f(x) =$. We read this notation "f of x". So for

387

the above example that was a function, instead of writing $y = 3x^2 - 5$, we could have written $f(x) = 3x^2 - 5$. It is important to point out that $f(x)$ does not mean f times x, it is nearly a notation that names the function with the first letter (function f) and then in parenthesis we are given information about what variables are in the function (variable x). The first letter can be anything we want it to be, often you will see $g(x)$ (read g of x).

World View Note: The concept of a function was first introduced by Arab mathematician Sharaf al-Din al-Tusi in the late 12th century

Once we know a relationship is a function, we may be interested in what values can be put into the equations. The values that are put into an equation (generally the x values) are called the **domain**. When finding the domain, often it is easier to consider what cannot happen in a given function, then exclude those values.

Example 507.

$$\text{Find the domain: } f(x) = \frac{3x-1}{x^2+x-6} \quad \text{With fractions, zero can't be in denominator}$$
$$x^2 + x - 6 \neq 0 \quad \text{Solve by factoring}$$
$$(x+3)(x-2) \neq 0 \quad \text{Set each factor not equal to zero}$$
$$x + 3 \neq 0 \text{ and } x - 2 \neq 0 \quad \text{Solve each equation}$$
$$\underline{-3 - 3} \qquad \underline{+2 + 2}$$
$$x \neq -3, 2 \qquad \text{Our Solution}$$

The notation in the previous example tells us that x can be any value except for -3 and 2. If x were one of those two values, the function would be undefined.

Example 508.

$$\text{Find the domain: } f(x) = 3x^2 - x \quad \text{With this equation there are no bad values}$$
$$\text{All Real Numbers or } \mathbb{R} \quad \text{Our Solution}$$

In the above example there are no real numbers that make the function undefined. This means any number can be used for x.

Example 509.

$$\text{Find the domain: } f(x) = \sqrt{2x-3} \quad \text{Square roots can't be negative}$$

$$2x - 3 \geqslant 0 \quad \text{Set up an inequality}$$
$$\underline{+3 + 3} \quad \text{Solve}$$
$$\frac{2x}{2} \geqslant \frac{3}{2}$$
$$x \geqslant \frac{3}{2} \quad \text{Our Solution}$$

The notation in the above example states that our variable can be $\frac{3}{2}$ or any number larger than $\frac{3}{2}$. But any number smaller would make the function undefined (without using imaginary numbers).

Another use of function notation is to easily plug values into functions. If we want to substitute a variable for a value (or an expression) we simply replace the variable with what we want to plug in. This is shown in the following examples.

Example 510.

$$f(x) = 3x^2 - 4x; \text{ find } f(-2) \quad \text{Substitute} -2 \text{ in for } x \text{ in the function}$$
$$f(-2) = 3(-2)^2 - 4(-2) \quad \text{Evaluate, exponents first}$$
$$f(-2) = 3(4) - 4(-2) \quad \text{Multiply}$$
$$f(-2) = 12 + 8 \quad \text{Add}$$
$$f(-2) = 20 \quad \text{Our Solution}$$

Example 511.

$$h(x) = 3^{2x-6}; \text{ find } h(4) \quad \text{Substitute } 4 \text{ in for } x \text{ in the function}$$
$$h(4) = 3^{2(4)-6} \quad \text{Simplify exponent, mutiplying first}$$
$$h(4) = 3^{8\ 6} \quad \text{Subtract in exponent}$$
$$h(4) = 3^2 \quad \text{Evaluate exponent}$$
$$h(4) = 9 \quad \text{Our Solution}$$

Example 512.

$$k(a) = 2|a + 4|; \text{ find } k(-7) \quad \text{Substitute} -7 \text{ in for } a \text{ in the function}$$
$$k(-7) = 2|-7 + 4| \quad \text{Add inside absolute values}$$
$$k(-7) = 2|-3| \quad \text{Evaluate absolute value}$$
$$k(-7) = 2(3) \quad \text{Multiply}$$

$$k(-7) = 6 \quad \text{Our Solution}$$

As the above examples show, the function can take many different forms, but the pattern to evaluate the function is always the same, replace the variable with what is in parenthesis and simplify. We can also substitute expressions into functions using the same process. Often the expressions use the same variable, it is important to remember each variable is replaced by whatever is in parenthesis.

Example 513.

$$\begin{aligned}
g(x) &= x^4 + 1; \text{find } g(3x) \quad &&\text{Replace } x \text{ in the function with } (3x) \\
g(3x) &= (3x)^4 + 1 \quad &&\text{Simplify exponet} \\
g(3x) &= 81x^4 + 1 \quad &&\text{Our Solution}
\end{aligned}$$

Example 514.

$$\begin{aligned}
p(t) &= t^2 - t; \text{find } p(t+1) \quad &&\text{Replace each } t \text{ in } p(t) \text{ with } (t+1) \\
p(t+1) &= (t+1)^2 - (t+1) \quad &&\text{Square binomial} \\
p(t+1) &= t^2 + 2t + 1 - (t+1) \quad &&\text{Distribute negative} \\
p(t+1) &= t^2 + 2t + 1 - t - 1 \quad &&\text{Combine like terms} \\
p(t+1) &= t^2 + t \quad &&\text{Our Solution}
\end{aligned}$$

It is important to become comfortable with function notation and how to use it as we transition into more advanced algebra topics.

10.1 Practice - Function Notation

Solve.

1) Which of the following is a function?

a)

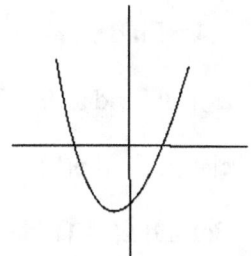

b)

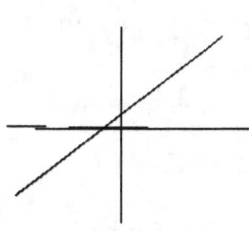

c)

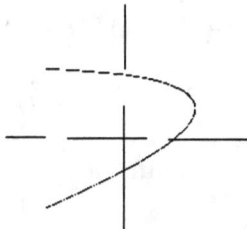

d)

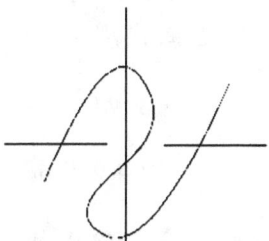

e) $y = 3x - 7$

f) $y^2 - x^2 = 1$

g) $\sqrt{y} + x = 2$

h) $x^2 + y^2 = 1$

Specify the domain of each of the following funcitons.

2) $f(x) = -5x + 1$

3) $f(x) = \sqrt{5 - 4x}$

4) $s(t) = \frac{1}{t^2}$

5) $f(x) = x^2 - 3x - 4$

6) $s(t) = \frac{1}{t^2 + 1}$

7) $f(x) = \sqrt{x - 16}$

8) $f(x) = \frac{-2}{x^2 \quad 3x \quad 4}$

9) $h(x) = \frac{\sqrt{3x \quad 12}}{x^2 \quad 25}$

10 $y(x) = \frac{x}{x^2 - 25}$

Evaluate each function.

11) $g(x) = 4x - 4$; Find $g(0)$

12) $g(n) = -3 \cdot 5^n$; Find $g(2)$

13) $f(x) = |3x + 1| + 1$; Find $f(0)$

14) $f(x) = x^2 + 4$; Find $f(-9)$

15) $f(n) = -2|-n-2| + 1$; Find $f(-$

16) $f(n) = n - 3$; Find $f(10)$

17) $f(t) = 3^t - 2$; Find $f(-2)$

18) $f(a) - 3^{a-1} - 3$; Find $f(2)$

19) $f(t) = |t + 3|$; Find $f(10)$

20) $w(x) = x^2 + 4x$; Find $w(-5)$

21) $w(n) = 4n + 3$; Find $w(2)$

22) $w(x) = -4x + 3$; Find $w(6)$

23) $w(n) = 2^{n+2}$; Find $w(-2)$

24) $p(x) = -|x| + 1$; Find $p(5)$

25) $p(n) = -3|n|$; Find $p(7)$

26) $k(a) = a + 3$; Find $k(-1)$

27) $p(t) = -t^3 + t$; Find $p(4)$

28) $k(x) = -2 \cdot 4^{2x\ 2}$; Find $k(2)$

29) $k(n) = |n - 1|$; Find $k(3)$

30) $p(t) = -2 \cdot 4^{2t+1} + 1$; Find $p(-2)$

31) $h(x) = x^3 + 2$; Find $h(-4x)$

32) $h(n) = 4n + 2$; Find $h(n + 2)$

33) $h(x) = 3x + 2$; Find $h(-1 + x)$

34) $h(a) = -3 \cdot 2^{a+3}$; Find $h(\frac{a}{4})$

35) $h(t) = 2|-3t - 1| + 2$; Find $h(n^2)$

36) $h(x) = x^2 + 1$; Find $h(\frac{x}{4})$

37) $g(x) = x + 1$; Find $g(3x)$

38) $h(t) = t^2 + t$; Find $h(t^2)$

39) $g(x) = 5^x$; Find $g(-3 - x)$

40) $h(n) = 5^{n\ 1} + 1$; Find $h(\frac{n}{2})$

Functions - Operations on Functions

Objective: Combine functions using sum, difference, product, quotient and composition of functions.

Several functions can work together in one larger function. There are 5 common operations that can be performed on functions. The four basic operations on functions are adding, subtracting, multiplying, and dividing. The notation for these functions is as follows.

$$\text{Addition} \quad (f+g)(x) = f(x) + g(x)$$
$$\text{Subtraction} \quad (f-g)(x) = f(x) - g(x)$$
$$\text{Multiplication} \quad (f \cdot g)(x) = f(x)g(x)$$
$$\text{Division} \quad \left(\frac{f}{g}\right)(x) = \frac{f(x)}{g(x)}$$

When we do one of these four basic operations we can simply evaluate the two functions at the value and then do the operation with both solutions

Example 515.

$$f(x) = x^2 - x - 2$$
$$g(x) = x + 1 \qquad \text{Evaluate } f \text{ and } g \text{ at } -3$$
$$\text{find } (f+g)(-3)$$

$$f(-3) = (-3)^2 - (-3) - 2 \qquad \text{Evaluate } f \text{ at } -3$$
$$f(-3) = 9 + 3 - 2$$
$$f(-3) = 10$$

$$g(-3) = (-3) + 1 \qquad \text{Evaluate } g \text{ at } -3$$
$$g(-3) = -2$$

$$f(-3) + g(-3) \qquad \text{Add the two functions together}$$
$$(10) + (-2) \qquad \text{Add}$$
$$8 \qquad \text{Our Solution}$$

The process is the same regardless of the operation being performed.

Example 516.

$$h(x) = 2x - 4$$
$$k(x) = -3x + 1 \qquad \text{Evaluate } h \text{ and } k \text{ at } 5$$
$$\text{Find } (h \cdot k)(5)$$

$$h(5) = 2(5) - 4 \qquad \text{Evaluate } h \text{ at } 5$$

393

$$h(5) = 10 - 4$$
$$h(5) = 6$$

$$k(5) = -3(5) + 1 \quad \text{Evaluate } k \text{ at } 5$$
$$k(5) = -15 + 1$$
$$k(5) = -14$$

$$h(5)k(5) \quad \text{Multiply the two results together}$$
$$(6)(-14) \quad \text{Multiply}$$
$$-84 \quad \text{Our Solution}$$

Often as we add, subtract, multiply, or divide functions, we do so in a way that keeps the variable. If there is no number to plug into the equations we will simply use each equation, in parenthesis, and simplify the expression.

Example 517.

$$f(x) = 2x - 4$$
$$g(x) = x^2 - x + 5 \quad \text{Write subtraction problem of functions}$$
$$\text{Find } (f - g)(x)$$

$$f(x) - g(x) \quad \text{Replace } f(x) \text{ with } (2x - 3) \text{ and } g(x) \text{ with } (x^2 - x + 5)$$
$$(2x - 4) - (x^2 - x + 5) \quad \text{Distribute the negative}$$
$$2x - 4 - x^2 + x - 5 \quad \text{Combine like terms}$$
$$-x^2 + 3x - 9 \quad \text{Our Solution}$$

The parenthesis are very important when we are replacing $f(x)$ and $g(x)$ with a variable. In the previous example we needed the parenthesis to know to distribute the negative.

Example 518.

$$f(x) = x^2 - 4x - 5$$
$$g(x) = x - 5$$
$$\text{Find } \left(\dfrac{f}{g}\right)(x) \quad \text{Write division problem of functions}$$

$$\frac{f(x)}{g(x)} \quad \text{Replace } f(x) \text{ with } (x^2 - 4x - 5) \text{ and } g(x) \text{ with } (x - 5)$$

$$\frac{(x^2 - 4x - 5)}{(x - 5)} \quad \text{To simplify the fraction we must first factor}$$

394

$$\frac{(x-5)(x+1)}{(x-5)} \qquad \text{Divide out common factor of } x-5$$

$$x+1 \qquad \text{Our Solution}$$

Just as we could substitute an expression into evaluating functions, we can substitute an expression into the operations on functions.

Example 519.

$$f(x) = 2x - 1$$
$$g(x) = x + 4 \qquad \text{Write as a sum of functions}$$
$$\text{Find } (f+g)(x^2)$$

$$f(x^2) + g(x^2) \qquad \text{Replace } x \text{ in } f(x) \text{ and } g(x) \text{ with } x^2$$
$$[2(x^2) - 1] + [(x^2) + 4] \qquad \text{Distribute the } + \text{ does not change the problem}$$
$$2x^2 - 1 + x^2 + 4 \qquad \text{Combine like terms}$$
$$3x^2 + 3 \qquad \text{Our Solution}$$

Example 520.

$$f(x) = 2x - 1$$
$$g(x) = x + 4 \qquad \text{Write as a product of functions}$$
$$\text{Find } (f \cdot g)(3x)$$

$$f(3x)\,g(3x) \qquad \text{Replace } x \text{ in } f(x) \text{ and } g(x) \text{ with } 3x$$
$$[2(3x) - 1][(3x) + 4] \qquad \text{Multiply our } 2(3x)$$
$$(6x - 1)(3x + 4) \qquad \text{FOIL}$$
$$18x^2 + 24x - 3x - 4 \qquad \text{Combine like terms}$$
$$18x^2 + 21x - 4 \qquad \text{Our Solution}$$

The fifth operation of functions is called composition of functions. A composition of functions is a function inside of a function. The notation used for composition of functions is:

$$(f \circ g)(x) = f(g(x))$$

To calculate a composition of function we will evaluate the inner function and substitute the answer into the outer function. This is shown in the following example.

Example 521.

$$a(x) = x^2 - 2x + 1$$
$$b(x) = x - 5 \qquad \text{Rewrite as } a \text{ function in function}$$
$$\text{Find } (a \circ b)(3)$$

$$a(b(3)) \qquad \text{Evaluate the inner function first, } b(3)$$
$$b(3) = (3) - 5 = -2 \qquad \text{This solution is put into } a, a(-2)$$
$$a(-2) = (-2)^2 - 2(-2) + 1 \qquad \text{Evaluate}$$
$$a(-2) = 4 + 4 + 1 \qquad \text{Add}$$
$$a(-2) = 9 \qquad \text{Our Solution}$$

We can also evaluate a composition of functions at a variable. In these problems we will take the inside function and substitute into the outside function.

Example 522.

$$f(x) = x^2 - x$$
$$g(x) = x + 3 \qquad \text{Rewrite as } a \text{ function in function}$$
$$\text{Find } (f \circ g)(x)$$

$$\begin{aligned}
f(g(x)) \qquad & \text{Replace } g(x) \text{ with } x + 3 \\
f(x + 3) \qquad & \text{Replace the variables in } f \text{ with } (x + 3) \\
(x + 3)^2 - (x + 3) \qquad & \text{Evaluate exponent} \\
(x^2 + 6x + 9) - (x + 3) \qquad & \text{Distribute negative} \\
x^2 + 6x + 9 - x - 3 \qquad & \text{Combine like terms} \\
x^2 + 5x + 6 \qquad & \text{Our Solution}
\end{aligned}$$

It is important to note that very rarely is $(f \circ g)(x)$ the same as $(g \circ f)(x)$ as the following example will show, using the same equations, but compositing them in the opposite direction.

Example 523.

$$f(x) = x^2 - x$$
$$g(x) = x + 3 \qquad \text{Rewrite as } a \text{ function in function}$$
$$\text{Find } (g \circ f)(x)$$

$$\begin{aligned}
g(f(x)) \qquad & \text{Replace } f(x) \text{ with } x^2 - x \\
g(x^2 - x) \qquad & \text{Replace the variable in } g \text{ with } (x^2 - x) \\
(x^2 - x) + 3 \qquad & \text{Here the parenthesis don}'t \text{ change the expression} \\
x^2 - x + 3 \qquad & \text{Our Solution}
\end{aligned}$$

World View Note: The term "function" came from Gottfried Wihelm Leibniz, a German mathematician from the late 17th century.

10.2 Practice - Operations on Functions

Perform the indicated operations.

1) $g(a) = a^3 + 5a^2$
   $f(a) = 2a + 4$
   Find $g(3) + f(3)$

2) $f(x) = -3x^2 + 3x$
   $g(x) = 2x + 5$
   Find $f(-4) \div g(-4)$

3) $g(a) = 3a + 3$
   $f(a) = 2a - 2$
   Find $(g + f)(9)$

4) $g(x) = 4x + 3$
   $h(x) = x^3 - 2x^2$
   Find $(g - h)(-1)$

5) $g(x) = x + 3$
   $f(x) = -x + 4$
   Find $(g - f)(3)$

6) $g(x) = -4x + 1$
   $h(x) = -2x - 1$
   Find $g(5) + h(5)$

7) $g(x) = x^2 + 2$
   $f(x) = 2x + 5$
   Find $(g - f)(0)$

8) $g(x) = 3x + 1$
   $f(x) = x^3 + 3x^2$
   Find $g(2) \cdot f(2)$

9) $g(t) = t - 3$
   $h(t) = -3t^3 + 6t$
   Find $g(1) + h(1)$

10) $f(n) = n - 5$
    $g(n) = 4n + 2$
    Find $(f + g)(-8)$

11) $h(t) = t + 5$
    $g(t) = 3t - 5$
    Find $(h \cdot g)(5)$

12) $g(a) = 3a - 2$
    $h(a) = 4a - 2$
    Find $(g + h)(-10)$

13) $h(n) = 2n - 1$
    $g(n) = 3n - 5$
    Find $h(0) \div g(0)$

14) $g(x) = x^2 - 2$
    $h(x) = 2x + 5$
    Find $g(-6) + h(-6)$

15) $f(a) = -2a - 4$
    $g(a) = a^2 + 3$
    Find $(\frac{f}{g})(7)$

16) $g(n) = n^2 - 3$
    $h(n) = 2n - 3$
    Find $(g - h)(n)$

17) $g(x) = -x^3 - 2$
    $h(x) = 4x$
    Find $(g - h)(x)$

18) $g(x) = 2x - 3$
    $h(x) = x^3 - 2x^2 + 2x$
    Find $(g - h)(x)$

19) $f(x) = -3x + 2$
    $g(x) = x^2 + 5x$
    Find $(f - g)(x)$

20) $g(t) = t - 4$
    $h(t) = 2t$
    Find $(g \cdot h)(t)$

21) $g(x) = 4x + 5$
    $h(x) = x^2 + 5x$
    Find $g(x) \cdot h(x)$

22) $g(t) = -2t^2 - 5t$
    $h(t) = t + 5$
    Find $g(t) \cdot h(t)$

23) $f(x) = x^2 - 5x$
$g(x) = x + 5$
Find $(f + g)(x)$

24) $f(x) = 4x - 4$
$g(x) = 3x^2 - 5$
Find $(f + g)(x)$

25) $g(n) = n^2 + 5$
$f(n) = 3n + 5$
Find $g(n) \div f(n)$

26) $f(x) = 2x + 4$
$g(x) = 4x - 5$
Find $f(x) - g(x)$

27) $g(a) = -2a + 5$
$f(a) = 3a + 5$
Find $\left(\frac{g}{f}\right)(a)$

28) $g(t) = t^3 + 3t^2$
$h(t) = 3t - 5$
Find $g(t) - h(t)$

29) $h(n) = n^3 + 4n$
$g(n) = 4n + 5$
Find $h(n) + g(n)$

30) $f(x) = 4x + 2$
$g(x) = x^2 + 2x$
Find $f(x) \div g(x)$

31) $g(n) = n^2 - 4n$
$h(n) = n - 5$
Find $g(n^2) \cdot h(n^2)$

32) $g(n) = n + 5$
$h(n) = 2n - 5$
Find $(g \cdot h)(-3n)$

33) $f(x) = 2x$
$g(x) = -3x - 1$
Find $(f + g)(-4 - x)$

34) $g(a) = -2a$
$h(a) = 3a$
Find $g(4n) \div h(4n)$

35) $f(t) = t^2 + 4t$
$g(t) = 4t + 2$
Find $f(t^2) + g(t^2)$

36) $h(n) = 3n - 2$
$g(n) = -3n^2 - 4n$
Find $h\left(\frac{n}{3}\right) \div g\left(\frac{n}{3}\right)$

37) $g(a) = a^3 + 2a$
$h(a) = 3a + 4$
Find $\left(\frac{g}{h}\right)(-x)$

38) $g(x) = -4x + 2$
$h(x) = x^2 - 5$
Find $g(x^2) + h(x^2)$

39) $f(n) = -3n^2 + 1$
$g(n) = 2n + 1$
Find $(f - g)\left(\frac{n}{3}\right)$

40) $f(n) = 3n + 4$
$g(n) = n^3 - 5n$
Find $f\left(\frac{n}{2}\right) - g\left(\frac{n}{2}\right)$

41) $f(x) = -4x + 1$
$g(x) = 4x + 3$
Find $(f \circ g)(9)$

42) $g(x) = x - 1$
Find $(g \circ g)(7)$

43) $h(a) = 3a + 3$
$g(a) = a + 1$
Find $(h \circ g)(5)$

44) $g(t) = t + 3$
$h(t) = 2t - 5$
Find $(g \circ h)(3)$

45) $g(x) = x + 4$
$h(x) = x^2 - 1$
Find $(g \circ h)(10)$

46) $f(a) = 2a - 4$
$g(a) = a^2 + 2a$
Find $(f \circ g)(-4)$

47) $f(n) = -4n + 2$
$g(n) = n + 4$
Find $(f \circ g)(9)$

48) $g(x) = 3x + 4$
$h(x) = x^3 + 3x$
Find $(g \circ h)(3)$

49) $g(x) = 2x - 4$
$h(x) = 2x^3 + 4x^2$
Find $(g \circ h)(3)$

50) $g(a) = a^2 + 3$
Find $(g \circ g)(-3)$

51) $g(x) = x^2 - 5x$
$h(x) = 4x + 4$
Find $(g \circ h)(x)$

52) $g(a) = 2a + 4$
$h(a) = -4a + 5$
Find $(g \circ h)(a)$

53) $f(a) = -2a + 2$
$g(a) = 4a$
Find $(f \circ g)(a)$

54) $g(t) = -t - 4$
Find $(g \circ g)(t)$

56) $f(n) = -2n^2 - 4n$
$g(n) = n + 2$
Find $(f \circ g)(n)$

55) $g(x) = 4x + 4$
$f(x) = x^3 - 1$
Find $(g \circ f)(x)$

57) $g(x) = -x + 5$
$f(x) = 2x - 3$
Find $(g \circ f)(x)$

58) $g(t) = t^3 - t$
$f(t) = 3t - 4$
Find $(g \circ f)(t)$

59) $f(t) = 4t + 3$
$g(t) = -4t - 2$
Find $(f \circ g)(t)$

60) $f(x) = 3x - 4$
$g(x) = x^3 + 2x^2$
Find $(f \circ g)(x)$

Functions - Inverse Functions

Objective: Identify and find inverse functions.

When a value goes into a function it is called the input. The result that we get when we evaluate the function is called the output. When working with functions sometimes we will know the output and be interested in what input gave us the output. To find this we use an inverse function. As the name suggests an inverse function undoes whatever the function did. If a function is named $f(x)$, the inverse function will be named $f^{-1}(x)$ (read "f inverse of x"). The negative one is not an exponent, but mearly a symbol to let us know that this function is the inverse of f.

World View Note: The notation used for functions was first introduced by the great Swiss mathematician, Leonhard Euler in the 18th century.

For example, if $f(x) = x + 5$, we could deduce that the inverse function would be $f^{-1}(x) = x - 5$. If we had an input of 3, we could calculate $f(3) = (3) + 5 = 8$. Our output is 8. If we plug this output into the inverse function we get $f^{-1}(8) = (8) - 5 = 3$, which is the original input.

Often the functions are much more involved than those described above. It may be difficult to determine just by looking at the functions if they are inverses. In order to test if two functions, $f(x)$ and $g(x)$ are inverses we will calculate the composition of the two functions at x. If f changes the variable x in some way, then g undoes whatever f did, then we will be back at x again for our final solution. In otherwords, if we simplify $(f \circ g)(x)$ the solution will be x. If it is anything but x the functions are not inverses.

Example 524.

Are $f(x) = \sqrt[3]{3x+4}$ and $g(x) = \dfrac{x^3 - 4}{3}$ inverses?   Caculate composition

$$f(g(x))$$ Replace $g(x)$ with $\dfrac{x^3 - 4}{3}$

$$f\left(\dfrac{x^3 - 4}{3}\right)$$ Substitute $\left(\dfrac{x^3 - 4}{3}\right)$ for variable in f

$$\sqrt[3]{3\left(\dfrac{x^3 - 4}{3}\right) + 4}$$ Divide out the $3's$

$$\sqrt[3]{x^3 - 4 + 4}$$ Combine like terms

$$\sqrt[3]{x^3}$$ Take cubed root

$$x$$ Simplified to x!

Yes, they are inverses!   Our Solution

Example 525.

Are $h(x) = 2x + 5$ and $g(x) = \dfrac{x}{2} - 5$ inverses?   Calculate composition

$$h(g(x))$$ Replace $g(x)$ with $\left(\dfrac{x}{2} - 5\right)$

$$h\left(\dfrac{x}{2} - 5\right)$$ Substitute $\left(\dfrac{x}{2} - 5\right)$ for variable in h

$$2\left(\dfrac{x}{2} - 5\right) + 5$$ Distrubte 2

$$x - 10 + 5$$ Combine like terms

$$x - 5$$ Did not simplify to x

No, they are not inverses   Our Solution

Example 526.

Are $f(x) = \dfrac{3x - 2}{4x + 1}$ and $g(x) = \dfrac{x + 2}{3 - 4x}$ inverses?   Calculate composition

$$f(g(x))$$ Replace $g(x)$ with $\left(\dfrac{x + 2}{3 - 4x}\right)$

$$f\left(\dfrac{x + 2}{3 - 4x}\right)$$ Substitute $\left(\dfrac{x + 2}{3 - 4x}\right)$ for variable in f

$$\dfrac{3\left(\frac{x+2}{3-4x}\right) - 2}{4\left(\frac{x+2}{3-4x}\right) + 1}$$ Distribute 3 and 4 into numerators

$$\frac{\frac{3x+6}{3-4x}-2}{\frac{4x+8}{3-4x}+1} \qquad \text{Multiply each term by LCD: } 3-4x$$

$$\frac{\frac{(3x+6)(3-4x)}{3-4x}-2(3-4x)}{\frac{(4x+8)(3-4x)}{3-4x}+1(3-4x)} \qquad \text{Reduce fractions}$$

$$\frac{3x+6-2(3-4x)}{4x+8+1(3-4x)} \qquad \text{Distribute}$$

$$\frac{3x+6-6+8x}{4x+8+3-4x} \qquad \text{Combine like terms}$$

$$\frac{11x}{11} \qquad \text{Divide out 11}$$

$$x \qquad \text{Simplified to } x!$$

$$\text{Yes, they are inverses} \qquad \text{Our Solution}$$

While the composition is useful to show two functions are inverses, a more common problem is to find the inverse of a function. If we think of x as our input and y as our output from a function, then the inverse will take y as an input and give x as the output. This means if we switch x and y in our function we will find the inverse! This process is called the switch and solve strategy.

Switch and solve strategy to find an inverse:

1. Replace $f(x)$ with y

2. Switch x and y's

3. Solve for y

4. Replace y with $f^{-1}(x)$

Example 527.

$$\begin{aligned}
\text{Find the inverse of } f(x) &= (x+4)^3-2 & &\text{Replace } f(x) \text{ with } y \\
y &= (x+4)^3-2 & &\text{Switch } x \text{ and } y \\
x &= (y+4)^3-2 & &\text{Solve for } y \\
\underline{+2 \qquad\qquad +2} & & &\text{Add 2 to both sides}
\end{aligned}$$

$$x + 2 = (y + 4)^3 \quad \text{Cube root both sides}$$
$$\sqrt[3]{x + 2} = y + 4 \quad \text{Subtract 4 from both sides}$$
$$\underline{\phantom{\sqrt[3]{x+2}} -4 \qquad -4}$$
$$\sqrt[3]{x + 2} - 4 = y \quad \text{Replace } y \text{ with } f^{-1}(x)$$
$$f^{-1}(x) = \sqrt[3]{x + 2} - 4 \quad \text{Our Solution}$$

Example 528.

$$\text{Find the inverse of } g(x) = \frac{2x - 3}{4x + 2} \quad \text{Replace } g(x) \text{ with } y$$

$$y = \frac{2x - 3}{4x + 2} \quad \text{Switch } x \text{ and } y$$

$$x = \frac{2y - 3}{4y + 2} \quad \text{Multiply by } (4y + 2)$$

$$x(4y + 2) = 2y - 3 \quad \text{Distribute}$$
$$4xy + 2x = 2y - 3 \quad \text{Move all } y's \text{ to one side, rest to other side}$$
$$\underline{-4xy + 3 \qquad -4xy + 3} \quad \text{Subtract } 4xy \text{ and add 3 to both sides}$$
$$2x + 3 = 2y - 4xy \quad \text{Factor out } y$$

$$\frac{2x + 3}{2 - 4x} = \frac{y(2 - 4x)}{2 - 4x} \quad \text{Divide by } 2 - 4x$$

$$\frac{2x + 3}{2 - 4x} = y \quad \text{Replace } y \text{ with } g^{-1}(x)$$

$$g^{-1}(x) = \frac{2x + 3}{2 - 4x} \quad \text{Our Solution}$$

In this lesson we looked at two different things, first showing functions are inverses by calculating the composition, and second finding an inverse when we only have one function. Be careful not to get them backwards. When we already have two functions and are asked to show they are inverses, we do not want to use the switch and solve strategy, what we want to do is calculate the inverse. There may be several ways to represent the same function so the switch and solve strategy may not look the way we expect and can lead us to conclude two functions are not inverses when they are in fact inverses.

10.3 Practice - Inverse Functions

State if the given functions are inverses.

1) $g(x) = -x^5 - 3$
   $f(x) = \sqrt[5]{-x - 3}$

2) $g(x) = \frac{4-x}{x}$
   $f(x) = \frac{4}{x}$

3) $f(x) = \frac{-x-1}{x-2}$
   $g(x) = \frac{2x+1}{-x-1}$

4) $h(x) = \frac{2}{x} \quad \frac{2x}{x}$
   $f(x) = \frac{2}{x+2}$

5) $g(x) = -10x + 5$
   $f(x) = \frac{x}{10} \quad \frac{5}{10}$

6) $f(x) = \frac{x-5}{10}$
   $h(x) = 10x + 5$

7) $f(x) = -\frac{2}{x+3}$
   $g(x) = \frac{3x+2}{x+2}$

8) $f(x) = \sqrt[5]{\frac{x+1}{2}}$
   $g(x) = 2x^5 - 1$

9) $g(x) = \sqrt[5]{\frac{x-1}{2}}$
   $f(x) = 2x^5 + 1$

10) $g(x) = \frac{8+9x}{2}$
    $f(x) = \frac{5x}{2} \quad \frac{9}{2}$

Find the inverse of each functions.

11) $f(x) = (x-2)^5 + 3$

12) $g(x) = \sqrt[3]{x+1} + 2$

13) $g(x) = \frac{4}{x+2}$

14) $f(x) = \frac{-3}{x} \quad \frac{}{3}$

15) $f(x) = \frac{2x}{x+2} \quad 2$

16) $g(x) = \frac{9+x}{3}$

17) $f(x) = \frac{10 \quad x}{5}$

18) $f(x) = \frac{5x \quad 15}{2}$

19) $g(x) = -(x-1)^3$

20) $f(x) = \frac{12-3x}{4}$

21) $f(x) = (x-3)^3$

22) $g(x) = \sqrt[5]{\frac{-x+2}{2}}$

23) $g(x) = \frac{x}{x-1}$

24) $f(x) = \frac{-3-2x}{x+3}$

25) $f(x) = \frac{x \quad 1}{x+1}$

26) $h(x) = \frac{x}{x+2}$

27) $g(x) = \frac{8 \quad 5x}{4}$

28) $g(x) = \frac{-x+2}{3}$

29) $g(x) = -5x + 1$

30) $f(x) = \frac{5x \quad 5}{4}$

31) $g(x) = -1 + x^3$

32) $f(x) = 3 - 2x^5$

33) $h(x) = \frac{4 \quad \sqrt[3]{4x}}{2}$

34) $g(x) = (x-1)^3 + 2$

35) $f(x) = \frac{x+1}{x+2}$

36) $f(x) = \frac{1}{x+1}$

37) $f(x) = \frac{7-3x}{x \quad 2}$

38) $f(x) = -\frac{3x}{4}$

39) $g(x) = -x$

40) $g(x) = \frac{-2x+1}{3}$

Functions - Exponential Functions

Objective: Solve exponential equations by finding a common base.

As our study of algebra gets more advanced we begin to study more involved functions. One pair of inverse functions we will look at are exponential functions and logarithmic functions. Here we will look at exponential functions and then we will consider logarithmic functions in another lesson. Exponential functions are functions where the variable is in the exponent such as $f(x) = a^x$. (It is important not to confuse exponential functions with polynomial functions where the variable is in the base such as $f(x) = x^2$).

World View Note One common application of exponential functions is population growth. According to the 2009 CIA World Factbook, the country with the highest population growth rate is a tie between the United Arab Emirates (north of Saudi Arabia) and Burundi (central Africa) at 3.69%. There are 32 countries with negative growth rates, the lowest being the Northern Mariana Islands (north of Australia) at -7.08%.

Solving exponetial equations cannot be done using the skill set we have seen in the past. For example, if $3^x = 9$, we cannot take the $x -$ root of 9 because we do not know what the index is and this doesn't get us any closer to finding x. However, we may notice that 9 is 3^2. We can then conclude that if $3^x = 3^2$ then $x = 2$. This is the process we will use to solve exponential functions. If we can re-write a problem so the bases match, then the exponents must also match.

Example 529.

$$
\begin{array}{ll}
5^{2x+1} = 125 & \text{Rewrite } 125 \text{ as } 5^3 \\
5^{2x+1} = 5^3 & \text{Same base, set exponents equal} \\
2x + 1 = 3 & \text{Solve} \\
\underline{-1 \quad -1} & \text{Subtract 1 from both sides} \\
\dfrac{2x}{2} = \dfrac{2}{2} & \text{Divide both sides by 2} \\
x = 1 & \text{Our Solution}
\end{array}
$$

Sometimes we may have to do work on both sides of the equation to get a common base. As we do so, we will use various exponent properties to help. First we will use the exponent property that states $(a^x)^y = a^{xy}$.

Example 530.

$$8^{3x} = 32 \qquad \text{Rewrite 8 as } 2^3 \text{ and 32 as } 2^5$$
$$(2^3)^{3x} = 2^5 \qquad \text{Multiply exponents 3 and } 3x$$
$$2^{9x} = 2^5 \qquad \text{Same base, set exponents equal}$$
$$9x = 5 \qquad \text{Solve}$$
$$\overline{9 \quad 9} \qquad \text{Divide both sides by 9}$$
$$x = \frac{5}{9} \qquad \text{Our Solution}$$

As we multiply exponents we may need to distribute if there are several terms involved.

Example 531.

$$27^{3x+5} = 81^{4x+1} \qquad \text{Rewrite 27 as } 3^3 \text{ and 81 as } 3^4 \, (9^2 \text{ would not be same base})$$
$$(3^3)^{3x+5} = (3^4)^{4x+1} \qquad \text{Multiply exponents } 3(3x+5) \text{ and } 4(4x+1)$$
$$3^{9x+15} = 3^{16x+4} \qquad \text{Same base, set exponents equal}$$
$$9x + 15 = 16x + 4 \qquad \text{Move variables to one side}$$
$$\underline{-9x \qquad\quad -9x} \qquad \text{Subtract } 9x \text{ from both sides}$$
$$15 = 7x + 4 \qquad \text{Subtract 4 from both sides}$$
$$\underline{-4 \qquad -4}$$
$$11 = 7x \qquad \text{Divide both sides by 7}$$
$$\overline{7 \quad 7}$$
$$\frac{11}{7} = x \qquad \text{Our Solution}$$

Another useful exponent property is that negative exponents will give us a reciprocal, $\frac{1}{a^n} = a^{-n}$

Example 532.

$$\left(\frac{1}{9}\right)^{2x} = 3^{7x-1} \qquad \text{Rewrite } \frac{1}{9} \text{ as } 3^{-2} \, (\text{negative exponet to flip})$$
$$(3^{-2})^{2x} = 3^{7x-1} \qquad \text{Multiply exponents } -2 \text{ and } 2x$$
$$3^{-4x} = 3^{7x-1} \qquad \text{Same base, set exponets equal}$$
$$-4x = 7x - 1 \qquad \text{Subtract } 7x \text{ from both sides}$$
$$\underline{-7x - 7x}$$
$$-11x = -1 \qquad \text{Divide by } -11$$
$$\overline{-11 \quad -11}$$
$$x = \frac{1}{11} \qquad \text{Our Solution}$$

If we have several factors with the same base on one side of the equation we can add the exponents using the property that states $a^x a^y = a^{x+y}$.

407

Example 533.

$$5^{4x} \cdot 5^{2x\,1} = 5^{3x+11} \qquad \text{Add exponents on left, combing like terms}$$
$$5^{6x-1} = 5^{3x+11} \qquad \text{Same base, set exponents equal}$$
$$6x - 1 = 3x + 11 \qquad \text{Move variables to one sides}$$
$$\underline{-3x \qquad -3x} \qquad \text{Subtract } 3x \text{ from both sides}$$
$$3x - 1 = 11 \qquad \text{Add 1 to both sides}$$
$$\underline{+1 \ +1}$$
$$3x = 12 \qquad \text{Divide both sides by 3}$$
$$\overline{3 \quad 3}$$
$$x = 4 \qquad \text{Our Solution}$$

It may take a bit of practice to get use to knowing which base to use, but as we practice we will get much quicker at knowing which base to use. As we do so, we will use our exponent properties to help us simplify. Again, below are the properties we used to simplify.

$$(a^x)^y = a^{xy} \quad \text{and} \quad \frac{1}{a^n} = a^{-n} \quad \text{and} \quad a^x a^y = a^{x+y}$$

We could see all three properties used in the same problem as we get a common base. This is shown in the next example.

Example 534.

$$16^{2x\,5} \cdot \left(\frac{1}{4}\right)^{3x+1} = 32 \cdot \left(\frac{1}{2}\right)^{x+3} \qquad \text{Write with } a \text{ common base of 2}$$
$$(2^4)^{2x\,5} \cdot (2^{\,2})^{3x+1} = 2^5 \cdot (2^{\,1})^{x+3} \qquad \text{Multiply exponents, distributing as needed}$$
$$2^{8x\,20} \cdot 2^{\,6x\,2} = 2^5 \cdot 2^{\,x\,3} \qquad \text{Add exponents, combining like terms}$$
$$2^{2x\,22} = 2^{\,x+2} \qquad \text{Same base, set exponents equal}$$
$$2x - 22 = -x + 2 \qquad \text{Move variables to one side}$$
$$\underline{+x \qquad\qquad +x} \qquad \text{Add } x \text{ to both sides}$$
$$3x - 22 = 2 \qquad \text{Add 22 to both sides}$$
$$\underline{+22 + 22}$$
$$3x = 24 \qquad \text{Divide both sides by 3}$$
$$\overline{3 \quad 3}$$
$$x = 8 \qquad \text{Our Solution}$$

All the problems we have solved here we were able to write with a common base. However, not all problems can be written with a common base, for example, $2 = 10^x$, we cannot write this problem with a common base. To solve problems like this we will need to use the inverse of an exponential function. The inverse is called a logarithmic function, which we will discuss in another secion.

10.4 Practice - Exponential Functions

Solve each equation.

1) $3^{1-2n} = 3^{1-3n}$

2) $4^{2x} = \frac{1}{16}$

3) $4^{2a} = 1$

4) $16^{-3p} = 64^{-3p}$

5) $\left(\frac{1}{25}\right)^{-k} = 125^{-2k-2}$

6) $625^{n-2} = \frac{1}{125}$

7) $6^{2m+1} = \frac{1}{36}$

8) $6^{2r-3} = 6^{r-3}$

9) $6^{3x} = 36$

10) $5^{2n} = 5^{-n}$

11) $64^b = 2^5$

12) $216^{-3v} = 36^{3v}$

13) $\left(\frac{1}{4}\right)^x = 16$

14) $27^{2n-1} = 9$

15) $4^{3a} = 4^3$

16) $4^{-3v} = 64$

17) $36^{3x} = 216^{2x+1}$

18) $64^{x+2} = 16$

19) $9^{2n+3} = 243$

20) $16^{2k} = \frac{1}{64}$

21) $3^{3x-2} = 3^{3x+1}$

22) $243^p = 27^{-3p}$

23) $3^{-2x} = 3^3$

24) $4^{2n} = 4^{2-3n}$

25) $5^{m+2} = 5^{-m}$

26) $625^{2x} = 25$

27) $\left(\frac{1}{36}\right)^{b-1} = 216$

28) $216^{2n} = 36$

29) $6^{2-2x} = 6^2$

30) $\left(\frac{1}{4}\right)^{3v-2} = 64^{1-v}$

31) $4 \cdot 2^{-3n-1} = \frac{1}{4}$

32) $\frac{216}{6^{2a}} = 6^{3a}$

33) $4^{3k-3} \cdot 4^{2-2k} = 16^{-k}$

34) $32^{2p-2} \cdot 8^p = \left(\frac{1}{2}\right)^{2p}$

35) $9^{-2x} \cdot \left(\frac{1}{243}\right)^{3x} = 243^{-x}$

36) $3^{2m} \cdot 3^{3m} = 1$

37) $64^{n-2} \cdot 16^{n+2} = \left(\frac{1}{4}\right)^{3n-1}$

38) $3^{2-x} \cdot 3^{3m} = 1$

39) $5^{-3n-3} \cdot 5^{2n} = 1$

40) $4^{3r} \cdot 4^{-3r} = \frac{1}{64}$

Functions - Logarithmic Functions

Objective: Convert between logarithms and exponents and use that relationship to solve basic logarithmic equations.

The inverse of an exponential function is a new function known as a logarithm. Lograithms are studied in detail in advanced algebra, here we will take an introductory look at how logarithms works. When working with radicals we found that there were two ways to write radicals. The expression $\sqrt[m]{a^n}$ could be written as $a^{\frac{n}{m}}$. Each form has its advantages, thus we need to be comfortable using both the radical form and the rational exponent form. Similarly an exponent can be written in two forms, each with its own advantages. The first form we are very familiar with, $b^x = a$, where b is the base, a can be thought of as our answer, and x is the exponent. The second way to write this is with a logarithm, $\log_b a = x$. The word "log" tells us that we are in this new form. The variables all still mean the same thing. b is still the base, a can still be thought of as our answer.

Using this idea the problem $5^2 = 25$ could also be written as $\log_5 25 = 2$. Both mean the same thing, both are still the same exponent problem, but just as roots can be written in radical form or rational exponent form, both our forms have their own advantages. The most important thing to be comfortable doing with logarithms and exponents is to be able to switch back and forth between the two forms. This is what is shown in the next few examples.

Example 535.

Write each exponential equation in logarithmic form

$$m^3 = 5 \quad \text{Identify base}, m, \text{answer}, 5, \text{and exponent } 3$$
$$\log_m 5 = 3 \quad \text{Our Solution}$$

$$7^2 = b \quad \text{Identify base}, 7, \text{answer}, b, \text{and exponent}, 2$$
$$\log_7 b = 2 \quad \text{Our Solution}$$

$$\left(\frac{2}{3}\right)^4 = \frac{16}{81} \quad \text{Identify base}, \frac{2}{3}, \text{answer}, \frac{16}{81}, \text{and exponent } 4$$
$$\log_{\frac{2}{3}} \frac{16}{81} = 4 \quad \text{Our Solution}$$

Example 536.

Write each logarithmic equation in exponential form

$$\log_4 16 = 2 \quad \text{Identify base}, 4, \text{answer}, 16, \text{and exponent}, 2$$
$$4^2 = 16 \quad \text{Our Solution}$$

$$\log_3 x = 7 \quad \text{Identify base, 3, answer, } x, \text{ and exponent, 7}$$
$$3^7 = x \quad \text{Our Solution}$$

$$\log_9 3 = \frac{1}{2} \quad \text{Identify base, 9, answer, 3, and exponent, } \frac{1}{2}$$
$$9^{\frac{1}{2}} = 3 \quad \text{Our Solution}$$

Once we are comfortable switching between logarithmic and exponential form we are able to evaluate and solve logarithmic expressions and equations. We will first evaluate logarithmic expressions. An easy way to evaluate a logarithm is to set the logarithm equal to x and change it into an exponential equation.

Example 537.

$$\text{Evaluate } \log_2 64 \quad \text{Set logarithm equal to } x$$
$$\log_2 64 = x \quad \text{Change to exponent form}$$
$$2^x = 64 \quad \text{Write as common base, } 64 = 2^6$$
$$2^x = 2^6 \quad \text{Same base, set exponents equal}$$
$$x = 6 \quad \text{Our Solution}$$

Example 538.

$$\text{Evaluate } \log_{125} 5 \quad \text{Set logarithm equal to } x$$
$$\log_{125} 5 = x \quad \text{Change to exponent form}$$
$$125^x = 5 \quad \text{Write as common base, } 125 = 5^3$$
$$(5^3)^x = 5 \quad \text{Multiply exponents}$$
$$5^{3x} = 5 \quad \text{Same base, set exponents equal } (5 = 5^1)$$
$$3x = 1 \quad \text{Solve}$$
$$\overline{3 \quad 3} \quad \text{Divide both sides by 3}$$
$$x = \frac{1}{3} \quad \text{Our Solution}$$

Example 539.

$$\text{Evaluate } \log_3 \frac{1}{27} \quad \text{Set logarithm equal to } x$$
$$\log_3 \frac{1}{27} = x \quad \text{Change to exponent form}$$
$$3^x = \frac{1}{27} \quad \text{Write as common base, } \frac{1}{27} = 3^{-3}$$
$$3^x = 3^{-3} \quad \text{Same base, set exponents equal}$$
$$x = -3 \quad \text{Our Solution}$$

World View Note: Dutch mathematician Adriaan Vlacq published a text in 1628 which listed logarithms calculated out from 1 to 100,000!

Solve equations with logarithms is done in a very similar way, we simply will change the equation into exponential form and try to solve the resulting equation.

Example 540.

$$\log_5 x = 2 \qquad \text{Change to exponential form}$$
$$5^2 = x \qquad \text{Evaluate exponent}$$
$$25 = x \qquad \text{Our Solution}$$

Example 541.

$$\log_2(3x + 5) = 4 \qquad \text{Change to exponential form}$$
$$2^4 = 3x + 5 \qquad \text{Evaluate exponent}$$
$$16 = 3x + 5 \qquad \text{Solve}$$
$$\underline{-5 \qquad\quad -5} \qquad \text{Subtract 5 from both sides}$$
$$\frac{11 = 3x}{3 \quad 3} \qquad \text{Divide both sides by 3}$$
$$\frac{11}{3} = x \qquad \text{Our Solution}$$

Example 542.

$$\log_x 8 = 3 \qquad \text{Change to exponential form}$$
$$x^3 = 8 \qquad \text{Cube root of both sides}$$
$$x = 2 \qquad \text{Our Solution}$$

There is one base on a logarithm that gets used more often than any other base, base 10. Similar to square roots not writting the common index of 2 in the radical, we don't write the common base of 10 in the logarithm. So if we are working on a problem with no base written we will always assume that base is base 10.

Example 543.

$$\log x = -2 \qquad \text{Rewrite as exponent, 10 is base}$$
$$10^{-2} = x \qquad \text{Evaluate, remember negative exponent is fraction}$$
$$\frac{1}{100} = x \qquad \text{Our Solution}$$

This lesson has introduced the idea of logarithms, changing between logs and exponents, evaluating logarithms, and solving basic logarithmic equations. In an advanced algebra course logarithms will be studied in much greater detail.

10.5 Practice - Logarithmic Functions

Rewrite each equation in exponential form.

1) $\log_9 81 = 2$

2) $\log_b a = -16$

3) $\log_7 \frac{1}{49} = -2$

4) $\log_{16} 256 = 2$

5) $\log_{13} 169 = 2$

6) $\log_{11} 1 = 0$

Rewrite each equations in logarithmic form.

7) $8^0 = 1$

8) $17^{-2} = \frac{1}{289}$

9) $15^2 = 225$

10) $144^{\frac{1}{2}} = 12$

11) $64^{\frac{1}{6}} = 2$

12) $19^2 = 361$

Evaluate each expression.

13) $\log_{125} 5$

14) $\log_5 125$

15) $\log_{343} \frac{1}{7}$

16) $\log_7 1$

17) $\log_4 16$

18) $\log_4 \frac{1}{64}$

19) $\log_6 36$

20) $\log_{36} 6$

21) $\log_2 64$

22) $\log_3 243$

Solve each equation.

23) $\log_5 x = 1$

24) $\log_8 k = 3$

25) $\log_2 x = -2$

26) $\log n = 3$

27) $\log_{11} k = 2$

28) $\log_4 p = 4$

29) $\log_9 (n+9) = 4$

30) $\log_{11} (x-4) = -1$

31) $\log_5 (-3m) = 3$

32) $\log_2 -8r = 1$

33) $\log_{11} (x+5) = -1$

34) $\log_7 -3n = 4$

35) $\log_4 (6b+4) = 0$

36) $\log_{11} (10v+1) = -1$

37) $\log_5 (-10x+4) = 4$

38) $\log_9 (7-6x) = -2$

39) $\log_2 (10-5a) = 3$

40) $\log_8 (3k-1) = 1$

Functions - Compound Interest

Objective: Calculate final account balances using the formulas for compound and continuous interest.

An application of exponential functions is compound interest. When money is invested in an account (or given out on loan) a certain amount is added to the balance. This money added to the balance is called interest. Once that interest is added to the balance, it will earn more interest during the next compounding period. This idea of earning interest on interest is called compound interest. For example, if you invest $100 at 10% interest compounded annually, after one year you will earn $10 in interest, giving you a new balance of $110. The next year you will earn another 10% or $11, giving you a new balance of $121. The third year you will earn another 10% or $12.10, giving you a new balance of $133.10. This pattern will continue each year until you close the account.

There are several ways interest can be paid. The first way, as described above, is compounded annually. In this model the interest is paid once per year. But interest can be compounded more often. Some common compounds include compounded semi-annually (twice per year), quarterly (four times per year, such as quarterly taxes), monthly (12 times per year, such as a savings account), weekly (52 times per year), or even daily (365 times per year, such as some student loans). When interest is compounded in any of these ways we can calculate the balance after any amount of time using the following formula:

$$\textbf{Compound Interest Formula: } A = P\left(1 + \frac{r}{n}\right)^{nt}$$

$$A = \text{Final Amount}$$
$$P = \text{Principle (starting balance)}$$
$$r = \text{Interest rate (as a decimal)}$$
$$n = \text{number of compounds per year}$$
$$t = \text{time (in years)}$$

Example 544.

If you take a car loan for $25000 with an interest rate of 6.5% compounded quarterly, no payments required for the first five years, what will your balance be at the end of those five years?

$P = 25000, r = 0.065, n = 4, t = 5$ Identify each variable

$A = 25000\left(1 + \dfrac{0.065}{4}\right)^{4 \cdot 5}$ Plug each value into formula, evaluate parenthesis

$A = 25000(1.01625)^{4 \cdot 5}$ Multiply exponents

$$A = 25000(1.01625)^{20} \quad \text{Evaluate exponent}$$
$$A = 25000(1.38041977...) \quad \text{Multiply}$$
$$A = 34510.49$$
$$\$34,510.49 \quad \text{Our Solution}$$

We can also find a missing part of the equation by using our techniques for solving equations.

Example 545.

What principle will amount to \$3000 if invested at 6.5% compounded weekly for 4 years?

$$A = 3000, r = 0.065, n = 52, t = 4 \quad \text{Identify each variable}$$
$$3000 = P\left(1 + \frac{0.065}{52}\right)^{52 \cdot 4} \quad \text{Evaluate parentheses}$$
$$3000 = P(1.00125)^{52 \cdot 4} \quad \text{Multiply exponent}$$
$$3000 = P(1.00125)^{208} \quad \text{Evaluate exponent}$$
$$\frac{3000}{1.296719528...} = \frac{P(1.296719528...)}{1.296719528...} \quad \text{Divide each side by } 1.296719528...$$
$$2313.53 = P \quad \text{Solution for } P$$
$$\$2313.53 \quad \text{Our Solution}$$

It is interesting to compare equal investments that are made at several different types of compounds. The next few examples do just that.

Example 546.

If \$4000 is invested in an account paying 3% interest compounded monthly, what is the balance after 7 years?

$$P = 4000, r = 0.03, n = 12, t = 7 \quad \text{Identify each variable}$$
$$A = 4000\left(1 + \frac{0.03}{12}\right)^{12 \cdot 7} \quad \text{Plug each value into formula, evaluate parentheses}$$
$$A = 4000(1.0025)^{12 \cdot 7} \quad \text{Multiply exponents}$$
$$A = 4000(1.0025)^{84} \quad \text{Evaluate exponent}$$
$$A = 4000(1.2333548) \quad \text{Multiply}$$
$$A = 4933.42$$
$$\$4933.42 \quad \text{Our Solution}$$

To investigate what happens to the balance if the compounds happen more often, we will consider the same problem, this time with interest compounded daily.

Example 547.

If \$4000 is invested in an account paying 3% interest compounded daily, what is the balance after 7 years?

$P = 4000, r = 0.03, n = 365, t = 7$ Identify each variable

$$A = 4000\left(1 + \frac{0.03}{365}\right)^{365 \cdot 7}$$ Plug each value into formula, evaluate parenthesis

$A = 4000(1.00008219...)^{365 \cdot 7}$ Multiply exponent

$A = 4000(1.00008219...)^{2555}$ Evaluate exponent

$A = 4000(1.23366741....)$ Multiply

$A = 4934.67$

\$4934.67   Our Solution

While this difference is not very large, it is a bit higher. The table below shows the result for the same problem with different compounds.

Compound	Balance
Annually	\$4919.50
Semi-Annually	\$4927.02
Quarterly	\$4930.85
Monthly	\$4933.42
Weekly	\$4934.41
Daily	\$4934.67

As the table illustrates, the more often interest is compounded, the higher the final balance will be. The reason is, because we are calculating compound interest or interest on interest. So once interest is paid into the account it will start earning interest for the next compound and thus giving a higher final balance. The next question one might consider is what is the maximum number of compounds possible? We actually have a way to calculate interest compounded an infinite number of times a year. This is when the interest is compounded continuously. When we see the word "continuously" we will know that we cannot use the first formula. Instead we will use the following formula:

Interest Compounded Continuously: $A = Pe^{rt}$

$A = $ Final Amount

$P = $ Principle (starting balance)

$e = $ a constant approximately $2.71828183....$

$r = $ Interest rate (written as a decimal)

$t = $ time (years)

The variable e is a constant similar in idea to pi (π) in that it goes on forever without repeat or pattern, but just as pi (π) naturally occurs in several geometry applications, so does e appear in many exponential applications, continuous interest being one of them. If you have a scientific calculator you probably have an e button (often using the 2nd or shift key, then hit ln) that will be useful in calculating interest compounded continuously.

World View Note: e first appeared in 1618 in Scottish mathematician's Napier's work on logarithms. However it was Euler in Switzerland who used the letter e first to represent this value. Some say he used e because his name begins with E. Others, say it is because exponent starts with e. Others say it is because Euler's work already had the letter a in use, so e would be the next value. Whatever the reason, ever since he used it in 1731, e became the natural base.

Example 548.

If \$4000 is invested in an account paying 3% interest compounded continuously, what is the balance after 7 years?

$$P = 4000, r = 0.03, t = 7 \quad \text{Identify each of the variables}$$
$$A = 4000e^{0.03 \cdot 7} \quad \text{Multiply exponent}$$
$$A = 4000e^{0.21} \quad \text{Evaluate } e^{0.21}$$
$$A = 4000(1.23367806...) \quad \text{Multiply}$$
$$A = 4934.71$$
$$\$4934.71 \quad \text{Our Solution}$$

Albert Einstein once said that the most powerful force in the universe is compound interest. Consider the following example, illustrating how powerful compound interest can be.

Example 549.

If you invest \$6.16 in an account paying 12% interest compounded continuously for 100 years, and that is all you have to leave your children as an inheritance, what will the final balance be that they will receive?

$$P = 6.16, r = 0.12, t = 100 \quad \text{Identify each of the variables}$$
$$A = 6.16e^{0.12 \cdot 100} \quad \text{Multiply exponent}$$
$$A = 6.16e^{12} \quad \text{Evaluate}$$
$$A = 6.16(162,544.79) \quad \text{Multiply}$$
$$A = 1,002,569.52$$
$$\$1,002,569.52 \quad \text{Our Solution}$$

In 100 years that one time investment of \$6.16 investment grew to over one million dollars! That's the power of compound interest!

10.6 Practice - Compound Interest

Solve

1) Find each of the following:
   a. $500 invested at 4% compounded annually for 10 years.
   b. $600 invested at 6% compounded annually for 6 years.
   c. $750 invested at 3% compounded annually for 8 years.
   d. $1500 invested at 4% compounded semiannually for 7 years.
   e. $900 invested at 6% compounded semiannually for 5 years.
   f. $950 invested at 4% compounded semiannually for 12 years.
   g. $2000 invested at 5% compounded quarterly for 6 years.
   h. $2250 invested at 4% compounded quarterly for 9 years.
   i. $3500 invested at 6% compounded quarterly for 12 years.

j. All of the above compounded continuously.

2) What principal will amount to $2000 if invested at 4% interest compounded semiannually for 5 years?

3) What principal will amount to $3500 if invested at 4% interest compounded quarterly for 5 years?

4) What principal will amount to $3000 if invested at 3% interest compounded semiannually for 10 years?

5) What principal will amount to $2500 if invested at 5% interest compounded semiannually for 7.5 years?

6) What principal will amount to $1750 if invested at 3% interest compounded quarterly for 5 years?

7) A thousand dollars is left in a bank savings account drawing 7% interest, compounded quarterly for 10 years. What is the balance at the end of that time?

8) A thousand dollars is left in a credit union drawing 7% compounded monthly. What is the balance at the end of 10 years?

9) $1750 is invested in an account earning 13.5% interest compounded monthly for a 2 year period. What is the balance at the end of 9 years?

10) You lend out $5500 at 10% compounded monthly. If the debt is repaid in 18 months, what is the total owed at the time of repayment?

11) A $10,000 Treasury Bill earned 16% compounded monthly. If the bill matured in 2 years, what was it worth at maturity?

12) You borrow $25000 at 12.25% interest compounded monthly. If you are unable to make any payments the first year, how much do you owe, excluding penalties?

13) A savings institution advertises 7% annual interest, compounded daily, How much more interest would you earn over the bank savings account or credit union in problems 7 and 8?

14) An 8.5% account earns continuous interest. If $2500 is deposited for 5 years, what is the total accumulated?

15) You lend $100 at 10% continuous interest. If you are repaid 2 months later, what is owed?

Functions - Trigonometric Functions

Objective: Solve for a missing side of a right triangle using trigonometric ratios.

There are six special functions that describe the relationship between the sides of a right triangle and the angles of the triangle. We will discuss three of the functions here. The three functions are called the sine, cosine, and tangent (the three others are cosecant, secant, and cotangent, but we will not need to use them here).

To the right is a picture of a right triangle. Based on which angle we are interested in on a given problem we will name the three sides in relationship to that angle. In the picture, angle A is the angle we will use to name the other sides. The longest side, the side opposite the right angle is always called the hypotenouse. The side across from the angle A is called the opposite side.

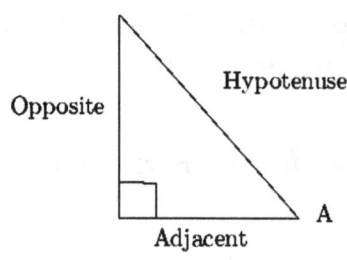

The third side, the side between our angle and the right angle is called the adjacent side. It is important to remember that the opposite and adjacent sides are named in relationship to the angle A or the angle we are using in a problem. If the angle had been the top angle, the opposite and adjacent sides would have been switched.

The three trigonometric funtions are functions taken of angles. When an angle goes into the function, the output is a ratio of two of the triangle sides. The ratios are as describe below:

$$\sin\theta = \frac{\text{opposite}}{\text{hypotenuse}} \qquad \cos\theta = \frac{\text{adjacent}}{\text{hypotenuse}} \qquad \tan\theta = \frac{\text{opposite}}{\text{adjacent}}$$

The "weird" variable θ is a greek letter, pronounced "theta" and is close in idea to our letter "t". Often working with triangles, the angles are repesented with Greek letters, in honor of the Ancient Greeks who developed much of Geometry. Some students remember the three ratios by remembering the word "SOH CAH TOA" where each letter is the first word of: "Sine: Opposite over Hypotenuse; Cosine: Adjacent over Hypotenuse; and Tangent: Opposite over Adjacent." Knowing how to use each of these relationships is fundamental to solving problems using trigonometry.

World View Note: The word "sine" comes from a mistranslation of the Arab word jayb

Example 550.

Using the diagram at right, find each of the following: $\sin\theta$, $\cos\theta$, $\tan\theta$, $\sin\alpha$, $\cos\alpha$, and $\tan\alpha$.

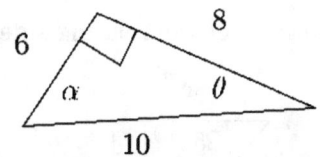

First we will find the three ratios of θ. The hypotenuse is 10, from θ, the opposite side is 6 and the adjacent side is 8. So we fill in the following:

$$\sin\theta = \frac{\text{opposite}}{\text{hypotenuse}} = \frac{6}{10} = \frac{3}{5}$$

$$\cos\theta = \frac{\text{adjacent}}{\text{hypotenuse}} = \frac{8}{10} = \frac{4}{5}$$

$$\tan\theta = \frac{\text{opposite}}{\text{adjacent}} = \frac{6}{8} = \frac{3}{4}$$

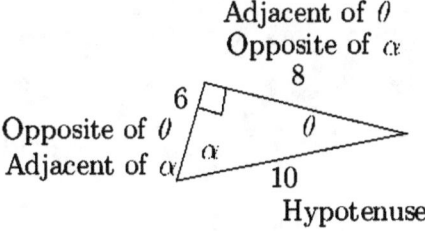

Adjacent of θ
Opposite of α

Opposite of θ
Adjacent of α

Hypotenuse

Now we will find the three ratios of α. The hypotenuse is 10, from α, the opposite side is 8 and the adjacent side is 6. So we fill in the following:

$$\sin\alpha = \frac{\text{opposite}}{\text{hypotenuse}} = \frac{8}{10} = \frac{4}{5}$$

$$\cos\alpha = \frac{\text{adjacent}}{\text{hypotenuse}} = \frac{6}{10} = \frac{3}{5}$$

$$\tan\alpha = \frac{\text{opposite}}{\text{adjacent}} = \frac{8}{6} = \frac{4}{3}$$

We can either use a trigonometry table or a calculator to find decimal values for sine, cosine, or tangent of any angle. We only put angle values into the trigonometric functions, never values for sides. Using either a table or a calculator, we can solve the next example.

Example 551.

$$\sin 42°\qquad \text{Use calculator or table}$$
$$0.669\qquad \text{Our Solution}$$

$$\tan 12°\qquad \text{Use calculator or table}$$
$$0.213\qquad \text{Our Solution}$$

$$\cos 18°\qquad \text{Use calculator or table}$$
$$0.951\qquad \text{Our Solution}$$

By combining the ratios together with the decimal approximations the calculator or table gives us, we can solve for missing sides of a triangle. The trick will be to determine which angle we are working with, naming the sides we are working with, and deciding which trig function can be used with the sides we have.

Example 552.

Find the measure of the missing side.

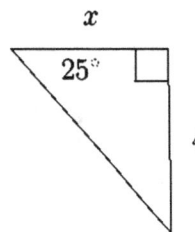

We will be using the angle marked 25°, from this angle, the side marked 4 is the opposite side and the side marked x is the adjacent side.

The trig ratio that uses the opposite and adjacent sides is tangent. So we will take the tangent of our angle.

$$\tan 25° = \frac{4}{x}$$ Tangent is opposite over adjacent

$$\frac{0.466}{1} = \frac{4}{x}$$ Evaluate $\tan 25°$, put over 1 so we have proportion

$$0.466x = 4$$ Find cross product
$$\overline{0.466} \quad \overline{0.466}$$ Divide both sides by 0.466
$$x = 8.58$$ Our Solution

Example 553.

Find the measure of the missing side.

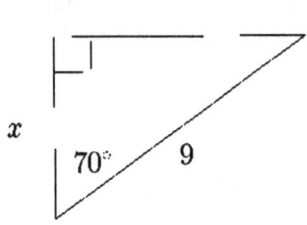

We will be using the angle marked 70°. From this angle, the x is the adjacent side and the 9 is the hypotenuse.

The trig ratio that uses adjacent and hypotenuse is the cosine. So we will take the cosine of our angle.

$$\cos 70° = \frac{x}{9}$$ Cosine is adjacent over hypotenuse

$$\frac{0.342}{1} = \frac{x}{9}$$ Evaluate $\cos 70°$, put over 1 so we have a proportion

$$3.08 = 1x$$ Find the cross product.
$$3.08 = x$$ Our Solution.

10.7 Practice - Trigonometric Functions

Find the value of each. Round your answers to the nearest ten-thousandth.

1) cos 71°

2) cos 23°

3) sin 75°

4) sin 50°

Find the value of the trig function indicated.

5) sin θ

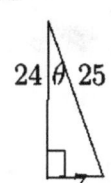

6) tan θ

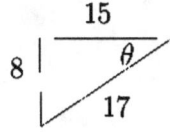

7) sin θ

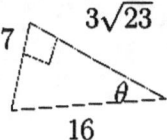

8) sin θ

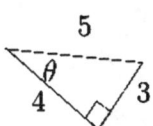

9) sin θ

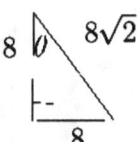

10) cos θ

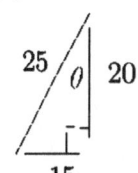

Find the measure of each side indicated. Round to the nearest tenth.

11)

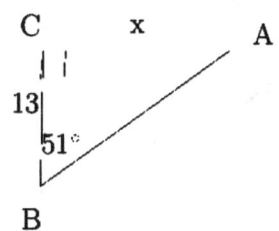

12)

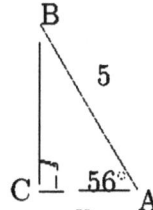

13)

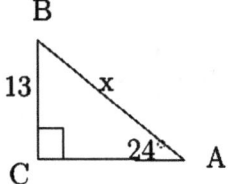

14)

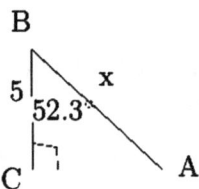

15)

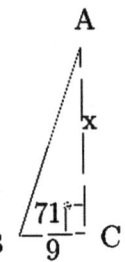

16)

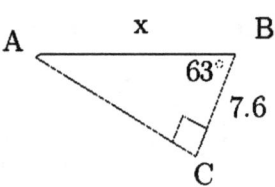

17)

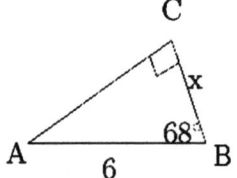

18)

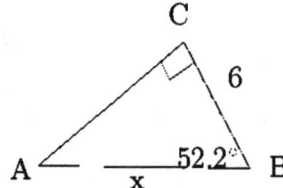

19)

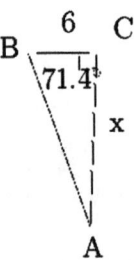

20)

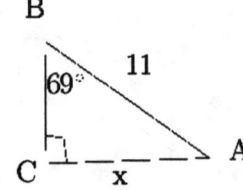

21)

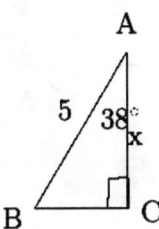

22)

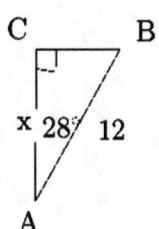

23)

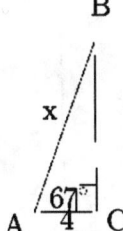

24)

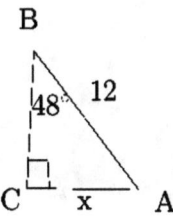

25)

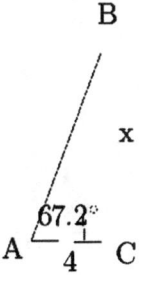

26)

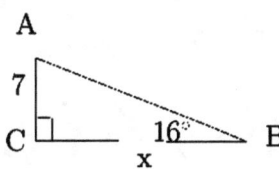

27)

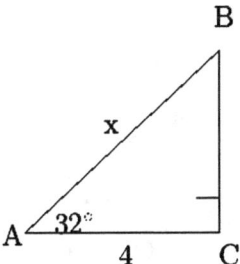

28)

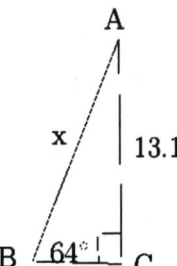

425

29)

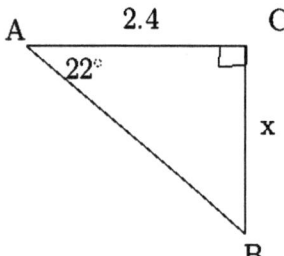

30)

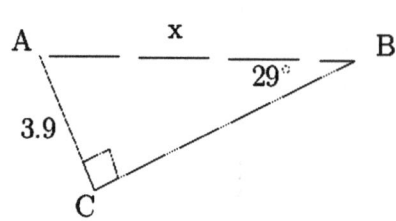

31)

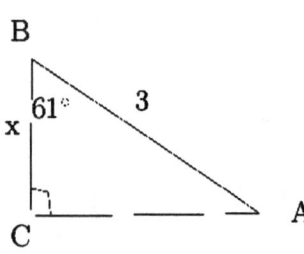

32)

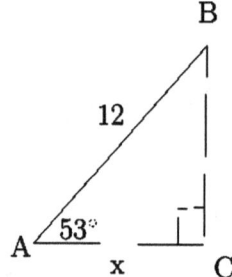

33)

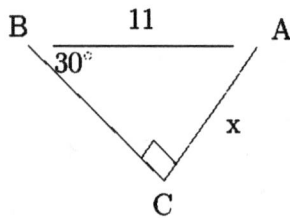

34)

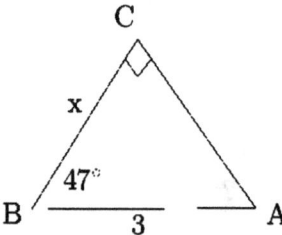

35)

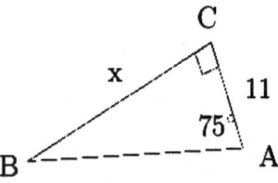

36)

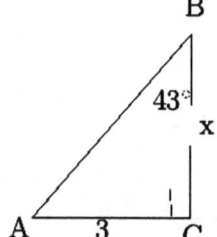

37)

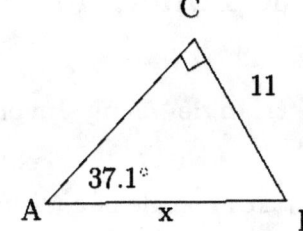

38)

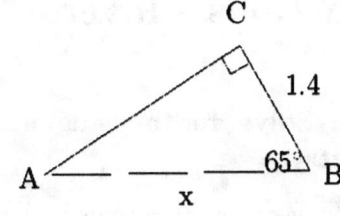

39)

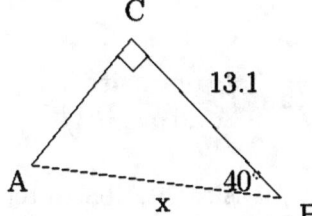

40)

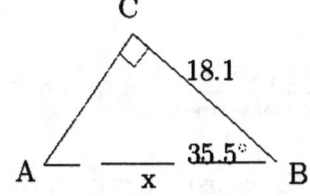

Functions - Inverse Trigonometric Functions

Objective: Solve for missing angles of a right triangle using inverse trigonometry.

We used a special function, one of the trig functions, to take an angle of a triangle and find the side length. Here we will do the opposite, take the side lengths and find the angle. Because this is the opposite operation, we will use the inverse function of each of the trig ratios we saw before. The notation we will use for the inverse trig functions will be similar to the inverse notation we used with functions.

$$\sin^{-1}\left(\frac{\textbf{opposite}}{\textbf{hypotenuse}}\right) = \theta \quad \cos^{-1}\left(\frac{\textbf{adjacent}}{\textbf{hypotenuse}}\right) = \theta \quad \tan^{-1}\left(\frac{\textbf{opposite}}{\textbf{adjacent}}\right) = \theta$$

Just as with inverse functions, the -1 is not an exponent, it is a notation to tell us that these are inverse functions. While the regular trig functions take angles as inputs, these inverse functions will always take a ratio of sides as inputs. We can calculate inverse trig values using a table or a calculator (usually pressing shift or 2nd first).

Example 554.

$$\begin{aligned}
\sin A &= 0.5 &&\text{We don't know the angle so we use an inverse trig function} \\
\sin^{-1}(0.5) &= A &&\text{Evaluate using table or calculator} \\
30^\circ &= A &&\text{Our Solution}
\end{aligned}$$

$$\begin{aligned}
\cos B &= 0.667 &&\text{We don't know the angle so we use an inverse trig function} \\
\cos^{-1}(0.667) &= B &&\text{Evaluate using table or calculator} \\
48^\circ &= B &&\text{Our Solution}
\end{aligned}$$

$$\begin{aligned}
\tan C &= 1.54 &&\text{We don't know the angle so we use an inverse trig function} \\
\tan^{-1}(1.54) &= C &&\text{Evaluate using table or calculator} \\
57^\circ &= C &&\text{Our Solution}
\end{aligned}$$

If we have two sides of a triangle, we can easily calculate their ratio as a decimal and then use one of the inverse trig functions to find a missing angle.

Example 555.

Find the indicated angle.

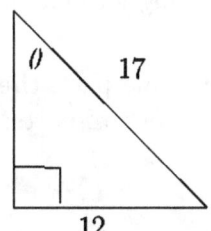

From angle θ the given sides are the opposite (12) and the hypotenuse (17).

The trig function that uses opposite and hypotenuse is the sine

Because we are looking for an angle we use the inverse sine

$\sin^{-1}\left(\dfrac{12}{17}\right)$ Sine is opposite over hyptenuse, use inverse to find angle

$\sin^{-1}(0.706)$ Evaluate fraction, take sine inverse using table or calculator

$\qquad\quad 45°$ Our Solution

Example 556.

Find the indicated angle

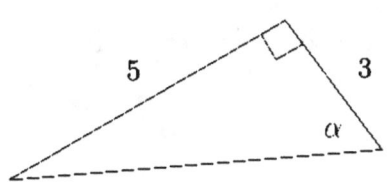

From the angle α, the given sides are the opposite (5) and the adjacent (3)

The trig function that uses opposite and adjacent is the tangent

As we are looking for an angle we will use the inverse tangent.

$\tan^{-1}\left(\dfrac{5}{3}\right)$ Tangent is opposite over adjacent. Use inverse to find angle

$\tan^{-1}(1.667)$ Evaluate fraction, take tangent inverse on table or calculator

$\qquad\quad 59°$ Our Solution

Using a combination of trig functions and inverse trig functions, if we are given two parts of a right triangle (two sides or a side and an angle), we can find all the other sides and angles of the triangle. This is called solving a triangle.

When we are solving a triangle, we can use trig ratios to solve for all the missing parts of it, but there are some properties from geometry that may be helpful along the way.

The angles of a triangle always add up to 180°, because we have a right triangle, 90° are used up in the right angle, that means there are another 90° left in the two acute angles. In other words, the smaller two angles will always add to 90, if we know one angle, we can quickly find the other by subtracting from 90.

Another trick is on the sides of the angles. If we know two sides of the right triangle, we can use the Pythagorean Theorem to find the third side. The Pythagorean Theorem states that if c is the hypotenuse of the triangle, and a and

b are the other two sides (legs), then we can use the following formula, $a^2 + b^2 = c^2$ to find a missing side.

Often when solving triangles we use trigonometry to find one part, then use the angle sum and/or the Pythagorean Theorem to find the other two parts.

Example 557.

Solve the triangle

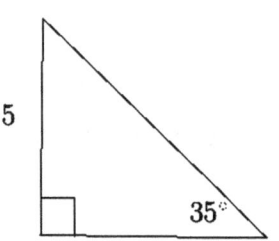

We have one angle and one side. We can use these to find either other side. We will find the other leg, the adjacent side to the 35° angle.

The 5 is the opposite side, so we will use the tangent to find the leg.

$$\tan 35° = \frac{5}{x}$$ Tangent is opposite over adjacent

$$\frac{0.700}{1} = \frac{5}{x}$$ Evaluate tangent, put it over one so we have a proportion

$$0.700x = 5$$ Find cross product

$$\overline{0.700 \quad 0.700}$$ Divide both sides by 0.700

$$x = 7.1$$ The missing leg.

$$a^2 + b^2 = c^2$$ We can now use pythagorean thorem to find hypotenuse, c

$$5^2 + 7.1^2 = c^2$$ Evaluate exponents

$$25 + 50.41 = c^2$$ Add

$$75.41 = c^2$$ Square root both sides

$$8.7 = c$$ The hypotenuse

$$90° - 35°$$ To find the missing angle we subtract from 90°

$$55°$$ The missing angle

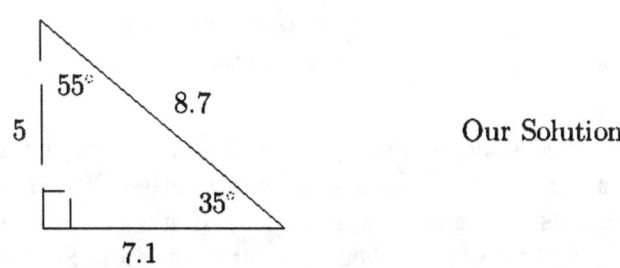

Our Solution

In the previous example, once we found the leg to be 7.1 we could have used the sine function on the 35° angle to get the hypotenuse and then any inverse trig

430

function to find the missing angle and we would have found the same answers. The angle sum and pythagorean theorem are just nice shortcuts to solve the problem quicker.

Example 558.

Solve the triangle

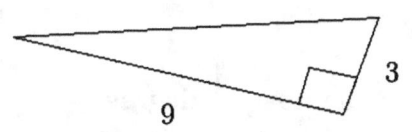

In this triangle we have two sides. We will first find the angle on the right side, adjacent to 3 and opposite from the 9.

Tangent uses opposite and adjacent

To find an angle we use the inverse tangent.

$$\tan^{-1}\left(\frac{9}{3}\right) \qquad \text{Evaluate fraction}$$
$$\tan^{-1}(3) \qquad \text{Evaluate tangent}$$
$$71.6^{\circ} \qquad \text{The angle on the right side}$$

$$90^{\circ} - 71.6^{\circ} \qquad \text{Subtract angle from } 90^{\circ} \text{ to get other angle}$$
$$18.4^{\circ} \qquad \text{The angle on the left side}$$

$$a^2 + b^2 = c^2 \qquad \text{Pythagorean theorem to find hypotenuse}$$
$$9^2 + 3^2 = c^2 \qquad \text{Evaluate exponents}$$
$$81 + 9 = c^2 \qquad \text{Add}$$
$$90 = c^2 \qquad \text{Square root both sides}$$
$$3\sqrt{10} \text{ or } 9.5 = c \qquad \text{The hypotenuse}$$

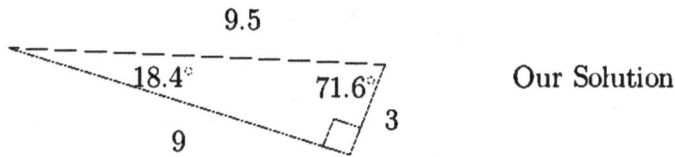

Our Solution

World View Note: Ancient Babylonian astronomers kept detailed records of the starts, planets, and eclipses using trigonometric ratios as early as 1900 BC!

10.8 Practice - Inverse Trigonometric Functions

Find each angle measure to the nearest degree.

1) sin Z = 0.4848

2) sin Y = 0.6293

3) sin Y = 0.6561

4) cos Y = 0.6157

Find the measure of the indicated angle to the nearest degree.

5)

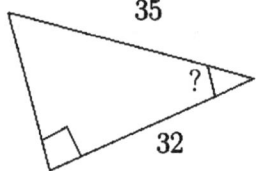

6)

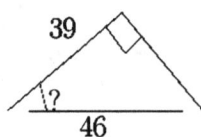

7)

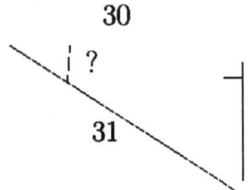

8)

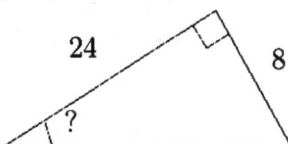

9)

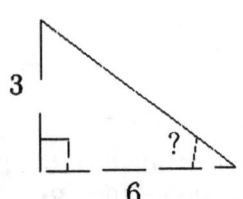

10)

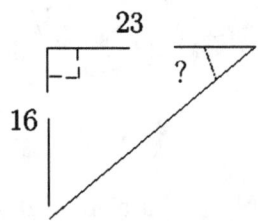

Find the measure of each angle indicated. Round to the nearest tenth.

11)

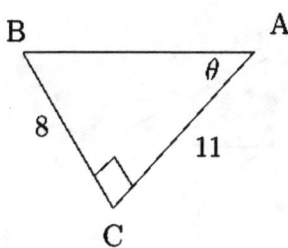

12)

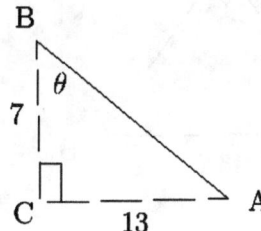

13)

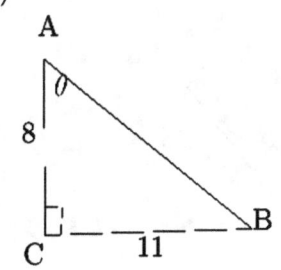

14)

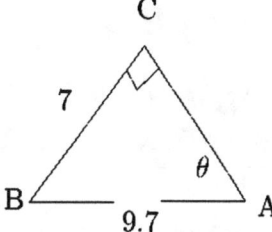

15)

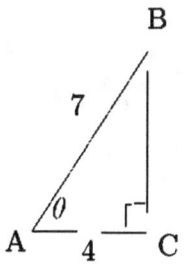

16)

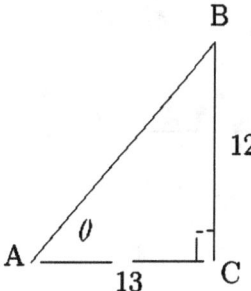

17)

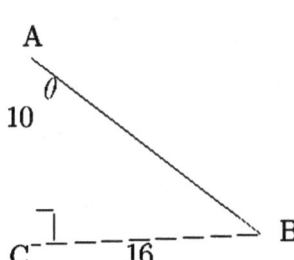

18)

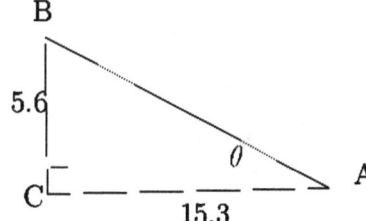

433

19)

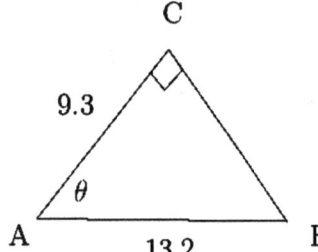

20)

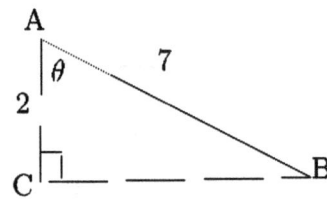

21)

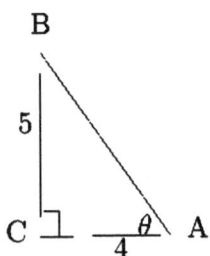

22)

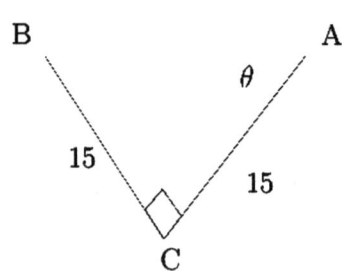

23)

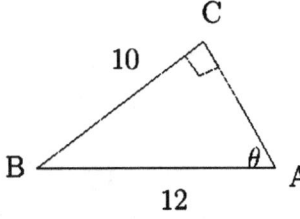

24)

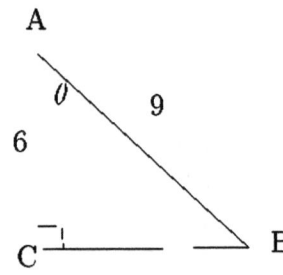

25)

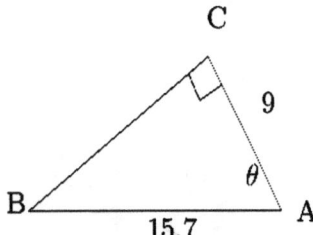

26)

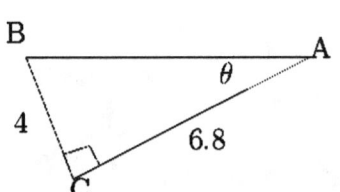

27)

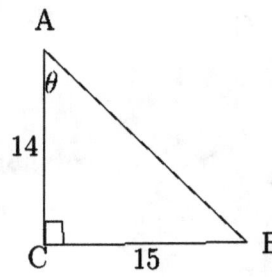

28)

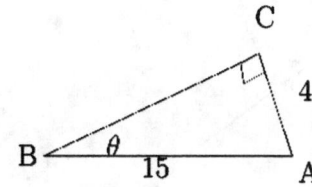

29)

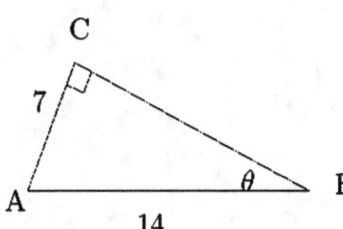

30)

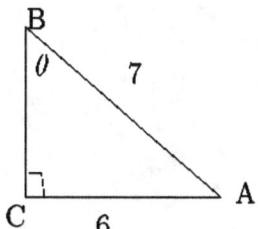

Solve each triangle. Round answers to the nearest tenth.

31)

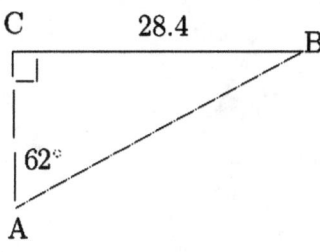

32)

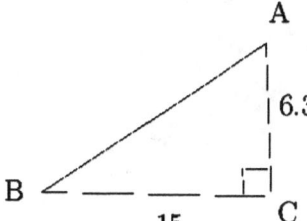

33)

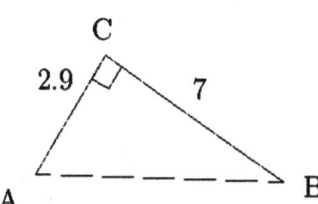

34)

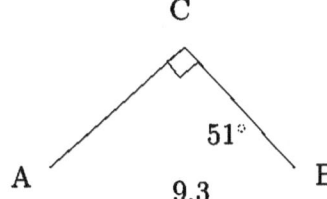

35)

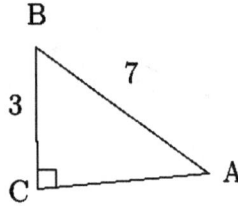

36)

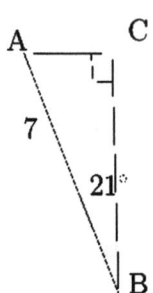

37)

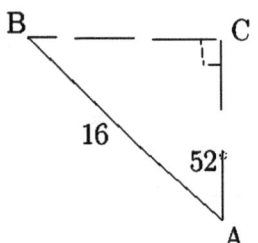

38)

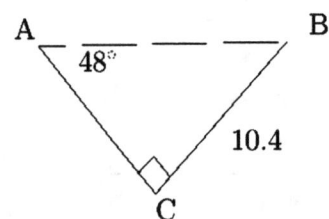

39)

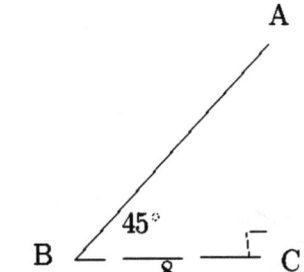

40)

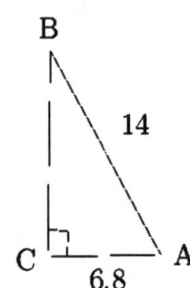

Answers - Chapter 0

0.1

Answers - Integers

1) -2	22) 0	43) -20
2) 5	23) 11	44) 27
3) 2	24) 9	45) -24
4) 2	25) -3	46) -3
5) -6	26) -4	47) 7
6) -5	27) -3	48) 3
7) 8	28) 4	49) 2
8) 0	29) 0	50) 5
9) -2	30) -8	51) 2
10) -5	31) -4	52) 9
11) 4	32) -35	53) 7
12) -7	33) -80	54) -10
13) 3	34) 14	55) 4
14) -9	35) 8	56) 10
15) -2	36) 6	57) -8
16) -9	37) -56	58) 6
17) -1	38) -6	59) -6
18) -2	39) -36	60) -9
19) -3	40) 63	
20) 2	41) -10	
21) -7	42) 4	

0.2

Answers - Fractions

1) $\frac{7}{2}$	4) $\frac{8}{3}$
2) $\frac{5}{4}$	5) $\frac{3}{2}$
3) $\frac{7}{5}$	6) $\frac{5}{4}$

7) $\frac{5}{4}$

8) $\frac{4}{3}$

9) $\frac{3}{2}$

10) $\frac{8}{3}$

11) $\frac{5}{2}$

12) $\frac{8}{7}$

13) $\frac{7}{2}$

14) $\frac{4}{3}$

15) $\frac{4}{3}$

16) $\frac{3}{2}$

17) $\frac{6}{5}$

18) $\frac{7}{6}$

19) $\frac{3}{2}$

20) $\frac{8}{7}$

21) 8

22) $\frac{5}{3}$

23) $-\frac{4}{9}$

24) $-\frac{2}{3}$

25) $-\frac{13}{4}$

26) $\frac{3}{4}$

27) $\frac{33}{20}$

28) $\frac{33}{56}$

29) 4

30) $\frac{18}{7}$

31) $\frac{1}{2}$

32) $-\frac{19}{20}$

33) 3

34) $-\frac{17}{15}$

35) $-\frac{7}{10}$

36) $\frac{5}{14}$

37) $-\frac{8}{7}$

38) $\frac{20}{21}$

39) $\frac{2}{9}$

40) $\frac{4}{3}$

41) $-\frac{21}{26}$

42) $\frac{25}{21}$

43) $-\frac{3}{2}$

44) $-\frac{5}{27}$

45) $\frac{40}{9}$

46) $-\frac{1}{10}$

47) $-\frac{45}{7}$

48) $\frac{13}{15}$

49) $\frac{4}{27}$

50) $\frac{32}{65}$

51) $\frac{1}{15}$

52) 1

53) -1

54) $-\frac{10}{7}$

55) $\frac{2}{7}$

56) 2

57) 3

58) $-\frac{31}{8}$

59) $\frac{37}{20}$

60) $-\frac{5}{3}$

61) $\frac{33}{20}$

62) $\frac{3}{7}$

63) $\frac{47}{56}$

64) $-\frac{7}{6}$

65) $\frac{2}{3}$

66) $-\frac{4}{3}$

67) 1

68) $\frac{7}{8}$

69) $\frac{19}{20}$

70) $-\frac{2}{5}$

71) $-\frac{145}{56}$

72) $-\frac{29}{15}$

73) $\frac{34}{7}$

74) $-\frac{23}{3}$

75) $-\frac{3}{8}$

76) $-\frac{2}{3}$

77) $-\frac{5}{24}$

78) $\frac{39}{14}$

79) $-\frac{5}{6}$

80) $\frac{1}{10}$

81) 2

82) $\frac{62}{21}$

0.3

Answers - Order of Operation

1) 24
2) -1
3) 5
4) 180
5) 4
6) 8
7) 1
8) 8
9) 6

10) -6
11) -10
12) -9
13) 20
14) -22
15) 2
16) 28
17) -40
18) -15

19) 3
20) 0
21) -18
22) -3
23) -4
24) 3
25) 2

0.4

Answers - Properties of Algebra

1) 7
2) 29
3) 1
4) 3
5) 23
6) 14
7) 25
8) 46
9) 7
10) 8
11) 5
12) 10
13) 1
14) 6
15) 1
16) 2
17) 36
18) 54

19) 7
20) 38
21) $r+1$
22) $-4x-2$
23) $2n$
24) $11b+7$
25) $15v$
26) $7x$
27) $-9x$
28) $-7a-1$
29) $k+5$
30) $-3p$
31) $-5x-9$
32) $-9-10n$
33) $-m$
34) $-5-r$
35) $10n+3$
36) $5b$

37) $-8x+32$
38) $24v+27$
39) $8n^2+72n$
40) $5-9a$
41) $-7k^2+42k$
42) $10x+20x^2$
43) $-6-36x$
44) $-2n-2$
45) $40m-8m^2$
46) $-18p^2+2p$
47) $-36x+9x^2$
48) $32n-8$
49) $-9b^2+90b$
50) $-4-28r$
51) $-40n-80n^2$
52) $16x^2-20x$
53) $14b+90$

54) $60v - 7$

55) $-3x + 8x^2$

56) $-89x + 81$

57) $-68k^2 - 8k$

58) $-19 - 90a$

59) $-34 - 49p$

60) $-10x + 17$

61) $10 - 4n$

62) $-30 + 9m$

63) $12x + 60$

64) $30r - 16r^2$

65) $-72n - 48 - 8n^2$

66) $-42b - 45 - 4b^2$

67) $79 - 79v$

68) $-8x + 22$

69) $-20n^2 + 80n - 42$

70) $-12 + 57a + 54a^2$

71) $-75 - 20k$

72) $-128x - 121$

73) $4n^2 - 3n - 5$

74) $2x^2 - 6x - 3$

75) $4p - 5$

76) $3x^2 + 7x - 7$

77) $-v^2 + 2v + 2$

78) $-7b^2 + 3b - 8$

79) $-4k^2 + 12$

80) $a^2 + 3a$

81) $3x^2 - 15$

82) $-n^2 + 6$

Answers - Chapter 1

1.1

Answers to One-Step Equations

1) 7

2) 11

3) -5

4) 4

5) 10

6) 6

7) -19

8) -6

9) 18

10) 6

11) -20

12) -7

13) -108

14) 5

15) -8

16) 4

17) 17

18) 4

19) 20

20) -208

21) 3

22) 16

23) -13

24) -9

25) 15

26) 8

27) -10

28) -204

29) 5

30) 2

31) -11

32) -14

33) 14

34) 1

35) -11

36) -15

37) -240

38) -135

39) -16

40) -380

1.2

Answers to Two-Step Equations

1) -4

2) 7

3) -14

4) -2

5) 10

6) -12

7) 0

8) 12

9) -10

10) -16

11) 14

12) -7

13) 4

14) -5

15) 16

16) -15

17) 7

18) 12

19) 9

20) 0

21) 11

22) -6

23) -10

24) 13

25) 1

26) 4

27) -9

28) 15

29) -6

30) 6

31) -16

32) -4

33) 8

34) -13

35) -2

36) 10

37) -12

38) 0

39) 12

40) -9

1.3

Answers to General Linear Equations

1) -3

2) 6

3) 7

4) 0

5) 1

6) 3

7) 5

8) -4

9) 0

10) 3

11) 1

12) All real numbers

13) 8

14) 1

15) -7

16) 0

17) 2

18) -3

19) -3

20) 3

21) 3

22) -1

23) -1

24) -1

25) 8

26) 0

27) -1

28) 5

29) -1

30) 1

31) -4

32) 0

33) -3

34) 0

35) 0

36) -2

37) -6

38) -3

39) 5

40) 6

41) 0

42) -2

43) No Solution

44) 0

45) 12   48) 1

46) All real numbers   49) -9

47) No Solution   50) 0

1.4

Answers to Solving with Fractions

1) $\frac{3}{4}$

2) $-\frac{4}{3}$

3) $\frac{6}{5}$

4) $\frac{1}{6}$

5) $-\frac{19}{6}$

6) $\frac{25}{8}$

7) $-\frac{7}{9}$

8) $-\frac{1}{3}$

9) -2

10) $\frac{3}{2}$

11) 0

12) $\frac{4}{3}$

13) $-\frac{3}{2}$

14) $\frac{1}{2}$

15) $-\frac{4}{3}$

16) 1

17) 0

18) $-\frac{5}{3}$

19) 1

20) 1

21) $\frac{1}{2}$

22) -1

23) -2

24) $-\frac{9}{4}$

25) 16

26) $-\frac{1}{2}$

27) $-\frac{5}{3}$

28) $-\frac{3}{2}$

29) $\frac{4}{3}$

30) $\frac{3}{2}$

1.5

Answers - Formulas

1. $b = \frac{c}{a}$

2. $h = gi$

3. $x = \frac{gb}{f}$

4. $y = \frac{pq}{3}$

5. $x = \frac{a}{3b}$

6. $y = \frac{cb}{dm}$

7. $m = \frac{E}{c^2}$

8. $D = \frac{ds}{S}$

9. $\pi = \frac{3V}{4r^3}$

10. $m = \frac{2E}{v_2}$

11. $c = b - a$

12. $x = g + f$

13. $y = \frac{cm + cn}{4}$

14. $r = \frac{k(a - 3)}{5}$

15. $D = \frac{12V}{\pi n}$

16. $k = \frac{F}{R - L}$

17. $n = \frac{P}{p - c}$

18. $L = S - 2B$

19. $D = TL + d$

20. $E_a = IR + Eg$

21. $L_o = \frac{L}{1 + at}$

22. $x = \frac{c - b}{a}$

23. $m = \frac{p - q}{2}$

24. $L = \frac{q + 6p}{6}$

25. $k = qr + m$

26. $T = \frac{R - b}{a}$

27. $v = \frac{16t^2 + h}{t}$

28. $h = \frac{s - \pi r^2}{\pi r}$

29. $Q_2 = \frac{Q_1 + PQ_1}{P}$

30. $r_1 = \dfrac{L - 2d - \pi r^2}{\pi}$

31. $T_1 = \dfrac{Rd - kAT_2}{kA}$

32. $v_2 = \dfrac{Pg + V_1^2}{V_1}$

33. $a = \dfrac{c - b}{x}$

34. $r = \dfrac{d}{t}$

35. $w = \dfrac{V}{\ell h}$

36. $h = \dfrac{3v}{\pi r^2}$

37. $a = \dfrac{c - 1}{b}$

38. $b = \dfrac{c - 1}{a}$

39. $t = \dfrac{5 + bw}{a}$

40. $w = \dfrac{at - s}{b}$

41. $x = \dfrac{c - bx}{x}$

42. $x = 3 - 5y$

43. $y = \dfrac{3 - x}{5}$

44. $x = \dfrac{7 - 2y}{3}$

45. $y = \dfrac{7 - 3x}{2}$

46. $a = \dfrac{7b + 4}{5}$

47. $b = \dfrac{5a - 4}{7}$

48. $x = \dfrac{8 + 5y}{4}$

49. $y = \dfrac{4x - 8}{5}$

50. $f = \dfrac{9c + 160}{5}$

1.6

Answers to Absolute Value Equations

1) $8, -8$

2) $7, -7$

3) $1, -1$

4) $2, -2$

5) $6, -\dfrac{29}{4}$

6) $\dfrac{38}{9}, -6$

7) $-2, -\dfrac{10}{3}$

8) $-3, 9$

9) $3, -\dfrac{39}{7}$

10) $\dfrac{16}{5}, -6$

11) $7, -\dfrac{29}{3}$

12) $-\dfrac{1}{3}, -1$

13) $-9, 15$

14) $3, -\dfrac{5}{3}$

15) $-2, 0$

16) $0, -2$

17) $-\dfrac{6}{7}, 0$

18) $-4, \dfrac{4}{3}$

19) $-\dfrac{17}{2}, \dfrac{7}{2}$

20) $-\dfrac{6}{5}, -2$

21) $-6, -8$

22) $6, -\dfrac{25}{3}$

23) $1, -\dfrac{13}{7}$

24) $7, -21$

25) $-2, 10$

26) $-\dfrac{7}{5}, 1$

27) $6, -\dfrac{16}{3}$

28) $\dfrac{2}{5}, 0$

29) $-\dfrac{13}{7}, 1$

30) $-3, 5$

31) $-\dfrac{4}{3}, -\dfrac{2}{7}$

32) $-6, \dfrac{2}{5}$

33) $7, \dfrac{1}{5}$

34) $-\dfrac{22}{5}, -\dfrac{2}{13}$

35) $-\dfrac{19}{22}, -\dfrac{11}{38}$

36) $0, -\dfrac{12}{5}$

1.7

Answers - Variation

1) $\dfrac{c}{a} = k$

2) $\dfrac{x}{yz} = k$

3) $wx = k$

4) $\dfrac{r}{s^2} = k$

5) $\dfrac{f}{xy} = k$

6) $jm^3 = k$

7) $\dfrac{h}{b} = k$

8) $\dfrac{x}{a\sqrt[2]{b}} = k$

9) $ab = k$

10) $\dfrac{a}{b} = 3$

11) $\dfrac{P}{rq} = 0.5$

12) $cd = 28$

13) $\dfrac{t}{u^2} = 0.67$

14) $\dfrac{e}{fg} = 4$

444

15) $wx^3 = 1458$

16) $\frac{h}{j} = 1.5$

17) $\frac{a}{x\sqrt[2]{y}} = 0.33$

18) $mn = 3.78$

19) 6 k

20) 5.3 k

21) 33.3 cm

22) 160 kg/cm^3

23) 241,920,000 cans

24) 3.5 hours

25) 4.29 dollars

26) 450 m

27) 40 kg

28) 5.7 hr

29) 40 lb

30) 100 N

31) 27 min

32) 1600 km

33) r = 36

34) 8.2 mph

35) 2.5 m

36) V = 100.5 cm^3

37) 6.25 km

38) I = 0.25

1.8

Answer Set - Number and Geometry

1) 11

2) 5

3) -4

4) 32

5) -13

6) 62

7) 16

8) $\frac{17}{4}$

9) 35, 36, 37

10) $-43, -42, -41$

11) $-14, -13, -12$

12) 52, 54

13) 61, 63, 65

14) 83, 85, 87

15) 9, 11, 13

16) 56, 56, 68

17) 64, 64, 52

18) 36, 36, 108

19) 30, 120, 30

20) 30, 90, 60

21) 40, 80, 60

22) 28, 84, 68

23) 24, 120, 36

24) 32, 96, 52

25) 25, 100, 55

26) 45, 30

27) 96, 56

28) 27, 49

29) 57, 83

30) 17, 31

31) 6000, 24000

32) 1000, 4000

33) 40, 200

34) 60, 180

35) 20, 200

36) 30, 15

37) 76, 532

38) 110, 880

39) 2500, 5000

40) 4, 8

41) 2, 4

42) 3, 5

43) 14, 16

44) 1644

45) 325, 950

1.9

Answers - Age Problems

1) 6, 16

2) 10, 40

3) 18, 38

4) 17, 40

5) 27, 31

6) 12, 48

7) 31, 36

8) 16, 32

9) 12, 20

10) 40, 16

11) 10, 6

12) 12, 8

13) 26

14) 8

15) 4

16) 3

17) 10, 20

18) 14

19) 9, 18

20) 15, 20

21) 50, 22

22) 12

23) 72, 16

24) 6

25) 37, 46

26) 15

27) 45

28) 14, 54

29) 8, 4

30) 16, 32

31) 10, 28

32) 12,20

33) 141, 67

34) 16, 40

35) 84, 52

36) 14, 42

37) 10

38) 10, 6

39) 38, 42

40) 5

1.10

Answers - Distance, Rate, and Time Problems

1) $1\frac{1}{3}$

2) $25\frac{1}{2}, 20\frac{1}{2}$

3) 3

4) 10

5) 30, 45

6) 3

7) $\frac{300}{13}$

8) 10

9) 7

10) 30

11) 150

12) 360

13) 8

14) 10

15) 2

16) 3

17) 48

18) 600

19) 6

20) 120

21) 36

22) 2

23) 570

24) 24, 18

25) 300

26) 8, 16

27) 56

28) 95, 120

29) 180

30) 105, 130

31) 2:15 PM

32) 200

33) $\frac{1}{3}$

34) 15

35) $\frac{27}{4}$

36) $\frac{1}{2}$

37) 3, 2

38) 90

Answers - Chapter 2

2.1

Answers - Points and Lines

1) B(4, −3) C(1, 2)   D(−1, 4)
   E(−5, 0)   F(2, −3) G(1, 3)
   H(−1, −4) I(−2, −1) J(0, 2)
   K(−4, 3)

2)

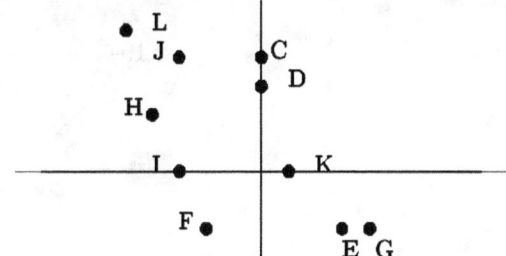

3)
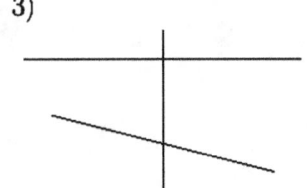

4)
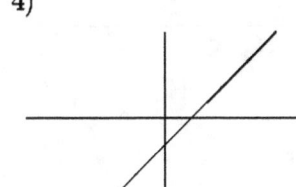

5)
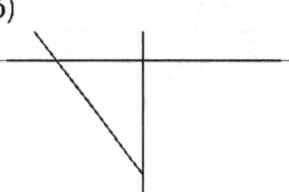

6)
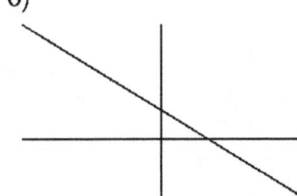

7)
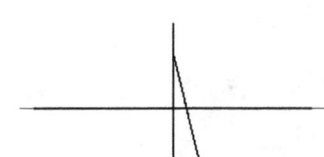

8)
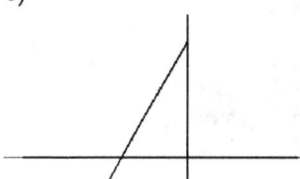

9)
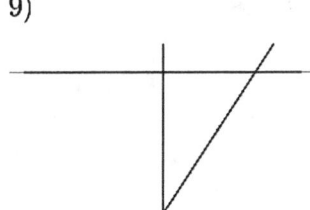

10)
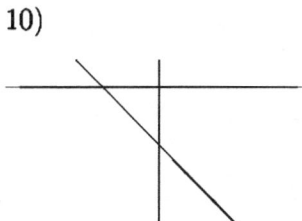

11)
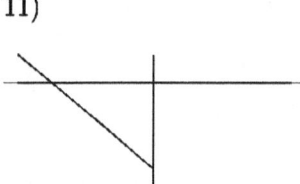

12)
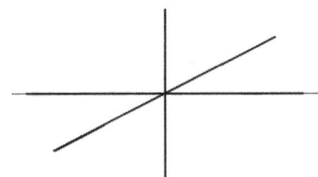

13)
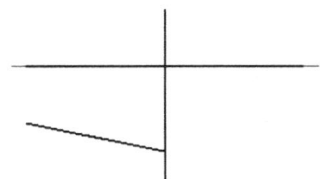

14)
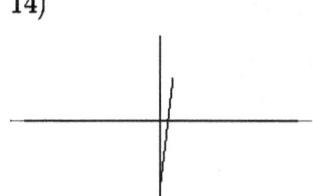

15)

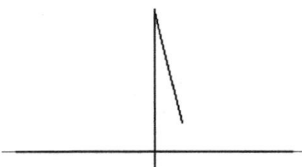

16)

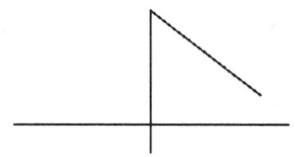

17)

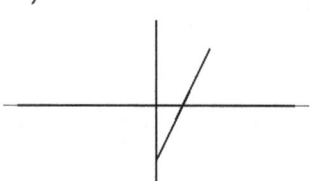

18)

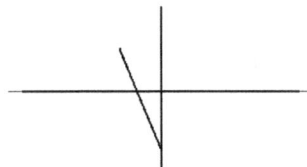

19)

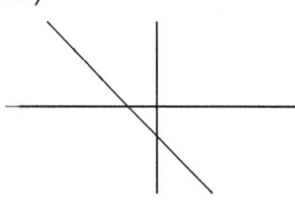

20)

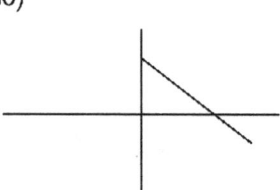

21)

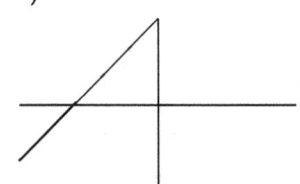

22)

2.2

Answers - Slope

1) $\frac{3}{2}$

2) 5

3) Undefined

4) $-\frac{1}{2}$

5) $\frac{5}{6}$

6) $-\frac{2}{3}$

7) -1

8) $\frac{5}{4}$

9) -1

10) 0

11) Undefined

12) $\frac{16}{7}$

13) $-\frac{17}{31}$

14) $-\frac{3}{2}$

15) $\frac{4}{3}$

16) $-\frac{7}{17}$

17) 0

18) $\frac{5}{11}$

19) $\frac{1}{2}$

20) $\frac{1}{16}$

21) $-\frac{11}{2}$

22) $-\frac{12}{31}$

23) Undefined

24) $\frac{24}{11}$

25) $-\frac{26}{27}$

26) $-\frac{19}{10}$

27) $-\frac{1}{3}$

28) $\frac{1}{16}$

29) $-\frac{7}{13}$

30) $\frac{2}{7}$

31) -5

32) 2

33) -8

34) 3

35) -5

36) 6

37) -4

38) 1

39) 2

40) 1

2.3

Answers - Slope-Intercept

1) $y = 2x + 5$

2) $y = -6x + 4$

3) $y = x - 4$

4) $y = -x - 2$

5) $y = -\frac{3}{4}x - 1$

6) $y = -\frac{1}{4}x + 3$

7) $y = \frac{1}{3}x + 1$

8) $y = \frac{2}{5}x + 5$

9) $y = -x + 5$

10) $y = -\frac{7}{2}x - 5$

11) $y = x - 1$

12) $y = -\frac{5}{3}x - 3$

13) $y = -4x$

14) $y = -\frac{3}{4}x + 2$

15) $y = -\frac{1}{10}x - \frac{37}{10}$

16) $y = \frac{1}{10}x - \frac{3}{10}$

17) $y = -2x - 1$

18) $y = \frac{6}{11}x + \frac{70}{11}$

19) $y = \frac{7}{3}x - 8$

20) $y = -\frac{4}{7}x + 4$

21) $x = -8$

22) $y = \frac{1}{7}x + 6$

23) $y = -x - 1$

24) $y = \frac{5}{2}x$

25) $y = 4x$

26) $y = -\frac{2}{3}x + 1$

27) $y = -4x + 3$

28) $x = 4$

2.4

29) $y = -\frac{1}{2}x + 1$

30) $y = \frac{6}{5}x + 4$

31)

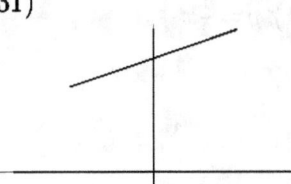

32)

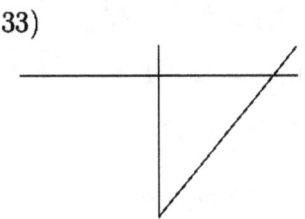

33)

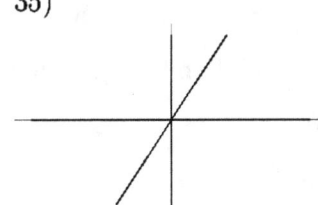

34)

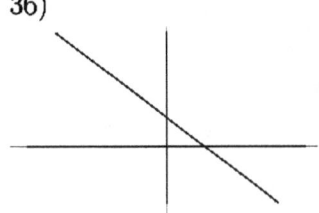

35)

36)

37)

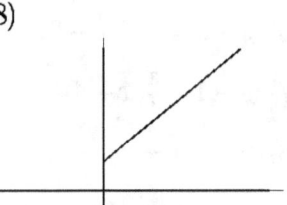

38)

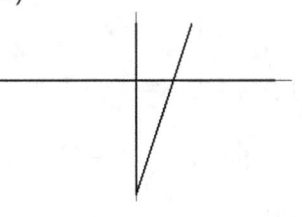

39)

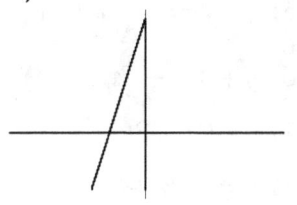

40)

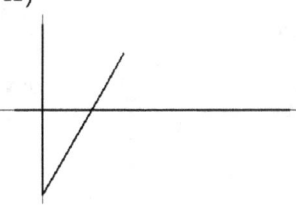

41)

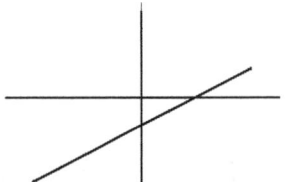

42)

Answers - Point-Slope Form

1) $x = 2$

2) $x = 1$

3) $y - 2 = \frac{1}{2}(x - 2)$

4) $y - 1 = -\frac{1}{2}(x - 2)$

5) $y + 5 = 9(x + 1)$

6) $y + 2 = -2(x - 2)$

7) $y - 1 = \frac{3}{4}(x + 4)$

8) $y + 3 = -2(x - 4)$

9) $y + 2 = -3x$

10) $y - 1 = 4(x + 1)$

11) $y + 5 = -\frac{1}{4}x$

12) $y - 2 = -\frac{5}{4}x$

13) $y + 3 = \frac{1}{5}(x + 5)$

14) $y + 4 = -\frac{2}{3}(x + 1)$

15) $y - 4 = -\frac{5}{4}(x + 1)$

16) $y + 4 = -\frac{3}{2}x(x - 1)$

17) $y = 2x - 3$

18) $y = -2x + 2$

19) $y = -\frac{3}{5}x + 2$

20) $y = -\frac{2}{3}x - \frac{10}{3}$

21) $y = \frac{1}{2}x + 3$

22) $y = -\frac{7}{4}x + 4$

23) $y = -\frac{3}{2}x + 4$

24) $y = -\frac{5}{2}x - 5$

25) $y = -\frac{2}{5}x - 5$

26) $y = \frac{7}{3}x - 4$

27) $y = x - 4$

28) $y = -3$

29) $x = -3$

30) $y = 2x - 1$

31) $y = -\frac{1}{2}x$

32) $y = \frac{6}{5}x - 3$

33) $y - 3 = -2(x + 4)$

34) $y = 3$

35) $y - 1 = \frac{1}{8}(x - 5)$

36) $y - 5 = -\frac{1}{8}(x + 4)$

37) $y + 2 = \frac{3}{2}(x + 4)$

38) $y - 1 = \frac{3}{8}(x + 4)$

39) $y - 5 = \frac{1}{4}(x - 3)$

40) $y + 4 = -(x + 1)$

41) $y + 3 = -\frac{8}{7}(x - 3)$

42) $y + 5 = -\frac{1}{4}(x + 1)$

43) $y = -\frac{3}{4}x - \frac{11}{4}$

44) $y = -\frac{1}{10}x - \frac{3}{2}$

45) $y = -\frac{8}{7}x - \frac{5}{7}$

46) $y = \frac{1}{2}x - \frac{3}{2}$

47) $y = -x + 5$

48) $y = \frac{1}{3}x + 1$

49) $y = -x + 2$

50) $y = x + 2$

51) $y = 4x + 3$

52) $y = \frac{3}{7}x + \frac{6}{7}$

2.5

Answers - Parallel and Perpendicular Lines

1) 2

2) $-\frac{2}{3}$

3) 4

4) $-\frac{10}{3}$

5) 1

6) $\frac{6}{5}$

7) -7

8) $-\frac{3}{4}$

9) 0

10) 2

11) 3

12) $-\frac{5}{4}$

13) -3

14) $-\frac{1}{3}$

15) 2

16) $-\frac{3}{8}$

17) $x = 2$

18) $y - 2 = \frac{7}{5}(x - 5)$

19) $y - 4 = \frac{9}{2}(x - 3)$

20) $y + 1 = -\frac{3}{4}(x - 1)$

21) $y - 3 = \frac{7}{5}(x - 2)$

22) $y - 3 = -3(x + 1)$

23) $x = 4$

24) $y - 4 = \frac{7}{5}(x - 1)$

25) $y + 5 = -(x - 1)$

26) $y + 2 = -2(x - 1)$

27) $y - 2 = \frac{1}{5}(x - 5)$

28) $y - 3 = -(x - 1)$

29) $y - 2 = -\frac{1}{4}(x - 4)$

30) $y + 5 = \frac{7}{3}(x + 3)$

31) $y + 2 = -3(x - 2)$

32) $y - 5 = -\frac{1}{2}(x + 2)$

33) $y = -2x + 5$

34) $y = \frac{3}{5}x + 5$

35) $y = -\frac{4}{3}x - 3$

36) $y = -\frac{5}{4}x - 5$

37) $y = -\frac{1}{2}x - 3$

38) $y = \frac{5}{2}x - 2$

39) $y = -\frac{1}{2}x - 2$

40) $y = \frac{3}{5}x - 1$

41) $y = x - 1$

42) $y = 2x + 1$

43) $y = 2$

44) $y = -\frac{2}{5}x + 1$

45) $y = -x + 3$

46) $y = -\frac{5}{2}x + 2$

47) $y = -2x + 5$

48) $y = \frac{3}{4}x + 4$

Answers - Chapter 3

3.1

Answers - Solve and Graph Inequalities

1) $(-5, \infty)$

2) $(-4, \infty)$

3) $(-\infty, -2]$

4) $(-\infty, 1]$

5) $(-\infty, 5]$

6) $(-5, \infty)$

7) $m < -2$

8) $m \leqslant 1$

9) $x \geqslant 5$

10) $a \leqslant -5$

11) $b > -2$

12) $x > 1$

13) $x \geqslant 110 : [110, \infty)$

14) $n \geqslant -26 : [-26, \infty)$

15) $r < 1 : (-\infty, 1)$

16) $m \leqslant -6 : (-\infty, -6]$

17) $n \geqslant -6 : [-6, \infty)$

18) $x < 6 : (-\infty, 6)$

19) $a < 12 : (-\infty, 12)$

20) $v \geqslant 1 : [1, \infty)$

21) $x \geqslant 11 : [11, \infty)$

22) $x \leqslant -18 : (-\infty, -18]$

23) $k > 19 : (19, \infty)$

24) $n \leqslant -10 : (-\infty, -10]$

25) $p < -1 : (-\infty, -1)$

26) $x \leqslant 20 : (-\infty, 20]$

27) $m \geqslant 2 : [2, \infty)$

28) $n \leqslant 5 : (-\infty, 5]$

29) $r > 8 : (8, \infty)$

30) $x \leqslant -3 : (-\infty, -3]$

31) $b > 1 : (1, \infty)$

32) $n \geqslant 0 : [0, \infty)$

33) $v < 0 : (-\infty, 0)$

34) $x > 2 : (2, \infty)$

35) No solution: $\bigcirc$

36) $n > 1 : (1, \infty)$

37) {All real numbers.} : $\mathbb{R}$

38) $p \leqslant 3 : (-\infty, 3]$

3.2

Answers - Compound Inequalities

1) $n \leqslant -9$ or $n \geqslant 2 : (-\infty, -9] \bigcup [2, \infty)$

2) $m \geqslant -4$ or $m < -5 : (-\infty, -5) \bigcup [-4, \infty)$

3) $x \geqslant 5$ or $x < -5 : (-\infty, -5) \bigcup [5, \infty)$

4) $r > 0$ or $r < -7 : (-\infty -7), \bigcup (0, \infty)$

5) $x < -7 : (-\infty, -7)$

6) $n < -7$ or $n > 8 : (-\infty -7), \bigcup (8, \infty)$

7) $-8 < v < 3 : (-8, 3)$

8) $-7 < x < 4 : (-7, 4)$

9) $b < 5 : (-\infty, 5)$

10) $-2 \leqslant n \leqslant 6 : [-2, 6]$

11) $-7 \leqslant a \leqslant 6 : [-7, 6]$

12) $v \geqslant 6 : [6, \infty)$

13) $-6 \leqslant x \leqslant -2 : [-6, -2]$

14) $-9 \leqslant x \leqslant 0 : [-9, 0]$

15) $3 < k \leqslant 4 : (3, 4]$

16) $-2 \leqslant n \leqslant 4 : [-2, 4]$

17) $-2 < x < 2 : (-2, 2)$

18) No solution : $\varnothing$

19) $-1 \leqslant m < 4 : [-1, 4)$

20) $r > 8$ or $r < -6 :: (-\infty, -6) \bigcup (8, \infty)$

21) No solution : $\bigcirc$

22) $x \leqslant 0$ or $x > 8 : (-\infty, 0] \bigcup (8, \infty)$

23) No solution : $\varnothing$

24) $n \geqslant 5$ or $n < 1 : (-\infty, 1) \bigcup [5, \infty)$

25) $5 \leqslant x < 19 : [5, 19)$

26) $n < -14 \, \text{or} \, n \geqslant 17 : (-\infty, -14) \bigcup [17, \infty)$

27) $1 \leqslant v \leqslant 8 : [1, 8]$

28) $a \leqslant 1 \, \text{or} \, a \geqslant 19 : (-\infty, 1] \bigcup [19, \infty)$

29) $k \geqslant 2 \, \text{or} \, k < -20 : (-\infty, -20) \bigcup [2, \infty)$

30) {All real numbers.} : $\mathbb{R}$

31) $-1 < x \leqslant 1 : (-1, 1]$

32) $m > 4 \, \text{or} \, m \leqslant -1 : (-\infty, -1] \bigcup (4, \infty)$

3.3

Answers - Absolute Value Inequalities

1) $-3, 3$

2) $-8, 8$

3) $-3, 3$

4) $-7, 1$

5) $-4, 8$

6) $-4, 20$

7) $-2, 4$

8) $-7, 1$

9) $-\frac{7}{3}, \frac{11}{3}$

10) $-7, 2$

11) $-3, 5$

12) $0, 4$

13) $1, 4$

14) $(-\infty, 5) \bigcup (5, \infty)$

15) $(-\infty, -\frac{5}{3}] \bigcup [\frac{5}{3}, \infty)$

16) $(-\infty, -1] \bigcup [9, \infty)$

17) $(-\infty, -6) \bigcup (0, \infty)$

18) $(-\infty, -1) \bigcup (5, \infty)$

19) $(-\infty, \frac{2}{3}) \bigcup (\frac{8}{3}, \infty)$

20) $(-\infty, 0) \bigcup (4, \infty)$

21) $(-\infty, -1] \bigcup [3, \infty)$

22) $[-\frac{4}{3}, 2]$

23) $(-\infty, -4] \bigcup [14, \infty)$

24) $(-\infty, -\frac{5}{2}] \bigcup [-\frac{3}{2}, \infty)$

25) $[1, 3]$

26) $[\frac{1}{2}, 1]$

27) $(-\infty, -4) \bigcup (-3, \infty)$

28) $[3, 7]$

29) $[1, \frac{3}{2}]$

30) $[-2, -\frac{4}{3}]$

31) $(-\infty, \frac{3}{2}) \bigcup (\frac{5}{2}, \infty)$

32) $(-\infty, -\frac{1}{2}) \bigcup (1, \infty)$

33) $[2, 4]$

34) $[-3, -2]$

Answers - Chapter 4

4.1

Answers - Graphing

1) $(-1, 2)$

2) $(-4, 3)$

3) $(-1, -3)$

4) $(-3, 1)$

5) No Solution

6) $(-2, -2)$

7) $(-3, 1)$

8) $(4, 4)$

9) $(-3, -1)$

10) No Solution

11) $(3, -4)$

12) $(4, -4)$

13) $(1, -3)$

14) $(-1, 3)$

15) $(3, -4)$

16) No Solution

17) $(2, -2)$

18) $(4, 1)$

19) $(-3, 4)$

20) $(2, -1)$

21) $(3, 2)$

22) $(-4, -4)$

23) $(-1, -1)$

24) $(2, 3)$

25) $(-1, -2)$

26) $(-4, -3)$

27) No Solution

28) $(-3, 1)$

29) $(4, -2)$

30) $(1, 4)$

4.2

Answers - Substitution

1) $(1, -3)$

2) $(-3, 2)$

3) $(-2, -5)$

4) $(0, 3)$

5) $(-1, -2)$

6) $(-7, -8)$

7) $(1, 5)$

8) $(-4, -1)$

9) $(3, 3)$

10) $(4, 4)$

11) $(2, 6)$

12) $(-3, 3)$

13) $(-2, -6)$

14) $(0, 2)$

15) $(1, -5)$

16) $(-1, 0)$

17) $(-1, 8)$

18) $(3, 7)$

19) $(2, 3)$

20) $(8, -8)$

21) $(1, 7)$

22) $(1, 7)$

23) $(-3, -2)$

24) $(1, -3)$

25) $(1, 3)$

26) $(2, 1)$

27) $(-2, 8)$

28) $(-4, 3)$

29) $(4, -3)$

30) $(-1, 5)$

31) $(0, 2)$

32) $(0, -7)$

33) $(0, 3)$

34) $(1, -4)$

35) $(4, -2)$

36) $(8, -3)$

37) $(2, 0)$

38) $(2, 5)$

39) $(-4, 8)$

40) $(2, 3)$

4.3

Answers - Addition/Elimination

1) $(-2, 4)$

2) $(2, 4)$

3) No solution

4) Infinite number of solutions

5) No solution

6) Infinite number of solutions

7) No solution

8) $(2, -2)$

9) $(-3, -5)$

10) $(-3, 6)$

11) $(-2, -9)$

12) $(1, -2)$

13) $(0, 4)$

14) $(-1, 0)$

15) $(8, 2)$

16) $(0, 3)$

17) $(4, 6)$

18) $(-6, -8)$

19) $(-2, 3)$

20) $(1, 2)$

21) $(0, -4)$

22) $(0, 1)$

23) $(-2, 0)$

24) $(2, -2)$

25) $(-1, -2)$

26) $(-3, 0)$

27) $(-1, -3)$

28) $(-3, 0)$

29) $(-8, 9)$

30) $(1, 2)$

31) $(-2, 1)$

32) $(-1, 1)$

33) $(0, 0)$

34) Infinite number of solutions

4.4

Answers - Three Variables

1) $(1, -1, 2)$

2) $(5, -3, 2)$

3) $(2, 3, -2)$

4) $(3, -2, 1)$

5) $(-2, -1, 4)$

6) $(-3, 2, 1)$

7) $(1, 2, 3)$

8) $\propto$ solutions

9) $(0, 0, 0)$

10) $\propto$ solutions

11) $(19, 0, -13)$

12) $\propto$ solutions

13) $(0, 0, 0)$

14) $\propto$ solutions

15) $(2, \frac{1}{2}, -2)$

16) $\propto$ solutions

17) $(-1, 2, 3)$

18) $(-1, 2, -2)$

19) $(0, 2, 1)$

20) no solution

21) $(10, 2, 3)$

22) no solution

23) $(2, 3, 1)$

24) $\propto$ solutions

25) no solutions

26) $(1, 2, 4)$

27) $(-25, 18, -25)$

28) $(\frac{2}{7}, \frac{3}{7}, -\frac{2}{7})$

29) $(1, -3, -2, -1)$

30) $(7, 4, 5, 6)$

31) $(1, -2, 4, -1)$

32) $(-3, -1, 0, 4)$

4.5

Answers - Value Problems

1) 33Q, 70D

2) 26 h, 8 n

3) 236 adult, 342 child

4) 9d, 12q

5) 9, 18

6) 7q, 4h

7) 9, 18

8) 25, 20

9) 203 adults, 226 child

10) 130 adults, 70 students

11) 128 card, 75 no card

12) 73 hotdogs, 58 hamburgers

13) 135 students, 97 non-students

14) 12d, 15q

15) 13n, 5d

16) 8 20¢, 32 25¢

17) 6 15¢, 9 25¢

18) 5

4.6

19) 13 d, 19 q

20) 28 q

21) 15 n, 20 d

22) 20 $1, 6 $5

23) 8 $20, 4 $10

24) 27

25) $12500 @ 12% $14500 @ 13%

26) $20000 @ 5% $30000 @ 7.5%

27) $2500 @ 10% $6500 @ 12%

28) $12400 @ 6% $5600 @ 9%

29) $4100 @ 9.5% $5900 @ 11%

30) $7000 @ 4.5% $9000 @ 6.5%

31) $1600 @ 4%; $2400 @ 8%

32) $3000 @ 4.6% $4500 @ 6.6%

33) $3500 @ 6%; $5000 @ 3.5%

34) $7000 @ 9% $5000 @ 7.5%

35) $6500 @ 8%; $8500 @ 11%

36) $12000 @ 7.25% $5500 @ 6.5%

37) $3000 @ 4.25%; $3000 @ 5.75%

38) $10000 @ 5.5% $4000 @ 9%

39) $7500 @ 6.8%; $3500 @ 8.2%

40) $3000 @ 11%; $24000 @ 7%

41) $5000 @ 12% $11000 @ 8%

42) 26n, 13d, 10q

43) 18, 4, 8

44) 20n, 15d, 10q

Answers - Mixture Problems

1) 2666.7

2) 2

3) 30

4) 1, 8

5) 8

6) 10

7) 20

8) 16

9) 17.25

10) 1.5

11) 10

12) 8

13) 9.6

14) 36

15) 40, 60

16) 30, 70

17) 40, 20

18) 40, 110

19) 20, 30

20) 100, 200

21) 40, 20

22) 10, 5

23) 250, 250

24) 21, 49

25) 20, 40

26) 2, 3

27) $56, 144$ 33) $440, 160$ 39) 10

28) $1.5, 3.5$ 34) 20 40) $30, 20$

29) 30 35) $35, 63$

30) 10 36) $3, 2$ 41) 75

31) $75, 25$ 37) 1.2 42) $20, 60$

32) $55, 20$ 38) 150 43) 25

Answers - Chapter 5

5.1

Answers to Exponent Properties

1) 4^9 17) 4^2 31) 64

2) 4^7 18) 3^4 32) $2a$

3) 2^4 19) 3 33) $\frac{y^3}{512x^{24}}$

4) 3^6 20) 3^3 34) $\frac{y^5 x^2}{2}$

5) $12m^2 n$ 21) m^2

6) $12x^3$ 22) $\frac{xy^3}{4}$ 35) $64m^{12}n^{12}$

7) $8m^6 n^3$ 23) $\frac{4x^2 y}{3}$ 36) $\frac{n^{10}}{2m}$

8) $x^3 y^6$ 24) $\frac{y^2}{4}$ 37) $2x^2 y$

9) 3^{12} 25) $4x^{10}y^{14}$ 38) $2y^2$

10) 4^{12} 26) $8u^{18}v^6$ 39) $2q^7 r^8 p$

11) 4^8 27) $2x^{17}y^{16}$ 40) $4x^2 y^4 z^2$

12) 3^6 28) $3uv$ 41) $x^4 y^{16} z^4$

13) $4u^6 v^4$ 29) $\frac{x^2 y}{6}$ 42) $256q^4 r^8$

14) $x^3 y^3$ 30) $\frac{4a^2}{3}$ 43) $4y^4 z$

15) $16a^{16}$

16) $16x^4 y^4$

5.2

Answers to Negative Exponents

1) $32x^8 y^{10}$ 3) $\frac{2a^{15}}{b^{11}}$

2) $\frac{32b^{13}}{a^2}$ 4) $2x^3 y^2$

5) $16x^4y^8$

6) 1

7) $y^{16}x^5$

8) $\frac{32}{m^5n^{15}}$

9) $\frac{2}{9y}$

10) $\frac{y^5}{2x^7}$

11) $\frac{1}{y^2x^3}$

12) $\frac{y^8x^5}{4}$

13) $\frac{u}{4v^6}$

14) $\frac{x^7y^2}{2}$

15) $\frac{u^2}{12v^5}$

16) $\frac{y}{2x^4}$

17) $\frac{2}{y^7}$

18) $\frac{a^{16}}{2b}$

19) $16a^{12}b^{12}$

20) $\frac{y^8x^4}{4}$

21) $\frac{1}{8m^4n^7}$

22) $2x^{16}y^2$

23) $16n^6m^4$

24) $\frac{2x}{y^3}$

25) $\frac{1}{x^{15}y}$

26) $4y^4$

27) $\frac{u}{2v}$

28) $4y^5$

29) 8

30) $\frac{1}{2u^3v^5}$

31) $2y^5x^4$

32) $\frac{a^3}{2b^3}$

33) $\frac{1}{x^2y^{11}z}$

34) $\frac{a^2}{8c^{10}b^{12}}$

35) $\frac{1}{h^3kj^6}$

36) $\frac{x^{30}z^6}{16y^4}$

37) $\frac{2b^{14}}{a^{12}c^7}$

38) $\frac{m^{14}q^8}{4p^4}$

39) $\frac{x^2}{y^4z^4}$

40) $\frac{mn^7}{p^5}$

5.3

Answers to Scientific Notation

1) 8.85×10^2

2) 7.44×10^{4}

3) 8.1×10^{2}

4) 1.09×10^0

5) 3.9×10^{2}

6) 1.5×10^4

7) 870000

8) 256

9) 0.0009

10) 50000

11) 2

12) 0.00006

13) 1.4×10^{-3}

14) 1.76×10^{-10}

15) 1.662×10^{6}

16) 5.018×10^6

17) 1.56×10^{3}

18) 4.353×10^8

19) 1.815×10^4

20) 9.836×10^{1}

21) 5.541×10^{5}

22) 6.375×10^{-4}

23) 3.025×10^{-9}

24) 1.177×10^{-16}

25) 2.887×10^{-6}

26) 6.351×10^{-21}

27) 2.405×10^{-20}

28) 2.91×10^{-2}

29) 1.196×10^{2}

30) 1.2×10^7

31) 2.196×10^{-2}

32) 2.52×10^3

33) 1.715×10^{14}

34) 8.404×10^1

35) 1.149×10^6

36) 3.939×10^9

37) 4.6×10^2

38) 7.474×10^3

39) 3.692×10^{7}

40) 1.372×10^3

41) 1.034×10^6

42) 1.2×10^6

5.4

Answers to Introduction to Polynomials

1) 3

2) 7

3) -10

4) -6

5) -7

6) 8

7) 5

8) -1

9) 12

10) -1

11) $3p^4 - 3p$

12) $-m^3 + 12m^2$

13) $-n^3 + 10n^2$

14) $8x^3 + 8x^2$

15) $5n^4 + 5n$

16) $2v^4 + 6$

17) $13p^3$

18) $-3x$

19) $3n^3 + 8$

20) $x^4 + 9x^2 - 5$

21) $2b^4 + 2b + 10$

22) $-3r^4 + 12r^2 - 1$

23) $-5x^4 + 14x^3 - 1$

24) $5n^4 - 4n + 7$

25) $7a^4 - 3a^2 - 2a$

26) $12v^3 + 3v + 3$

27) $p^2 + 4p - 6$

28) $3m^4 - 2m + 6$

29) $5b^3 + 12b^2 + 5$

30) $-15n^4 + 4n - 6$

31) $n^3 - 5n^2 + 3$

32) $-6x^4 + 13x^3$

33) $-12n^4 + n^2 + 7$

34) $9x^2 + 10x^2$

35) $r^4 - 3r^3 + 7r^2 + 1$

36) $10x^3 - 6x^2 + 3x - 8$

37) $9n^4 + 2n^3 + 6n^2$

38) $2b^4 - b^3 + 4b^2 + 4b$

39) $-3b^4 + 13b^3 - 7b^2 - 11b + 19$

40) $12n^4 - n^3 - 6n^2 + 10$

41) $2x^4 - x^3 - 4x + 2$

42) $3x^4 + 9x^2 + 4x$

5.5

Answers to Multiply Polynomials

1) $6p - 42$

2) $32k^2 + 16k$

3) $12x + 6$

4) $18n^3 + 21n^2$

5) $20m^5 + 20m^4$

6) $12r - 21$

7) $32n^2 + 80n + 48$

8) $2x^2 - 7x - 4$

9) $56b^2 - 19b - 15$

10) $4r^2 + 40r + 64$

11) $8x^2 + 22x + 15$

12) $7n^2 + 43n - 42$

13) $15v^2 - 26v + 8$

14) $6a^2 - 44a - 32$

15) $24x^2 - 22x - 7$

16) $20x^2 - 29x + 6$

17) $30x^2 - 14xy - 4y^2$

18) $16u^2 + 10uv - 21v^2$

19) $3x^2 + 13xy + 12y^2$

20) $40u^2 - 34uv - 48v^2$

21) $56x^2 + 61xy + 15y^2$

22) $5a^2 - 7ab - 24b^2$

23) $6r^3 - 43r^2 + 12r - 35$

24) $16x^3 + 44x^2 + 44x + 40$

25) $12n^3 - 20n^2 + 38n - 20$

26) $8b^3 - 4b^2 - 4b - 12$

27) $36x^3 - 24x^2y + 3xy^2 + 12y^3$

28) $21m^3 + 4m^2n - 8n^3$

29) $48n^4 - 16n^3 + 64n^2 - 6n + 36$

30) $14a^4 + 30a^3 - 13a^2 - 12a + 3$

31) $15k^4 + 24k^3 + 48k^2 + 27k + 18$

32) $42u^4 + 76u^3v + 17u^2v^2 - 18v^4$

33) $18x^2 - 15x - 12$

34) $10x^2 - 55x + 60$

35) $24x^2 - 18x - 15$

36) $16x^2 - 44x - 12$

37) $7x^2 - 49x + 70$

38) $40x^2 - 10x - 5$

39) $96x^2 - 6$

40) $36x^2 + 108x + 81$

5.6

Answers to Multiply Special Products

1) $x^2 - 64$

2) $a^2 - 16$

3) $1 - 9p^2$

4) $x^2 - 9$

5) $1 - 49n^2$

6) $64m^2 - 25$

7) $25n^2 - 64$

8) $4r^2 - 9$

9) $16x^2 - 64$

10) $b^2 - 49$

11) $16y^2 - x^2$

12) $49a^2 - 49b^2$

13) $16m^2 - 64n^2$

14) $9y^2 - 9x^2$

15) $36x^2 - 4y^2$

16) $1 + 10n + 25n^2$

17) $a^2 + 10a + 25$

18) $v^2 + 8v + 16$

19) $x^2 - 16x + 64$

20) $1 - 12n + 36n^2$

21) $p^2 + 14p + 49$

22) $49k^2 - 98k + 49$

23) $49 - 70n + 25n^2$

24) $16x^2 - 40x + 25$

25) $25m^2 - 80m + 64$

26) $9a^2 + 18ab + 9b^2$

27) $25x^2 + 70xy + 49y^2$

28) $16m^2 - 8mn + n^2$

29) $4x^2 + 8xy + 4y^2$

30) $64x^2 + 80xy + 25y^2$

31) $25 + 20r + 4r^2$

32) $m^2 - 14m + 49$

33) $4 + 20x + 25x^2$

34) $64n^2 - 49$

35) $16v^2 - 49$

36) $b^2 - 16$

37) $n^2 - 25$

38) $49x^2 + 98x + 49$

39) $16k^2 + 16k + 4$

40) $9a^2 - 64$

5.7

Answers to Divide Polynomials

1) $5x + \frac{1}{4} + \frac{1}{2x}$

2) $\frac{5x^3}{9} + 5x^2 + \frac{4x}{9}$

3) $2n^3 + \frac{n^2}{10} + 4n$

4) $\frac{3k^2}{8} + \frac{k}{2} + \frac{1}{4}$

5) $2x^3 + 4x^2 + \frac{x}{2}$

6) $\frac{5p^3}{4} + 4p^2 + 4p$

7) $n^2 + 5n + \frac{1}{5}$

8) $\frac{m^2}{3} + 2m + 3$

9) $x - 10 + \frac{9}{x+8}$

10) $r + 6 + \frac{1}{r\ 9}$

11) $n + 8 - \frac{8}{n+5}$

12) $b - 3 - \frac{5}{b\ 7}$

13) $v + 8 - \frac{9}{v-10}$

14) $x - 3 - \frac{5}{x+7}$

15) $a + 4 - \frac{6}{a-8}$

16) $x - 6 - \frac{2}{x-4}$

17) $5p + 4 + \frac{3}{9p+4}$

18) $8k - 9 - \frac{1}{3k\ 1}$

19) $x - 3 + \frac{3}{10x\ 2}$

20) $n + 3 + \frac{3}{n+4}$

21) $r - 1 + \frac{2}{4x+3}$

22) $m + 4 + \frac{1}{m\ 1}$

23) $n + 2$

24) $x - 4 + \frac{4}{2x+3}$

25) $9b + 5 - \frac{5}{3b+8}$

26) $v + 3 - \frac{5}{3v\ 9}$

27) $x - 7 - \frac{7}{4x\ 5}$

28) $n - 7 - \frac{3}{4n+5}$

29) $a^2 + 8a - 7 - \frac{6}{a+7}$

30) $8k^2 - 2k - 4 + \frac{5}{k-8}$

31) $x^2 - 4x - 10 - \frac{1}{x+4}$

32) $x^2 - 8x + 7$

33) $3n^2 - 9n - 10 - \frac{8}{n+6}$

34) $k^2 - 3k - 9 - \frac{5}{k\ 1}$

35) $x^2 - 7x + 3 + \frac{1}{x+7}$

36) $n^2 + 9n - 1 + \frac{3}{2n+3}$

37) $p^2 + 4p - 1 + \frac{4}{9p+9}$

38) $m^2 - 8m + 7 - \frac{7}{8m+7}$

39) $r^2 + 3r - 4 - \frac{8}{r-4}$

40) $x^2 + 3x - 7 + \frac{5}{2x+6}$

41) $6n^2 - 3n - 3 + \frac{5}{2n+3}$

42) $6b^2 + b + 9 + \frac{3}{4b-7}$

43) $v^2 - 6v + 6 + \frac{1}{4v+3}$

Answers - Chapter 6

6.1

Answers - Greatest Common Factor

1) $9 + 8b^2$

2) $x - 5$

3) $5(9x^2 - 5)$

4) $1 + 2n^2$

5) $7(8 - 5p)$

6) $10(5x - 8y)$

7) $7ab(1 - 5a)$

8) $9x^2y^2(3y^3 - 8x)$

9) $3a^2b(-1 + 2ab)$

10) $4x^3(2y^2 + 1)$

11) $-5x^2(1 + x + 3x^2)$

12) $8n^5(-4n^4 + 4n + 5)$

13) $10(2x^4 - 3x + 3)$

14) $3(7p^6 + 10p^2 + 9)$

15) $4(7m^4 + 10m^3 + 2)$

16) $2x(-5x^3 + 10x + 6)$

17) $5(6b^9 + ab - 3a^2)$

18) $3y^2(9y^5 + 4x + 3)$

19) $-8a^2b(6b + 7a + 7a^3)$

20) $5(6m^6 + 3mn^2 - 5)$

21) $5x^3y^2z(4x^5z + 3x^2 + 7y)$

22) $3(p + 4q - 5q^2r^2)$

23) $10(5x^2y + y^2 + 7xz^2)$

24) $10y^4z^3(3x^5 + 5z^2 - x)$

25) $5q(6pr - p + 1)$

26) $7b(4 + 2b + 5b^2 + b^4)$

27) $3(-6n^5 + n^3 - 7n + 1)$

28) $3a^2(10a^6 + 2a^3 + 9a + 7)$

29) $10x^{11}(-4 - 2x + 5x^2 - 5x^3)$

30) $4x^2(-6x^4 - x^2 + 3x + 1)$

31) $4mn(-8n^7 + m^5 + 3n^3 + 4)$

32) $2y^7(-5 + 3y^3 - 2xy^3 - 4xy)$

6.2

Answers - Grouping

1) $(8r^2 - 5)(5r - 1)$

2) $(5x^2 - 8)(7x - 2)$

3) $(n^2 - 3)(3n - 2)$

4) $(2v^2 - 1)(7v + 5)$

5) $(3b^2 - 7)(5b + 7)$

6) $(6x^2 + 5)(x - 8)$

7) $(3x^2 + 2)(x + 5)$

8) $(7p^2 + 5)(4p + 3)$

9) $(7x^2 - 4)(5x - 4)$

10) $(7n^2 - 5)(n + 3)$

11) $(7x + 5)(y - 7)$

12) $(7r^2 + 3)(6r - 7)$

13) $(8x + 3)(4y + 5x)$

14) $(3a + b^2)(5b - 2)$

15) $(8x + 1)(2y - 7)$

16) $(m + 5)(3n - 8)$

17) $(2x + 7y^2)(y - 4x)$

18) $(m - 5)(5n + 2)$

19) $(5x - y)(8y + 7)$

20) $(8x - 1)(y + 7)$

21) $(4u + 3)(8v - 5)$

22) $2(u + 3)(2v + 7u)$

23) $(5x + 6)(2y + 5)$

24) $(4x - 5y^2)(6y - 5)$

25) $(3u - 7)(v - 2u)$

26) $(7a - 2)(8b - 7)$

27) $(2x + 1)(8y - 3x)$

6.3

Answers - Trinomials where a = 1

1) $(p + 9)(p + 8)$

2) $(x - 8)(x + 9)$

3) $(n - 8)(n - 1)$

4) $(x - 5)(x + 6)$

5) $(x + 1)(x - 10)$

6) $(x + 5)(x + 8)$

7) $(b+8)(b+4)$

8) $(b-10)(b-7)$

9) $(x-7)(x+10)$

10) $(x-3)(x+6)$

11) $(n-5)(n-3)$

12) $(a+3)(a-9)$

13) $(p+6)(p+9)$

14) $(p+10)(p-3)$

15) $(n-8)(n-7)$

16) $(m-5n)(m-10n)$

17) $(u-5v)(u-3v)$

18) $(m+5n)(m-8n)$

19) $(m+4n)(m-2n)$

20) $(x+8y)(x+2y)$

21) $(x-9y)(x-2y)$

22) $(u-7v)(u-2v)$

23) $(x-3y)(x+4y)$

24) $(x+5y)(x+9y)$

25) $(x+6y)(x-2y)$

26) $4(x+7)(x+6)$

27) $5(a+10)(a+2)$

28) $5(n-8)(n-1)$

29) $6(a-4)(a+8)$

30) $5(v-1)(v+5)$

31) $6(x+2y)(x+y)$

32) $5(m^2+6mn-18n^2)$

33) $6(x+9y)(x+7y)$

34) $6(m-9n)(m+3n)$

6.4

Answers - Trinomials where a $\neq$ 1

1) $(7x-6)(x-6)$

2) $(7n-2)(n-6)$

3) $(7b+1)(b+2)$

4) $(7v+4)(v-4)$

5) $(5a+7)(a-4)$

6) Prime

7) $(2x-1)(x-2)$

8) $(3r+2)(r-2)$

9) $(2x+5)(x+7)$

10) $(7x-6)(x+5)$

11) $(2b-3)(b+1)$

12) $(5k-6)(k-4)$

13) $(5k+3)(k+2)$

14) $(3r+7)(r+3)$

15) $(3x-5)(x-4)$

16) $(3u-2v)(u+5v)$

17) $(3x+2y)(x+5y)$

18) $(7x+5y)(x-y)$

19) $(5x-7y)(x+7y)$

20) $(5u-4v)(u+7v)$

21) $3(2x+1)(x-7)$

22) $2(5a+3)(a-6)$

23) $3(7k+6)(k-5)$

24) $3(7n-6)(n+3)$

25) $2(7x-2)(x-4)$

26) $(r+1)(4r-3)$

27) $(x+4)(6x+5)$

28) $(3p+7)(2p-1)$

29) $(k-4)(4k-1)$

30) $(r-1)(4r+7)$

31) $(x+2y)(4x+y)$

32) $2(2m^2+3mn+3n^2)$

33) $(m-3n)(4m+3n)$

34) $2(2x^2-3xy+15y^2)$

35) $(x+3y)(4x+y)$

36) $3(3u+4v)(2u-3v)$

37) $2(2x+7y)(3x+5y)$

38) $4(x+3y)(4x+3y)$

39) $4(x-2y)(6x-y)$

40) $2(3x+2y)(2x+7y)$

6.5

Answers - Factoring Special Products

1) $(r+4)(r-4)$

2) $(x+3)(x-3)$

3) $(v+5)(v-5)$

4) $(x+1)(x-1)$

5) $(p+2)(p-2)$

6) $(2v+1)(2v-1)$

7) $(3k+2)(3k-2)$

8) $(3a+1)(3a-1)$

9) $3(x+3)(x-3)$

10) $5(n+2)(n-2)$

11) $4(2x+3)(2x-3)$

12) $5(25x^2+9y^2)$

13) $2(3a+5b)(3a-5b)$

14) $4(m^2+16n^2)$

15) $(a-1)^2$

16) $(k+2)^2$

17) $(x+3)^2$

18) $(n-4)^2$

19) $(x-3)^2$

20) $(k-2)^2$

21) $(5p-1)^2$

22) $(x+1)^2$

23) $(5a+3b)^2$

24) $(x+4y)^2$

25) $(2a-5b)^2$

26) $2(3m-2n)^2$

27) $2(2x-3y)^2$

28) $5(2x+y)^2$

29) $(2-m)(4+2m+m^2)$

30) $(x+4)(x^2-4x+16)$

31) $(x-4)(x^2+4x+16)$

32) $(x+2)(x^2-2x+4)$

33) $(6-u)(36+6u+u^2)$

34) $(5x-6)(25x^2+30x+36)$

35) $(5a-4)(25a^2+20a+16)$

36) $(4x-3)(16x^2+12x+9)$

37) $(4x+3y)(16x^2-12xy+9y^2)$

38) $4(2m-3n)(4m^2+6mn+9n^2)$

39) $2(3x+5y)(9x^2-15xy+25y^2)$

40) $3(5m+6n)(25m^2-30mn+36n^2)$

41) $(a^2+9)(a+3)(a-3)$

42) $(x^2+16)(x+4)(x-4)$

43) $(4+z^2)(2+z)(2-z)$

44) $(n^2+1)(n+1)(n-1)$

45) $(x^2+y^2)(x+y)(x-y)$

46) $(4a^2+b^2)(2a+b)(2a-b)$

47) $(m^2+9b^2)(m+3b)(m-3b)$

48) $(9c^2+4d^2)(3c+2d)(3c-2d)$

6.6

Answers - Factoring Strategy

1) $3(2a+5y)(4z-3h)$

2) $(2x-5)(x-3)$

3) $(5u-4v)(u-v)$

4) $4(2x+3y)^2$

5) $2(-x+4y)(x^2+4xy+16y^2)$

6) $5(4u-x)(v-3u^2)$

7) $n(5n-3)(n+2)$

8) $x(2x+3y)(x+y)$

9) $2(3u-2)(9u^2+6u+4)$

10) $2(3-4x)(9+12x+16x^2)$

464

11) $n(n-1)$

12) $(5x+3)(x-5)$

13) $(x-3y)(x-y)$

14) $5(3u-5v)^2$

15) $(3x+5y)(3x-5y)$

16) $(x-3y)(x^2+3xy+9y^2)$

17) $(m+2n)(m-2n)$

18) $3(2a+n)(2b-3)$

19) $4(3b^2+2x)(3c-2d)$

20) $3m(m+2n)(m-4n)$

21) $2(4+3x)(16-12x+9x^2)$

22) $(4m+3n)(16m^2-12mn+9n^2)$

23) $2x(x+5y)(x-2y)$

24) $(3a+x^2)(c+5d^2)$

25) $n(n+2)(n+5)$

26) $(4m-n)(16m^2+4mn+n^2)$

27) $(3x-4)(9x^2+12x+16)$

28) $(4a+3b)(4a-3b)$

29) $x(5x+2)$

30) $2(x-2)(x-3)$

31) $3k(k-5)(k-4)$

32) $2(4x+3y)(4x-3y)$

33) $(m-4x)(n+3)$

34) $(2k+5)(k-2)$

35) $(4x-y)^2$

36) $v(v+1)$

37) $3(3m+4n)(3m-4n)$

38) $x^2(x+4)$

39) $3x(3x-5y)(x+4y)$

40) $3n^2(3n-1)$

41) $2(m-2n)(m+5n)$

42) $v^2(2u-5v)(u-3v)$

6.7

Answers - Solve by Factoring

1) $7, -2$

2) $-4, 3$

3) $1, -4$

4) $-\frac{5}{2}, 7$

5) $-5, 5$

6) $4, -8$

7) $2, -7$

8) $-5, 6$

9) $-\frac{5}{7}, -3$

10) $-\frac{7}{8}, 8$

11) $-\frac{1}{5}, 2$

12) $-\frac{1}{2}, 2$

13) $4, 0$

14) $8, 0$

15) $1, 4$

16) $4, 2$

17) $\frac{3}{7}, -8$

18) $-\frac{1}{7}, -8$

19) $\frac{4}{7}, -3$

20) $\frac{1}{4}, 3$

21) $-4, -3$

22) $8, -4$

23) $8, -2$

24) $4, 0$

25) $\frac{8}{3}, -5$

26) $-\frac{1}{2}, \frac{5}{3}$

27) $-\frac{3}{7}, -3$

28) $-\frac{4}{3}, -3$

29) $-4, 1$

30) $2, -3$

31) $-7, 7$

32) $-4, -6$

33) $-\frac{5}{2}, -8$

34) $-\frac{6}{5}, -7$

35) $\frac{4}{5}, -6$ 36) $\frac{5}{3}, -2$

Answers - Chapter 7

7.1

Answers - Reduce Rational Expressions

1) 3

2) $\frac{1}{3}$

3) $-\frac{1}{5}$

4) undefined

5) $\frac{1}{2}$

6) 6

7) -10

8) $0, 2$

9) $-\frac{5}{2}$

10) $0, -10$

11) 0

12) $-\frac{10}{3}$

13) -2

14) $0, -\frac{1}{2}$

15) $-8, 4$

16) $0, \frac{1}{7}$

17) $\frac{7x}{6}$

18) $\frac{3}{n}$

19) $\frac{3}{5a}$

20) $\frac{7}{8k}$

21) $\frac{4}{x}$

22) $\frac{9x}{2}$

23) $\frac{3m-4}{10}$

24) $\frac{10}{9n^2(9n+4)}$

25) $\frac{10}{2p+1}$

26) $\frac{1}{9}$

27) $\frac{1}{x+7}$

28) $\frac{7m+3}{9}$

29) $\frac{8x}{7(x+1)}$

30) $\frac{7r+8}{8r}$

31) $\frac{n+6}{n-5}$

32) $\frac{b+6}{b+7}$

33) $\frac{9}{v-10}$

34) $\frac{3(x-3)}{5x+4}$

35) $\frac{2x-7}{5x-7}$

36) $\frac{k-8}{k+4}$

37) $\frac{3a-5}{5a+2}$

38) $\frac{9}{p+2}$

39) $\frac{2n-1}{9}$

40) $\frac{3x-5}{5(x+2)}$

41) $\frac{2(m+2)}{5m-3}$

42) $\frac{9r}{5(r+1)}$

43) $\frac{2(x-4)}{3x-4}$

44) $\frac{5b-8}{5b+2}$

45) $\frac{7n-4}{4}$

46) $\frac{5(v+1)}{3v+1}$

47) $\frac{(n-1)^2}{6(n+1)}$

48) $\frac{7x-6}{(3x+4)(x+1)}$

49) $\frac{7a+9}{2(3a-2)}$

50) $\frac{2(2k+1)}{9(k-1)}$

7.2

Answers - Multipy and Divide

1) $4x^2$

2) $\frac{14}{3}$

3) $\frac{63}{10n}$

4) $\frac{63}{10m}$

466

5) $\frac{3x^2}{2}$

6) $\frac{5p}{2}$

7) $5m$

8) $\frac{7}{10}$

9) $\frac{r\ 6}{r+10}$

10) $x+4$

11) $\frac{2}{3}$

12) $\frac{9}{b\ 5}$

13) $\frac{x\ 10}{7}$

14) $\frac{1}{v-10}$

15) $x+1$

16) $\frac{a+10}{a-6}$

17) 5

18) $\frac{p-10}{p-4}$

19) $\frac{3}{5}$

20) $\frac{x+10}{x+4}$

21) $\frac{4(m-5)}{5m^2}$

22) 7

23) $\frac{x+3}{4}$

24) $\frac{n-9}{n+7}$

25) $\frac{b+2}{8b}$

26) $\frac{v-9}{5}$

27) $-\frac{1}{n\ 6}$

28) $\frac{x+1}{x-3}$

29) $\frac{1}{a+7}$

30) $\frac{7}{8(k+3)}$

31) $\frac{x\ 4}{x+3}$

32) $\frac{9(x+6)}{10}$

33) $9m^2(m+10)$

34) $\frac{10}{9(n+6)}$

35) $\frac{p+3}{6(p+8)}$

36) $\frac{x\ 8}{x+7}$

37) $\frac{5b}{b+5}$

38) $n+3$

39) $r-8$

40) $\frac{18}{5}$

41) $\frac{3}{2}$

42) $\frac{1}{a+2b}$

43) $\frac{1}{x+2}$

44) $\frac{3(x\ 2)}{4(x+2)}$

7.3

Answers - Least Common Denominators

1) 18

2) a^2

3) ay

4) $20xy$

5) $6a^2c^3$

6) 12

7) $2x-8$

8) x^2-2x-3

9) x^2-x-12

10) $x^2-11x+30$

11) $12a^4b^5$

12) $25x^3y^5z$

13) $x\,(x-3)$

14) $4(x-2)$

15) $(x+2)(x-4)$

16) $x(x-7)(x+1)$

17) $(x+5)(x-5)$

18) $(x-3)^2(x+3)$

19) $(x+1)(x+2)(x+3)$

20) $(x-2)(x-5)(x+3)$

21) $\dfrac{6a^4}{10a^3b^2}, \dfrac{2b}{10a^3b^2}$

22) $\dfrac{3x^2+6x}{(x-4)(x+2)}, \dfrac{2x-8}{(x-4)(x+2)}$

23) $\dfrac{x^2+4x+4}{(x-3)(x+2)}, \dfrac{x^2-6x+9}{(x-3)(x+2)}$

24) $\dfrac{5}{x(x-6)}, \dfrac{2x-12}{x(x-6)}, \dfrac{-3x}{x(x-6)}$

25) $\dfrac{x^2-4x}{(x-4)^2(x+4)}, \dfrac{3x^2+12x}{(x-4)^2(x+4)}$

26) $\dfrac{5x+1}{(x-5)(x+2)}, \dfrac{4x+8}{(x-5)(x+2)}$

27) $\dfrac{x^2+7x+6}{(x-6)(x+6)^2}, \dfrac{2x^2-9x-18}{(x-6)(x+6)^2}$

28) $\dfrac{3x^2+4x+1}{(x-4)(x+3)(x+1)}, \dfrac{2x^2-8x}{(x-4)(x+3)(x+1)}$

29) $\dfrac{4x}{(x-3)(x+2)}, \dfrac{x^2+4x+4}{(x-3)(x+2)}$

30) $\dfrac{3x^2+15x}{(x-4)(x-2)(x+5)}, \dfrac{x^2-4x+4}{(x-4)(x-2)(x+5)}, \dfrac{5x-20}{(x-4)(x-2)(x+5)}$

7.4

Answers - Add and Subtract

1) $\dfrac{6}{a+3}$

2) $x-4$

3) $t+7$

4) $\dfrac{a+4}{a+6}$

5) $\dfrac{x+6}{x-5}$

6) $\dfrac{3x+4}{x^2}$

7) $\dfrac{5}{24r}$

8) $\dfrac{7x+3y}{x^2y^2}$

9) $\dfrac{15t+16}{18t^3}$

10) $\dfrac{5x+9}{24}$

11) $\dfrac{a+8}{4}$

12) $\dfrac{5a^2+7a-3}{9a^2}$

13) $\dfrac{-7x-13}{4x}$

14) $\dfrac{c^2+cd-d^2}{c^2d^2}$

15) $\dfrac{3y^2-3xy-6x^2}{2x^2y^2}$

16) $\dfrac{4x}{x^2-1}$

17) $\dfrac{-z^2+5z}{z^2-1}$

18) $\dfrac{11x+15}{4x(x+5)}$

19) $\dfrac{14-3x}{x^2-4}$

20) $\dfrac{x^2-x}{x^2-25}$

21) $\dfrac{4t-5}{4(t-3)}$

22) $\dfrac{2x+10}{(x+3)^2}$

23) $\dfrac{6-20x}{15x(x+1)}$

24) $\dfrac{9a}{4(a-5)}$

25) $\dfrac{t^2+2ty-y^2}{y^2-t^2}$

26) $\dfrac{2x^2-10x+25}{x(x-5)}$

27) $\dfrac{x-3}{(x+3)(x+1)}$

28) $\dfrac{2x+3}{(x-1)(x+4)}$

29) $\dfrac{x-8}{(x+8)(x+6)}$

30) $\dfrac{2x-5}{(x-3)(x-2)}$

31) $\dfrac{5x+12}{x^2+5x+6}$

32) $\dfrac{4x+1}{(x+1)(x-2)}$

33) $\dfrac{2x+4}{x^2+4x+3}$

34) $\dfrac{2x+7}{x^2+5x+6}$

35) $\dfrac{2x-8}{x^2-5x-14}$

36) $\dfrac{3x^2+7x+4}{3(x+2)(2-x)}$

37) $\dfrac{a-2}{a^2-9}$

38) $\dfrac{2}{y^2-y}$

39) $\dfrac{z-3}{2z-1}$

40) $\dfrac{2}{r+s}$

41) $\dfrac{5(x-1)}{(x+1)(x+3)}$

42) $\dfrac{5x+5}{x^2+2x-15}$

43) $\dfrac{-(x-29)}{(x-3)(x+5)}$

44) $\dfrac{5x-10}{x^2+5x+4}$

7.5

Answers - Complex Fractions

1) $\dfrac{x}{x-1}$

2) $\dfrac{1-y}{y}$

3) $\dfrac{a}{a+2}$

4) $\dfrac{5-a}{a}$

5) $-\dfrac{a-1}{a+1}$

6) $\dfrac{b^3+2b-b-2}{8b}$

7) $\dfrac{2}{5}$

8) $\dfrac{4}{5}$

9) $-\dfrac{1}{2}$

10) $-\dfrac{1}{2}$

11) $\dfrac{x^2-x-1}{x^2+x+1}$

12) $\dfrac{2a^2-3a+3}{4a^2-2a}$

13) $\dfrac{x}{3}$

14) $3x+2$

15) $\dfrac{4b(a-b)}{a}$

16) $\dfrac{x+2}{x-1}$

17) $\dfrac{x-5}{x+9}$

18) $-\dfrac{(x-3)(x+5)}{4x^2-5x+4}$

19) $\dfrac{1}{3x+8}$

20) $\dfrac{1}{x+4}$

21) $\dfrac{x-2}{x+2}$

22) $\dfrac{x-7}{x+5}$

23) $\dfrac{x-3}{x+4}$

24) $\dfrac{2a-2}{3a-4}$

25) $-\dfrac{b-2}{2b+3}$

26) $\dfrac{x+y}{x-y}$

27) $\dfrac{a-3b}{a+3b}$

28) $-\dfrac{2x}{x^2+1}$

29) $-\dfrac{2}{y}$

30) x^2-1

31) $\dfrac{y-x}{xy}$

32) $\dfrac{x^2-xy+y^2}{y-x}$

469

33) $\dfrac{x^2 + y^2}{xy}$

35) $\dfrac{(1 - 3x)^2}{x^2(x+3)(x-3)}$

34) $\dfrac{2x}{2x+1}$ $\dfrac{1}{}$

36) $\dfrac{x+y}{xy}$

7.6

Answers - Proportions

1) $\dfrac{40}{3} = a$

2) $n = \dfrac{14}{3}$

3) $k = \dfrac{12}{7}$

4) $x = 16$

5) $x = \dfrac{3}{2}$

6) $n = 34$

7) $m = 21$

8) $x = \dfrac{79}{8}$

9) $p = 49$

10) $n = 25$

11) $b = -\dfrac{40}{3}$

12) $r = \dfrac{36}{5}$

13) $x = \dfrac{5}{2}$

14) $n = \dfrac{32}{5}$

15) $a = \dfrac{6}{7}$

16) $v = -\dfrac{16}{7}$

17) $v = \dfrac{69}{5}$

18) $n = \dfrac{61}{3}$

19) $x = \dfrac{38}{3}$

20) $k = \dfrac{73}{3}$

21) $x = -8, 5$

22) $x = -7, 5$

23) $m = -7, 8$

24) $x = -3, 9$

25) $p = -7, -2$

26) $n = -6, 9$

27) $n = -1$

28) $n = -4, -1$

29) $x = -7, 1$

30) $x = -1, 3$

31) \$9.31

32) 16

33) 2.5 in

34) 12.1 ft

35) 39.4 ft

36) 3.1 in

37) T: 38, V: 57

38) J: 4 hr, S: 14 hr

39) \$8

40) C: 36 min,
    K: 51 min

7.7

Answers - Solving Rational Equations

1) $-\dfrac{1}{2}, \dfrac{2}{3}$

2) $-3, 1$

3) 3

4) $-1, 4$

5) 2

6) $\dfrac{1}{3}$

7) -1

8) $-\dfrac{1}{3}$

9) -5

10) $-\dfrac{7}{15}$

11) $-5, 0$

12) $5, 10$

13) $\dfrac{16}{3}, 5$

14) $2, 13$

15) -8

16) 2

17) $-\dfrac{1}{5}, 5$

18) $-\dfrac{9}{5}, 1$

19) $\dfrac{3}{2}$

20) 10

21) 0, 5

22) $-2, \frac{5}{3}$

23) 4, 7

24) -1

25) $\frac{2}{3}$

26) $\frac{1}{2}$

27) $\frac{3}{10}$

28) 1

29) $-\frac{2}{3}$

30) -1

31) $\frac{13}{4}$

32) 1

33) -10

34) $\frac{7}{4}$

7.8

Answers - Dimensional Analysis

1) 12320 yd

2) 0.0073125 T

3) 0.0112 g

4) 135,000 cm

5) 6.1 mi

6) 0.5 yd^2

7) 0.435 km^2

8) 86,067,200 ft^2

9) 6,500,000 m^3

10) 239.58 cm^3

11) 0.0072 yd^3

12) 5.13 ft/sec

13) 6.31 mph

14) 104.32 mi/hr

15) 111 m/s

16) 2,623,269,600 km/yr

17) 11.6 lb/in^2

18) 63,219.51 km/hr^2

19) 32.5 mph; 447 yd/oz

20) 6.608 mi/hr

21) 17280 pages/day; 103.4 reams/month

22) 2,365,200,000 beats/lifetime

23) 1.28 g/L

24) $3040

25) 56 mph; 25 m/s

26) 148.15 yd^3; 113 m^3

27) 3630 ft^2, 522,720 in^2

28) 350,000 pages

29) 15,603,840,000 ft^3/week

30) 621,200 mg; 1.42 lb

Answers - Chapter 8

8.1

Answers - Square Roots

1) $7\sqrt{5}$

2) $5\sqrt{5}$

3) 6

4) 14

5) $2\sqrt{3}$

6) $6\sqrt{2}$

7) $6\sqrt{3}$

8) $20\sqrt{2}$

9) $48\sqrt{2}$

10) $56\sqrt{2}$

11) $-112\sqrt{2}$

12) $-21\sqrt{7}$

13) $8\sqrt{3n}$

14) $7\sqrt{7b}$

15) $14v$

16) $10n\sqrt{n}$

17) $6x\sqrt{7}$

8.2

18) $10a\sqrt{2a}$

19) $-10k^2$

20) $-20p^2\sqrt{7}$

21) $-56x^2$

22) $-16\sqrt{2n}$

23) $-30\sqrt{m}$

24) $32p\sqrt{7}$

25) $3xy\sqrt{5}$

26) $6b^2a\sqrt{2a}$

27) $4xy\sqrt{xy}$

28) $16a^2b\sqrt{2}$

29) $8x^2y^2\sqrt{5}$

30) $16m^2n\sqrt{2n}$

31) $24y\sqrt{5x}$

32) $56\sqrt{2mn}$

33) $35xy\sqrt{5y}$

34) $12xy$

35) $-12u\sqrt{5uv}$

36) $-30y^2x\sqrt{2x}$

37) $-48x^2z^2y\sqrt{5}$

38) $30a^2c\sqrt{2b}$

39) $8j^2\sqrt{5hk}$

40) $-4yz\sqrt{2xz}$

41) $-12p\sqrt{6mn}$

42) $-32p^2m\sqrt{2q}$

Answers - Higher Roots

1) $5\sqrt[3]{5}$

2) $5\sqrt[3]{3}$

3) $5\sqrt[3]{6}$

4) $5\sqrt[3]{2}$

5) $5\sqrt[3]{7}$

6) $2\sqrt[3]{3}$

7) $-8\sqrt[4]{6}$

8) $-16\sqrt[4]{3}$

9) $12\sqrt[4]{7}$

10) $6\sqrt[4]{3}$

11) $-2\sqrt[4]{7}$

12) $15\sqrt[4]{3}$

13) $3\sqrt[4]{8a^2}$

14) $2\sqrt[4]{4n^3}$

15) $2\sqrt[5]{7n^3}$

16) $-2\sqrt[5]{3x^3}$

17) $2p\sqrt[5]{7}$

18) $2x\sqrt[6]{4}$

19) $-6\sqrt[7]{7r}$

20) $-16b\sqrt[7]{3b}$

21) $4v^2\sqrt[3]{6v}$

22) $20a^2\sqrt[3]{2}$

23) $-28n^2\sqrt[3]{5}$

24) $-8n^2$

25) $-3xy\sqrt[3]{5x^2}$

26) $4uv\sqrt[3]{u^2}$

27) $-2xy\sqrt[3]{4xy}$

28) $10ab\sqrt[3]{ab^2}$

29) $4xy^2\sqrt[3]{4x}$

30) $3xy^2\sqrt[3]{7}$

31) $-21xy^2\sqrt[3]{3y}$

32) $-8y^2\sqrt[3]{7x^2y^2}$

33) $10v^2\sqrt[3]{3u^2v^2}$

34) $-40\sqrt[3]{6xy}$

35) $-12\sqrt[3]{3ab^2}$

36) $9y\sqrt[3]{5x}$

37) $-18m^2np^2\sqrt[3]{2m^2p}$

38) $-12mpq\sqrt[4]{5p^3}$

39) $18xy^4\sqrt[4]{8xy^3z^2}$

40) $-18ab^2\sqrt[4]{5ac}$

41) $14hj^2k^2\sqrt[4]{8h^2}$

472

42) $-18xz\sqrt[4]{4x^3yz^3}$

8.3

Answers - Adding Radicals

1) $6\sqrt{5}$

2) $-3\sqrt{6}-5\sqrt{3}$

3) $-3\sqrt{2}+6\sqrt{5}$

4) $-5\sqrt{6}-\sqrt{3}$

5) $-5\sqrt{6}$

6) $-3\sqrt{3}$

7) $3\sqrt{6}+5\sqrt{5}$

8) $-\sqrt{5}+\sqrt{3}$

9) $-8\sqrt{2}$

10) $-6\sqrt{6}+9\sqrt{3}$

11) $-3\sqrt{6}+\sqrt{3}$

12) $-2\sqrt{5}-6\sqrt{6}$

13) $-2\sqrt{2}$

14) $8\sqrt{5}-\sqrt{3}$

15) $5\sqrt{2}$

16) $-9\sqrt{3}$

17) $-3\sqrt{6}-\sqrt{3}$

18) $3\sqrt{2}+3\sqrt{6}$

19) $-12\sqrt{2}+2\sqrt{5}$

20) $-3\sqrt{2}$

21) $-4\sqrt{6}+4\sqrt{5}$

22) $-\sqrt{5}-3\sqrt{6}$

23) $8\sqrt{6}-9\sqrt{3}+4\sqrt{2}$

24) $-\sqrt{6}-10\sqrt{3}$

25) $2\sqrt[3]{2}$

26) $6\sqrt[3]{5}-3\sqrt[3]{3}$

27) $-\sqrt[4]{3}$

28) $10\sqrt[4]{4}$

29) $\sqrt[4]{2}-3\sqrt[4]{3}$

30) $5\sqrt[4]{6}+2\sqrt[4]{4}$

31) $6\sqrt[4]{3}-3\sqrt[4]{4}$

32) $-6\sqrt[4]{3}+2\sqrt[4]{6}$

33) $2\sqrt[4]{2}+\sqrt[4]{3}+6\sqrt[4]{4}$

34) $-2\sqrt[4]{3}-9\sqrt[4]{5}-3\sqrt[4]{2}$

35) $\sqrt[5]{6}-6\sqrt[5]{2}$

36) $\sqrt[7]{3}-6\sqrt[7]{6}+3\sqrt[7]{5}$

37) $4\sqrt[5]{5}-4\sqrt[5]{6}$

38) $-11\sqrt[7]{2}-2\sqrt[7]{5}$

39) $-4\sqrt[6]{4}-6\sqrt[6]{5}-4\sqrt[6]{2}$

8.4

Answers - Multiply and Divide Radicals

1) $-48\sqrt{5}$

2) $-25\sqrt{6}$

3) $6m\sqrt{5}$

4) $-25r^2\sqrt{2r}$

5) $2x^2\sqrt[3]{x}$

6) $6a^2\sqrt[3]{5a}$

7) $2\sqrt{3}+2\sqrt{6}$

8) $5\sqrt{2}+2\sqrt{5}$

9) $-45\sqrt{5} - 10\sqrt{15}$

10) $45\sqrt{5} + 10\sqrt{15}$

11) $25n\sqrt{10} + 10\sqrt{5}$

12) $5\sqrt{3} - 9\sqrt{5v}$

13) $-2 - 4\sqrt{2}$

14) $16 - 9\sqrt{3}$

15) $15 - 11\sqrt{5}$

16) $30 + 8\sqrt{3} + 5\sqrt{15} + 4\sqrt{5}$

17) $6a + a\sqrt{10} + 6a\sqrt{6} + 2a\sqrt{15}$

18) $-4p\sqrt{10} + 50\sqrt{p}$

19) $63 + 32\sqrt{3}$

20) $-10\sqrt{m} + 25\sqrt{2} + \sqrt{2m} - 5$

21) $\frac{\sqrt{3}}{25}$

22) $\frac{\sqrt{15}}{4}$

23) $\frac{1}{20}$

24) 2

8.5

25) $\frac{\sqrt{15}}{3}$

26) $\frac{\sqrt{10}}{15}$

27) $\frac{4\sqrt{3}}{9}$

28) $\frac{4\sqrt{5}}{5}$

29) $\frac{5\sqrt{3xy}}{12y^2}$

30) $\frac{4\sqrt{3x}}{16xy^2}$

31) $\frac{\sqrt{6p}}{3}$

32) $\frac{2\sqrt{5n}}{5}$

33) $\frac{\sqrt[3]{10}}{5}$

34) $\frac{\sqrt[3]{15}}{4}$

35) $\frac{\sqrt[3]{10}}{8}$

36) $\frac{\sqrt[4]{8}}{8}$

37) $\frac{5\sqrt[4]{10r^2}}{2}$

38) $\frac{\sqrt[4]{4n^2}}{mn}$

Answers - Rationalize Denominators

1) $\frac{4 + 2\sqrt{3}}{3}$

2) $\frac{4 + \sqrt{3}}{12}$

3) $\frac{2 + \sqrt{3}}{5}$

4) $\frac{\sqrt{3}\ 1}{4}$

5) $\frac{2\sqrt{13} - 5\sqrt{65}}{52}$

6) $\frac{\sqrt{85} + 4\sqrt{17}}{68}$

7) $\frac{\sqrt{6}\ 9}{3}$

8) $\frac{\sqrt{30} - 2\sqrt{3}}{18}$

9) $\frac{15\sqrt{5}\ 5\sqrt{2}}{43}$

10) $\frac{5\sqrt{3} + 20\sqrt{5}}{77}$

11) $\frac{10 - 2\sqrt{2}}{23}$

12) $\frac{2\sqrt{3} + \sqrt{2}}{2}$

13) $\frac{12\ 9\sqrt{3}}{11}$

14) $-2\sqrt{2} - 4$

15) $3 - \sqrt{5}$

16) $\frac{\sqrt{5}\ \sqrt{3}}{2}$

17) $1 + \sqrt{2}$

474

18) $\frac{16\sqrt{3}+4\sqrt{5}}{43}$

19) $\sqrt{2}-1$

20) $3+2\sqrt{3}$

21) $\sqrt{2}$

22) $\sqrt{2}$

23) $\sqrt{a}$

24) $3-2\sqrt{2}$

25) $\sqrt{a}$

26) $\frac{1}{3}$

27) $4-2\sqrt{3}+2\sqrt{6}-3\sqrt{2}$

28) $\frac{2\sqrt{5}\;2\sqrt{15}+\sqrt{3}+3}{2}$

29) $\frac{a^2\;2a\sqrt{b}+b}{a^2\;b}$

30) $\sqrt{a}-\sqrt{b}$

31) $3\sqrt{2}+2\sqrt{3}$

32) $\frac{a\sqrt{b}+b\sqrt{a}}{a-b}$

33) $\frac{a\sqrt{b}+b\sqrt{a}}{ab}$

34) $\frac{24\;4\sqrt{6}+9\sqrt{2}\;3\sqrt{3}}{15}$

35) $\frac{-1+\sqrt{5}}{4}$

36) $\frac{2\sqrt{5}-5\sqrt{2}-10+5\sqrt{10}}{30}$

37) $\frac{5\sqrt{2}+10\;\sqrt{3}+\sqrt{6}}{5}$

38) $\frac{8+3\sqrt{6}}{10}$

8.6

Answers - Rational Exponents

1) $(\sqrt[5]{m})^3$

2) $\frac{1}{(\sqrt[4]{10r})^3}$

3) $(\sqrt{7x})^3$

4) $\frac{1}{(\sqrt[3]{6b})^4}$

5) $(6x)^{\frac{3}{2}}$

6) $v^{\frac{1}{2}}$

7) $n^{-\frac{7}{4}}$

8) $(5a)^{\frac{1}{2}}$

9) 4

10) 2

11) 8

12) $\frac{1}{1000}$

13) $x^{\frac{4}{3}}y^{\frac{5}{2}}$

14) $\frac{4}{v^{\frac{1}{3}}}$

15) $\frac{1}{a^{\frac{1}{2}}b^{\frac{1}{2}}}$

16) 1

17) $\frac{1}{3a^2}$

18) $\frac{y^{\frac{25}{12}}}{x^{\frac{5}{6}}}$

19) $u^2 v^{\frac{11}{2}}$

20) 1

21) $y^{\frac{1}{2}}$

22) $\frac{v^2}{u^{\frac{7}{2}}}$

23) $\frac{b^{\frac{7}{4}}a^{\frac{3}{4}}}{3}$

24) $\frac{2y^{\frac{17}{6}}}{x^{\frac{7}{4}}}$

25) $\frac{3y^{\frac{1}{12}}}{2}$

26) $\frac{a^{\frac{3}{2}}}{2b^{\frac{1}{4}}}$

27) $\frac{m^{\frac{35}{8}}}{n^{\frac{7}{6}}}$

28) $\frac{1}{y^{\frac{5}{4}}x^{\frac{3}{2}}}$

29) $\frac{1}{n^{\frac{3}{4}}}$

30) $\frac{y^{\frac{1}{3}}}{x^{\frac{1}{4}}}$

31) $x\,y^{\frac{4}{3}}$

32) $\frac{x^{\frac{4}{3}}}{y^{\frac{10}{3}}}$

33) $\frac{u^{\frac{1}{2}}}{v^{\frac{3}{4}}}$

34) $x^{\frac{15}{4}} y^{\frac{17}{4}}$

8.7

Answers - Radicals of Mixed Index

1) $\sqrt[4]{4x^2y^3}$

2) $\sqrt{3xy^3}$

3) $\sqrt[6]{8x^2y^3z^4}$

4) $\dfrac{\sqrt{5}}{2x}$

5) $\dfrac{\sqrt[3]{36xy}}{3y}$

6) $\sqrt[5]{x^3y^4z^2}$

7) $\sqrt[4]{x^2y^3}$

8) $\sqrt[5]{8x^4y^2}$

9) $\sqrt[4]{x^3y^2z}$

10) $\sqrt{5y}$

11) $\sqrt[3]{2xy^2}$

12) $\sqrt[4]{3x^2y^3}$

13) $\sqrt[6]{5400}$

14) $\sqrt[12]{300125}$

15) $\sqrt[6]{49x^3y^2}$

16) $\sqrt[15]{27y^5z^5}$

17) $\sqrt[6]{x^3(x-2)^2}$

18) $\sqrt[4]{3x(y+4)^2}$

19) $\sqrt[10]{x^9y^7}$

20) $\sqrt[10]{4a^9b^9}$

21) $\sqrt[12]{x^{11}y^{10}}$

22) $\sqrt[20]{a^{18}b^{17}}$

23) $\sqrt[20]{a^{18}b^{17}c^{14}}$

24) $\sqrt[30]{x^{22}y^{11}z^{27}}$

25) $a\sqrt[4]{a}$

26) $x\sqrt{x}$

27) $b\sqrt[10]{b^9}$

28) $a\sqrt[12]{a^5}$

29) $xy\sqrt[6]{xy^5}$

30) $a\sqrt[10]{ab^7}$

31) $3a^2b\sqrt[4]{ab}$

32) $2xy^2\sqrt[6]{2x^5y}$

33) $x\sqrt[12]{59049xy^{11}z^{10}}$

34) $a^2b^2c^2\sqrt[6]{a^2bc^2}$

35) $9a^2(b+1)\sqrt[6]{243a^5(b+1)^5}$

36) $4x(y+z)^3\sqrt[6]{2x(y+z)}$

37) $\sqrt[12]{a^5}$

38) $\sqrt[15]{x^7}$

39) $\sqrt[12]{x^2y^5}$

40) $\dfrac{\sqrt[15]{a^7b^{11}}}{b}$

41) $\sqrt[10]{ab^9c^7}$

42) $yz\sqrt[10]{xy^8z^3}$

43) $\sqrt[20]{(3x-1)^3}$

44) $\sqrt[12]{(2+5x)^5}$

45) $\sqrt[15]{(2x+1)^4}$

46) $\sqrt[12]{(5-3x)}$

476

Answers - Complex Numbers

1) $11 - 4i$

2) $-4i$

3) $-3 + 9i$

4) $-1 - 6i$

5) $-3 - 13i$

6) $5 - 12i$

7) $-4 - 11i$

8) $-3 - 6i$

9) $-8 - 2i$

10) $13 - 8i$

11) 48

12) 24

13) 40

14) 32

15) -49

16) $28 - 21i$

17) $11 + 60i$

18) $-32 - 128i$

19) $80 - 10i$

20) $36 - 36i$

21) $27 + 38i$

22) $-28 + 76i$

23) $44 + 8i$

24) $16 - 18i$

25) $-3 + 11i$

26) $-1 + 13i$

27) $9i + 5$

28) $\frac{3i \quad 2}{3}$

29) $\frac{10i - 9}{6}$

30) $\frac{4i + 2}{3}$

31) $\frac{3i - 6}{4}$

32) $\frac{5i + 9}{9}$

33) $10i + 1$

34) $-2i$

35) $\frac{40i + 4}{101}$

36) $\frac{9i \quad 45}{26}$

37) $\frac{56 + 48i}{85}$

38) $\frac{4 - 6i}{13}$

39) $\frac{70 + 49i}{149}$

40) $\frac{36 + 27i}{50}$

41) $\frac{30i \quad 5}{37}$

42) $\frac{48i - 56}{85}$

43) $9i$

44) $3i\sqrt{5}$

45) $-2\sqrt{5}$

46) $-2\sqrt{6}$

47) $\frac{1 + i\sqrt{3}}{2}$

48) $\frac{2 + i\sqrt{2}}{2}$

49) $2 - i$

50) $\frac{3 + 2i\sqrt{2}}{2}$

51) i

52) $-i$

53) 1

54) 1

55) -1

56) i

57) -1

58) $-i$

Answers - Chapter 9

Answers - Solving with Radicals

1) 3

2) 3

3) $1, 5$

4) no solution

5) ± 2

6) 3

7) $\frac{1}{4}$

8) no solution

9) 5

10) 7

11) 6

12) 46

13) 5

14) 21

15) $-\frac{3}{2}$ 16) $-\frac{7}{3}$

9.2

Answers - Solving with Exponents

1) $\pm 5\sqrt{3}$

2) -2

3) $\pm 2\sqrt{2}$

4) 3

5) $\pm 2\sqrt{6}$

6) $-3, 11$

7) -5

8) $\frac{1}{5}, -\frac{3}{5}$

9) -1

10) $\frac{1 \pm 3\sqrt{2}}{2}$

11) $65, -63$

12) 5

13) -7

14) $-\frac{11}{2}, \frac{5}{2}$

15) $\frac{11}{2}, -\frac{5}{2}$

16) $-\frac{191}{64}$

17) $-\frac{3}{8}, -\frac{5}{8}$

18) $\frac{9}{8}$

19) $\frac{5}{4}$

20) No Solution

21) $-\frac{34}{3}, -10$

22) 3

23) $-\frac{17}{2}$

24) No Solutoin

9.3

Answers - Complete the Square

1) $225; (x-15)^2$

2) $144; (a-12)^2$

3) $324; (m-18)^2$

4) $289; (x-17)^2$

5) $\frac{225}{4}; (x-\frac{15}{2})^2$

6) $\frac{1}{324}; (r-\frac{1}{18})^2$

7) $\frac{1}{4}; (y-\frac{1}{2})^2$

8) $\frac{289}{4}; (p-\frac{17}{2})^2$

9) $11, 5$

10) $4+2\sqrt{7}, 4-2\sqrt{7}$

11) $4+i\sqrt{29}, 4-i\sqrt{29}$

12) $-1+i\sqrt{42}, -1-i\sqrt{42}$

13) $\frac{-2+i\sqrt{38}}{2}, \frac{-2-i\sqrt{38}}{2}$

14) $\frac{3+2i\sqrt{33}}{3}, \frac{3-2i\sqrt{33}}{3}$

15) $\frac{5+i\sqrt{215}}{5}, \frac{5-i\sqrt{215}}{5}$

16) $\frac{-4+3\sqrt{2}}{4}, \frac{-4-3\sqrt{2}}{4}$

17) $-5+\sqrt{86}, -5-\sqrt{86}$

18) $8+2\sqrt{29}, 8-2\sqrt{29}$

19) $9, 7$

20) $9, -1$

21) $-1+i\sqrt{21}, -1-i\sqrt{21}$

22) $1, -3$

23) $\frac{3}{2}, -\frac{7}{2}$

24) $3, -1$

25) $-5+2i, -5-2i$

26) $7+\sqrt{85}, 7-\sqrt{85}$

27) $7, 3$

28) $4, -14$

29) $1+i\sqrt{2}, 1-i\sqrt{2}$

30) $\frac{5+i\sqrt{105}}{5}, \frac{5-i\sqrt{105}}{5}$

31) $\frac{4+i\sqrt{110}}{2}, \frac{4\ i\sqrt{110}}{2}$

32) $1, -3$

33) $4 + i\sqrt{39}, 4 - i\sqrt{39}$

34) $-1. -7$

35) $7, 1$

36) $2, -6$

37) $\frac{6 + i\sqrt{258}}{6}, \frac{6 \ i\sqrt{258}}{6}$

38) $\frac{-6 + i\sqrt{111}}{3}, \frac{-6 - i\sqrt{111}}{3}$

39) $\frac{5 + i\sqrt{130}}{5}, \frac{5 - i\sqrt{130}}{5}$

40) $2, -4$

41) $\frac{5 + i\sqrt{87}}{2}, \frac{5 \ i\sqrt{87}}{2}$

42) $\frac{7 + \sqrt{181}}{2}, \frac{7 \ \sqrt{181}}{2}$

43) $\frac{3 + i\sqrt{271}}{7}, \frac{3 - i\sqrt{271}}{7}$

44) $\frac{-1 + 2i\sqrt{6}}{2}, \frac{-1 - 2i\sqrt{6}}{2}$

45) $\frac{7 + i\sqrt{139}}{2}, \frac{7 \ i\sqrt{139}}{2}$

46) $\frac{5 + 3i\sqrt{7}}{2}, \frac{5 - 3i\sqrt{7}}{2}$

47) $\frac{12}{5}, -4$

48) $\frac{1 + i\sqrt{511}}{4}, \frac{1 - i\sqrt{511}}{4}$

49) $\frac{9 + \sqrt{21}}{2}, \frac{9 - \sqrt{21}}{2}$

50) $\frac{1 + i\sqrt{163}}{2}, \frac{1 - i163}{2}$

51) $\frac{5 + i\sqrt{415}}{8}, \frac{5 \ i\sqrt{415}}{8}$

52) $\frac{11 + i\sqrt{95}}{6}, \frac{11 - i\sqrt{95}}{6}$

53) $\frac{5 + i\sqrt{191}}{2}, \frac{5 - i\sqrt{191}}{2}$

54) $8, 7$

55) $1, -\frac{5}{2}$

56) $3, -\frac{3}{2}$

9.4

Answers - Quadratic Formula

1) $\frac{i\sqrt{6}}{2}, -\frac{i\sqrt{6}}{2}$

2) $\frac{i\sqrt{6}}{3}, -\frac{i\sqrt{6}}{3}$

3) $2 + \sqrt{5}, 2 - \sqrt{5}$

4) $\frac{\sqrt{6}}{6}, -\frac{\sqrt{6}}{6}$

5) $\frac{\sqrt{6}}{2}, -\frac{\sqrt{6}}{2}$

6) $\frac{1 + i\sqrt{29}}{5}, \frac{1 \ i\sqrt{29}}{5}$

7) $1, -\frac{1}{3}$

8) $\frac{1 + \sqrt{31}}{2}, \frac{1 \ \sqrt{31}}{2}$

9) $3, -3$

10) $i\sqrt{2}, -i\sqrt{2}$

11) $3, 1$

12) $-1 + i, -1 - i$

13) $\frac{3 + i\sqrt{55}}{4}, \frac{3 \ i\sqrt{55}}{4}$

14) $\frac{-3 + i\sqrt{159}}{12}, \frac{-3 - i\sqrt{159}}{12}$

15) $\frac{-3 + \sqrt{141}}{6}, \frac{-3 - \sqrt{141}}{6}$

16) $\sqrt{3}, -\sqrt{3}$

17) $\frac{3 + \sqrt{401}}{14}, \frac{3 \ \sqrt{401}}{14}$

18) $\frac{5 + \sqrt{137}}{8}, \frac{5 \ \sqrt{137}}{8}$

19) $2, -5$

20) $5, -9$

21) $\frac{-1 + i\sqrt{3}}{2}, \frac{-1 - i\sqrt{3}}{2}$

22) $3, -\frac{1}{3}$

23) $\frac{7}{2}, -7$

24) $\frac{-3 + i\sqrt{3}}{3}, \frac{-3 - i\sqrt{3}}{3}$

25) $\frac{7 + 3\sqrt{21}}{10}, \frac{7 \ 3\sqrt{21}}{10}$

26) $\frac{5 + \sqrt{337}}{12}, \frac{5 \ \sqrt{337}}{12}$

27) $\frac{-3 + i\sqrt{247}}{16}, \frac{-3 - i\sqrt{247}}{16}$

479

28) $\frac{3+\sqrt{33}}{6}, \frac{3}{6} \frac{\sqrt{33}}{6}$

29) $-1, -\frac{3}{2}$

30) $2\sqrt{2}, -2\sqrt{2}$

31) $4, -4$

32) $2, -4$

33) $4, -9$

34) $\frac{2+3i\sqrt{5}}{7}, \frac{2-3i\sqrt{5}}{7}$

35) $6, -\frac{9}{2}$

36) $\frac{5+i\sqrt{143}}{14}, \frac{5-i\sqrt{143}}{14}$

37) $\frac{3+\sqrt{345}}{14}, \frac{3}{14} \frac{\sqrt{345}}{14}$

38) $\frac{\sqrt{6}}{2}, -\frac{\sqrt{6}}{2}$

39) $\frac{\sqrt{26}}{2}, -\frac{\sqrt{26}}{2}$

40) $\frac{-1+\sqrt{141}}{10}, \frac{-1-\sqrt{141}}{10}$

9.5

Answers - Build Quadratics from Roots

NOTE: There are multiple answers for each problem. Try checking your answers because your answer may also be correct.

1) $x^2 - 7x + 10 = 0$

2) $x^2 - 9x + 18 = 0$

3) $x^2 - 22x + 40 = 0$

4) $x^2 - 14x + 13 = 0$

5) $x^2 - 8x + 16 = 0$

6) $x^2 - 9x = 0$

7) $x^2 = 0$

8) $x^2 + 7x + 10 = 0$

9) $x^2 - 7x - 44 = 0$

10) $x^2 - 2x - 3 = 0$

11) $16x^2 - 16x + 3 = 0$

12) $56x^2 - 75x + 25 = 0$

13) $6x^2 - 5x + 1 = 0$

14) $6x^2 - 7x + 2 = 0$

15) $7x^2 - 31x + 12 = 0$

16) $9x^2 - 20x + 4 = 0$

17) $18x^2 - 9x - 5 = 0$

18) $6x^2 - 7x - 5 = 0$

19) $9x^2 + 53x - 6 = 0$

20) $5x^2 + 2x = 0$

21) $x^2 - 25 = 0$

22) $x^2 - 1 = 0$

23) $25x^2 - 1 = 0$

24) $x^2 - 7 = 0$

25) $x^2 - 11 = 0$

26) $x^2 - 12 = 0$

27) $16x^2 - 3 = 0$

28) $x^2 + 121 = 0$

29) $x^2 + 13 = 0$

30) $x^2 + 50 = 0$

31) $x^2 - 4x - 2 = 0$

32) $x^2 + 6x + 7 = 0$

33) $x^2 - 2x + 10 = 0$

34) $x^2 + 4x + 20 = 0$

35) $x^2 - 12x + 39 = 0$

36) $x^2 + 18x + 86 = 0$

37) $4x^2 + 4x - 5 = 0$

38) $9x^2 - 12x + 29 = 0$

39) $64x^2 - 96x + 38 = 0$

40) $4x^2 + 8x + 19 = 0$

9.6

Answers - Quadratic in Form

1) $\pm 1, \pm 2$

2) $\pm 2, \pm \sqrt{5}$

3) $\pm i, \pm 2\sqrt{2}$

4) $\pm 5, \pm 2$

5) $\pm 1, \pm 7$

6) $\pm 3, \pm 1$

7) $\pm 3, \pm 4$

8) $\pm 6, \pm 2$

9) $\pm 2, \pm 4$

10) $2, 3, -1 \pm i\sqrt{3}, \frac{3 \pm 3i\sqrt{3}}{2}$

11) $-2, 3, 1 \pm i\sqrt{3}, \frac{3 \pm i\sqrt{3}}{2}$

12) $\pm \sqrt{6}, \pm 2i$

13) $\frac{\pm 2i\sqrt{3}}{3}, \frac{\sqrt{6}}{2}$

14) $\frac{1}{4}, -\frac{1}{3}$

15) $-125, 343$

16) $-\frac{5}{4}, \frac{1}{5}$

17) $1, -\frac{1}{2}, \frac{1 \pm i\sqrt{3}}{4}, \frac{1 = i\sqrt{3}}{2}$

18) $\pm 2, \pm \sqrt{3}$

19) $\pm i, \pm \sqrt{3}$

20) $\pm i\sqrt{5}, \pm i\sqrt{2}$

21) $\pm \sqrt{2}, \pm \frac{\sqrt{2}}{2}$

22) $\pm i, \frac{\pm 6}{2}$

23) $\pm 1, \pm 2\sqrt{2}$

24) $2, \sqrt[3]{2}, -1 \pm i\sqrt{3}, \frac{-\sqrt[3]{2} \quad i\sqrt[6]{108}}{2}$

25) $1, \frac{1}{2}, \frac{1 \pm i\sqrt{3}}{4}, \frac{1 \pm i\sqrt{3}}{2}$

26) $\frac{1}{2}, -1, \frac{1 \pm i\sqrt{3}}{4}, \frac{1 = i\sqrt{3}}{2}$

27) $\pm 1, \pm i, \pm 2, \pm 2i$

28) $6, 0$

29) $-(b+3), 7-b$

30) -4

31) $-4, 6$

32) $8, -1$

33) $-2, 10$

34) $2, -6$

35) $-1, 11$

36) $\frac{5}{2}, 0$

37) $4, -\frac{4}{3}$

38) $\pm \sqrt{6}, \pm \sqrt{2}$

39) $\pm 1, -\frac{1}{3}, \frac{5}{3}$

40) $0, \pm 1, -2$

41) $\frac{511}{3}, -\frac{1339}{24}$

42) $-3, \pm 2, 1$

43) $\pm 1, -3$

44) $-3, -1, \frac{3}{2}, -\frac{1}{2}$

45) $\pm 1, -\frac{1}{2}, \frac{3}{2}$

46) $1, 2, \frac{1}{3}, -\frac{2}{3}$

9.7

Answers - Rectangles

1) 6 m x 10 m

2) 5

3) 40 yd x 60 yd

4) 10 ft x 18 ft

5) 6 x 10

6) 20 ft x 35 ft

7) 6" x 6"

8) 6 yd x 7 yd

9) 4 ft x 12 ft

10) 1.54 in

11) 3 in

12) 10 ft

13) 1.5 yd

14) 6 m x 8 m

15) 7 x 9

16) 1 in

17) 10 rods

18) 2 in

19) 15 ft

20) 20 ft

21) 1.25 in

22) 23.16 ft

23) 17.5 ft

24) 25 ft

25) 3 ft

26) 1.145 in

9.8

Answers - Teamwork

1) 4 and 6
2) 6 hours
3) 2 and 3
4) 2.4
5) C = 4, J = 12
6) 1.28 days
7) $1\frac{1}{3}$ days
8) 12 min
9) 8 days
10) 15 days

11) 2 days
12) $4\frac{4}{9}$ days
13) 9 hours
14) 12 hours
15) 16 hours
16) $7\frac{1}{2}$ min
17) 15 hours
18) 18 min
19) $5\frac{1}{4}$ min
20) 3.6 hours

21) 24 min
22) 180 min or 3 hrs
23) Su = 6, Sa = 12
24) 3 hrs and 12 hrs
25) P = 7, S = $17\frac{1}{2}$
26) 15 and 22.5 min
27) A = 21, B = 15
28) 12 and 36 min

9.9

Answers - Simultaneous Product

1) $(2, 36), (-18, -4)$
2) $(-9, -20), (-40, -\frac{9}{2})$
3) $(10, 15), (-90, -\frac{5}{3})$
4) $(8, 15), (-10, -12)$
5) $(5, 9), (18, 2.5)$
6) $(13, 5), (-20, -\frac{13}{4})$

7) $(45, 2), (-10, -9)$
8) $(16, 3), (-6, -8)$
9) $(1, 12), (-3, -4)$
10) $(20, 3), (5, 12)$
11) $(45, 1), (-\frac{5}{3}, -27)$
12) $(8, 10), (-10, -8)$

9.10

Answers - Revenue and Distance

1) 12
2) $4
3) 24
4) 55
5) 20
6) 30
7) 25 @ $18
8) 12 @ $6

9) 60 mph, 80 mph
10) 60, 80
11) 6 km/hr
12) 200 km/hr
13) 56, 76
14) 3.033 km/hr
15) 12 mph, 24 mph
16) 30 mph, 40 mph

17) r = 5
18) 36 mph
19) 45 mph
20) 40 mph, 60 mph
21) 20 mph
22) 4 mph

9.11

Answers - Graphs of Quadratics

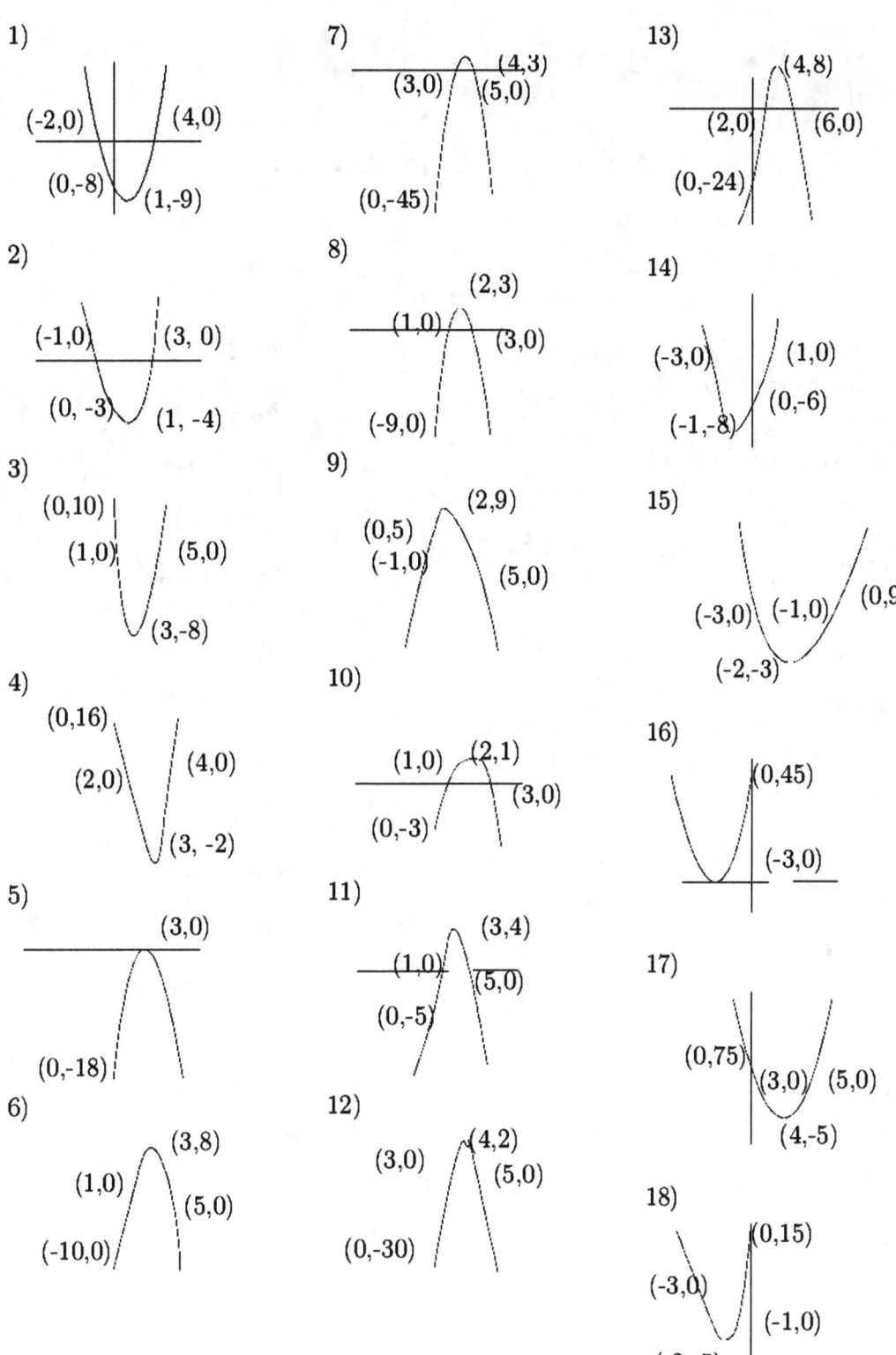

1)
(-2,0) (4,0)
(0,-8) (1,-9)

2)
(-1,0) (3, 0)
(0, -3) (1, -4)

3)
(0,10)
(1,0) (5,0)
(3,-8)

4)
(0,16)
(2,0) (4,0)
(3, -2)

5)
(3,0)
(0,-18)

6)
(3,8)
(1,0) (5,0)
(-10,0)

7)
(4,3)
(3,0) (5,0)
(0,-45)

8)
(2,3)
(1,0) (3,0)
(-9,0)

9)
(2,9)
(0,5)
(-1,0) (5,0)

10)
(1,0) (2,1)
(3,0)
(0,-3)

11)
(3,4)
(1,0) (5,0)
(0,-5)

12)
(4,2)
(3,0) (5,0)
(0,-30)

13)
(4,8)
(2,0) (6,0)
(0,-24)

14)
(-3,0) (1,0)
(0,-6)
(-1,-8)

15)
(-3,0) (-1,0) (0,9)
(-2,-3)

16)
(0,45)
(-3,0)

17)
(0,75)
(3,0) (5,0)
(4,-5)

18)
(0,15)
(-3,0)
(-1,0)
(-2,-5)

19)

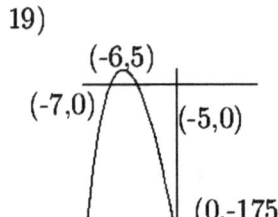

20)

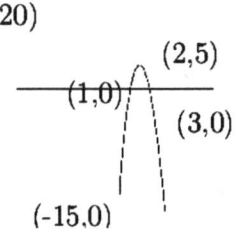

Answers - Chapter 10

10.1

Answers - Function Notation

1) a. yes b. yes c. no
   d. no e. yes f. no
   g. yes h. no

2) all real numbers

3) $x \leqslant \frac{5}{4}$

4) $t \neq 0$

5) all real numbers

6) all real numbers

7) $x \geqslant 16$

8) $x \neq -1, 4$

9) $x \geqslant 4, x \neq 5$

10) $x \neq \pm 5$

11) -4

12) $-\frac{3}{25}$

13) 2

14) 85

15) -7

16) 7

17) $-\frac{17}{9}$

18) -6

19) 13

20) 5

21) 11

22) -21

23) 1

24) -4

25) -21

26) 2

27) -60

28) -32

29) 2

30) $\frac{31}{32}$

31) $-64x^3 + 2$

32) $4n + 10$

33) $-1 + 3x$

34) $-3 \cdot 2^{\frac{12+a}{4}}$

35) $2\left| -3n^2 - 1 \right| + 2$

36) $1 + \frac{1}{16}x^2$

37) $3x + 1$

38) $t^4 + t^2$

39) 5^{-3-x}

40) $5^{\frac{2+n}{2}} + 1$

10.2

Answers - Operations on Functions

1) 82

2) 20

3) 46

4) 2

5) 5

6) -30

7) -3

8) 140

9) 1

10) -43

11) 100

12) -74

13) $\frac{1}{5}$

14) 27

15) $-\frac{9}{26}$

16) $n^2 - 2n$

17) $-x^3 - 4x - 2$

18) $-x^3 + 2x^2 - 3$

19) $-x^2 - 8x + 2$

20) $2t^2 - 8t$

21) $4x^3 + 25x^2 + 25x$

22) $-2t^3 - 15t^2 - 25t$

23) $x^2 - 4x + 5$

24) $3x^2 + 4x - 9$

25) $\frac{n^2 + 5}{3n + 5}$

26) $-2x + 9$

27) $\frac{2a + 5}{3a + 5}$

28) $t^3 + 3t^2 - 3t + 5$

29) $n^3 + 8n + 5$

30) $\frac{4x + 2}{x^2 + 2x}$

31) $n^6 - 9n^4 + 20n^2$

32) $18n^2 - 15n - 25$

33) $x + 3$

34) $-\frac{2}{3}$

35) $t^4 + 8t^2 + 2$

36) $\frac{3n - 6}{-n^2 - 4n}$

37) $\frac{-x^3 - 2x}{-3x + 4}$

38) $x^4 - 4x^2 - 3$

39) $\frac{-n^2 - 2n}{3}$

40) $\frac{32 + 23n}{8} \quad n^3$

41) -155

42) 5

43) 21

44) 4

45) 103

46) 12

47) -50

48) 112

49) 176

50) 147

51) $16x^2 + 12x - 4$

52) $-8a + 14$

53) $-8a + 2$

54) t

55) $4x^3$

56) $-2n^2 - 12n - 16$

57) $-2x + 8$

58) $27t^3 - 108t^2 + 141t - 60$

59) $-16t - 5$

60) $3x^3 + 6x^2 - 4$

10.3

Answers - Inverse Functions

1) Yes   2) No

3) Yes

4) Yes

5) No

6) Yes

7) No

8) Yes

9) Yes

10) No

11) $f^{-1}(x)=\sqrt[5]{x-3}+2$

12) $g^{-1}(x)=(x-2)^3-1$

13) $g^{-1}(x)=\frac{4\ 2x}{x}$

14) $f^{-1}(x)=\frac{-3+3x}{x}$

15) $f^{-1}(x)=\frac{-2x-2}{x+2}$

16) $g^{-1}(x)=3x-9$

17) $f^{-1}(x)=-5x+10$

18) $f^{-1}(x)=\frac{15+2x}{5}$

19) $g^{-1}(x)=-\sqrt[3]{x}+1$

20) $f^{-1}(x)=\frac{-4x+12}{3}$

21) $f^{-1}(x)=\sqrt[3]{x}+3$

22) $g^{-1}(x)=-2x^5+2$

23) $g^{-1}(x)=\frac{x}{x\ 1}$

24) $f^{-1}(x)=\frac{3x\ 3}{x+2}$

25) $f^{-1}(x)=\frac{x\ 1}{x-1}$

26) $h^{-1}(x)=\frac{-2x}{x\ 1}$

27) $g^{-1}(x)=\frac{4x+8}{5}$

28) $g^{-1}(x)=-3x+2$

29) $g^{-1}(x)=\frac{x+1}{5}$

30) $f^{-1}(x)=\frac{5+4x}{5}$

31) $g^{-1}(x)=\sqrt[3]{x+1}$

32) $f^{-1}(x)=\sqrt[5]{\frac{x+3}{2}}$

33) $h^{-1}(x)=\frac{(-2x+4)^3}{4}$

34) $g^{-1}(x)=\sqrt[3]{x-2}+1$

35) $f^{-1}(x)=\frac{-2x+1}{x-1}$

36) $f^{-1}(x)=\frac{-1-x}{x}$

37) $f^{-1}(x)=\frac{2x+7}{x+3}$

38) $f^{-1}(x)=-\frac{4x}{3}$

39) $g^{-1}(x)=-x$

40) $g^{-1}(x)=\frac{-3x+1}{2}$

10.4

Answers - Exponential Functions

1) 0

2) -1

3) 0

4) 0

5) $-\frac{3}{4}$

6) $-\frac{5}{4}$

7) $-\frac{3}{2}$

8) 0

9) $-\frac{2}{3}$

10) 0

11) $\frac{5}{6}$

12) 0

13) -2

14) $-\frac{5}{6}$

15) 1

16) -1

17) No solution

18) $-\frac{4}{3}$

19) $-\frac{1}{4}$

20) $-\frac{3}{4}$

21) No solution

22) 0

23) $-\frac{3}{2}$

24) $\frac{2}{5}$

25) -1

26) $\frac{1}{4}$

27) $-\frac{1}{2}$

28) $\frac{1}{3}$

29) 0

30) No solution

31) 1

32) 3

33) $\frac{1}{3}$

34) $\frac{2}{3}$

35) 0

36) 0

37) $\frac{3}{8}$

38) -1

39) -3

40) No solution

10.5

Answers - Logarithmic Functions

1) $9^2 = 81$

2) $b^{-16} = a$

3) $7^{-2} = \frac{1}{49}$

4) $16^2 = 256$

5) $13^2 = 169$

6) $11^0 = 1$

7) $\log_8 1 = 0$

8) $\log_{17} \frac{1}{289} = -2$

9) $\log_{15} 225 = 2$

10) $\log_{144} 12 = \frac{1}{2}$

11) $\log_{64} 2 = \frac{1}{6}$

12) $\log_{19} 361 = 2$

13) $\frac{1}{3}$

14) 3

15) $-\frac{1}{3}$

16) 0

17) 2

18) -3

19) 2

20) $\frac{1}{2}$

21) 6

22) 5

23) 5

24) 512

25) $\frac{1}{4}$

26) 1000

27) 121

28) 256

29) 6552

30) $\frac{45}{11}$

31) $-\frac{125}{3}$

32) $-\frac{1}{4}$

33) $-\frac{54}{11}$

34) $-\frac{2401}{3}$

35) $-\frac{1}{2}$

36) $-\frac{1}{11}$

37) $-\frac{621}{10}$

38) $\frac{283}{243}$

39) $\frac{2}{5}$

40) 3

10.6

Answers - Interest Rate Problems

1)

a. 740.12; 745.91

b. 851.11; 859.99

c. 950.08; 953.44

d. 1979.22; 1984.69

e. 1209.52; 1214.87

f. 1528.02; 1535.27

g. 2694.70; 2699.72

h. 3219.23; 3224.99

i. 7152.17; 7190.52

2) 1640.70

3) 2868.41

4) 2227.41

5) 1726.16

6) 1507.08

7) 2001.60

8) 2009.66

9) 2288.98

10) 6386.12

11) 13742.19

12) 28240.43

13) 12.02; 3.96

14) 3823.98

15) 101.68

10.7

Answers - Trigonometric Functions

1) 0.3256	14) 8.2	28) 14.6
2) 0.9205	15) 26.1	29) 1
3) 0.9659	16) 16.8	30) 8
4) 0.7660	17) 2.2	31) 1.5
5) $\frac{7}{25}$	18) 9.8	32) 7.2
6) $\frac{8}{15}$	19) 17.8	33) 5.5
7) $\frac{7}{16}$	20) 10.3	34) 2
8) $\frac{3}{5}$	21) 3.9	35) 41.1
9) $\frac{\sqrt{2}}{2}$	22) 10.6	36) 3.2
10) $\frac{4}{5}$	23) 10.2	37) 18.2
11) 16.1	24) 8.9	38) 3.3
12) 2.8	25) 9.5	39) 17.1
13) 32	26) 24.4	40) 22.2
	27) 4.7	

10.8

Answers - Inverse Trigonometric Functions

1) 29°	11) 36°
2) 39°	12) 61.7°
3) 41°	13) 54°
4) 52°	14) 46.2°
5) 24°	15) 55.2°
6) 32°	16) 42.7°
7) 15°	17) 58°
8) 18°	18) 20.1°
9) 27°	19) 45.2°
10) 35°	20) 73.4°

21) 51.3°

22) 45°

23) 56.4°

24) 48.2°

25) 55°

26) 30.5°

27) 47°

28) 15.5°

29) 30°

30) 59°

31) $m\angle B = 28°, b = 15.1, c = 32.2$

32) $m\angle B = 22.8°, m\angle A = 67.2°,$
$c = 16.3$

33) $m\angle B = 22.5°, m\angle A = 67.5°, c = 7.6$

34) $m\angle A = 39°, b = 7.2, a = 5.9$

35) $m\angle B = 64.6°, m\angle A = 25.4°, b = 6.3$

36) $m\angle A = 69°, b = 2.5, a = 6.5$

37) $m\angle B = 38°, b = 9.9, a = 12.6$

38) $m\angle B = 42°, b = 9.4, c = 14$

39) $m\angle A = 45°, b = 8, c = 11.3$

40) $m\angle B = 29.1°, m\angle A = 60.9°,$
$a = 12.2$

www.ingramcontent.com/pod-product-compliance
Lightning Source LLC
Chambersburg PA
CBHW080756180526
45168CB00006B/2226